Lecture Notes in Mathematics

Edited by A. Dold and B. Eckmann

1333

A. S. Kechris D. A. Martin
J. R. Steel (Eds.)

Cabal Seminar 81–85

Proceedings, Caltech-UCLA Logic Seminar 1981–85

Springer-Verlag
Berlin Heidelberg New York London Paris Tokyo

Editors

Alexander S. Kechris
Department of Mathematics, California Institute of Technology
Pasadena, California 91125, USA

Donald A. Martin
John R. Steel
Department of Mathematics, University of California at Los Angeles
Los Angeles, CA 90024, USA

Mathematics Subject Classification (1980): 03EXX, 03DXX, 04-XX

ISBN 3-540-50020-0 Springer-Verlag Berlin Heidelberg New York
ISBN 0-387-50020-0 Springer-Verlag New York Berlin Heidelberg

Printing and binding: Druckhaus Beltz, Hemsbach/Bergstr.
2146/3140-543210

שלשה המה נפלאו ממני וארבע לא ידעתים

-Proverbs 30:18

INTRODUCTION

This is the fourth volume of the proceedings of the Caltech-UCLA Logic Seminar. Except for the first paper, which contains very recent results, the papers of this volume deal with some of the material discussed in the period 1981-85. There is also an Appendix with the second list of Victoria Delfino problems distributed during the 5th Very Informal Gathering of logicians at UCLA in January 1984.

Los Angeles
June 1987

Alexander S. Kechris
Donald A. Martin
John R. Steel

TABLE OF CONTENTS

The Strength of Borel Wadge Determinacy

Alain Louveau and Jean Saint-Raymond
Equipe d'Analyse
Université Paris VI
4, Place Jussieu
75230 Paris, Cedex 05

One of the nice consequences of Martin's theorem that Borel games are determined is the so-called Borel Wadge Determinacy, the determinacy of all games $G(A,B)$ of the following kind: Player I produces $\alpha \in \omega^\omega$, player II produces $\beta \in \omega^\omega$ and II wins $G(A,B)$ if $\alpha \in A \leftrightarrow \beta \in B$, whenever A and B are Borel subsets of ω^ω. Borel Wadge Determinacy allows to get a complete description of all classes of Borel sets, i.e. of all families $\Gamma \subseteq \mathbf{\Delta}^1_1$ which are continuously closed.

Borel Determinacy is not an elementary statement (Friedman), as its proof requires the existence of uncountably many cardinals; so the question arises of the strength of Borel Wadge Determinacy. In case of $\mathbf{\Pi}^1_1$ Wadge determinacy, it is known from work of Harrington and Steel that it is equivalent to the full $\mathbf{\Pi}^1_1$ determinacy, and Steel conjectured in [6] that the same result should be true at the Borel level. However, lightface results of Louveau [1] about Borel Wadge classes seemed to indicate Borel Wadge determinacy might be much weaker. In this paper, we settle the question by showing that Borel Wadge Determinacy–and even stronger forms of it that do not follow a priori from Borel Determinacy–are provable in (a weak subsystem of) second order arithmetics.

Our main result is proved in section 3. It is a sequel to two earlier papers: First, Louveau's paper [1] which analyzes the Borel Wadge classes in a way which does not depend too heavily on Borel Wadge determinacy. We quickly review in the first section the material from [1] that we need. The second source is our joint paper [2], where we prove the particular instances of Borel Wadge Determinacy which correspond to the Baire classes ($\mathbf{\Sigma}^0_\xi$ and $\mathbf{\Pi}^0_\xi$), and introduce the main device for the general proof, a specific way of associating to a closed game another closed game, that we called the ramification method. We present the material from [2] that we need in section 2. Both papers are rather long and technical, so the information we provide in sections 1 and 2 is a bit sketchy. In particular, the existence of ramifications will be used as a black box here, and we will also leave to the reader the verification that the results from Louveau [1] we use do not depend on Borel Wadge Determinacy.

The main consequence of Borel Wadge determinacy is Wadge's lemma which asserts that any Borel set in ω^ω which is not in a class Γ always generates by continuous preimages the dual class $\check{\Gamma}$ (of complements of sets in Γ). In section 4 we show how Wadge's lemma extends to arbitrary Polish (even Suslin) spaces in place of ω^ω, even if no game is available in this general context. This part is much more topological in nature, and uses some transfer methods and selection results for continuous functions which might be of interest in other contexts. Finally in section 5 we develop a notion of Hurewicz test for a class Γ, in order to extend to all Borel Wadge classes the well known theorem of Hurewicz which characterizes among Borel sets those which are not Polish as those which contain a relatively closed set homeomorphic to $\boldsymbol{Q}$. Most results in sections 4 and 5 are again sequels of our paper [2].

Section 1. Descriptions of Borel Wadge Classes. A family Γ of subsets of ω^ω is a *class* if it is closed under continuous preimages, and a *Wadge class* if it is generated by one set $A \subseteq \omega^\omega$. If moreover A is Borel, it is a Borel Wadge class.

Using as main tool Borel Wadge Determinacy, Wadge analyzed in his thesis [7] all Borel Wadge classes. Relying heavily on Wadge's work, Louveau proposed in [1] an inductive construction of all Borel Wadge classes in terms of certain Boolean operations. Fortunately, these works do not depend too much on Borel Wadge determinacy: Although it is unclear (at this point) that the analysis in [1] exhausts all Borel Wadge classes, one can still show directly that it almost does, in a precise sense (given by Theorem 4 below). And part of our proof of Borel Wadge Determinacy will in fact consist in showing that the analysis is exhaustive.

Let us first introduce some notations and definitions.

DEFINITION 1. Let Γ be a class of subsets of ω^ω

(i) If $A \subseteq \omega^\omega$, we let $\check{A} = \omega^\omega - A$, and we let $\check{\Gamma} = \{\check{A} : A \in \Gamma\}$ be the *dual* class of Γ. We also set $\Delta(\Gamma) = \Gamma \cap \check{\Gamma}$, the *ambiguous* class of Γ. We say that Γ is *self-dual* if $\Gamma = \check{\Gamma}(= \Delta(\Gamma))$.

We also define the ordering $<$ between classes by

$$\Gamma < \Gamma' \iff \Gamma \subseteq \Delta(\Gamma').$$

(ii) $PU(\Gamma)$ is the class of all A's $\subseteq \omega^\omega$ of form $A = \cup_n (A_n \cap C_n)$, where $A_n \in \Gamma$ and $(C_n)_{n\in\omega}$ is a partition of ω^ω in clopen sets (PU stands for "partitioned union"). We will use this operation mainly in two cases: If Γ is a non self-dual class, we set $\Gamma^+ = PU(\Gamma \cup \check{\Gamma})$. And if $\langle \Gamma_n : n \in \omega \rangle$ is a $<$-increasing sequence of classes, $\langle \Gamma_n \rangle^+ = PU(\cup_n \Gamma_n)$.

DEFINITION 2. Let Γ, Γ' be classes, ξ, η ordinals ≥ 1.

(a) $A \in D_\eta(\mathbf{\Sigma}^\circ_\xi) \leftrightarrow A = \cup\{A_\theta - \cup_{\theta'<\theta} A_{\theta'} : \theta < \eta,\ \theta$ of a different parity than $\eta\}$ for some increasing sequence $\langle A_\theta : \theta < \eta\rangle$ of $\mathbf{\Sigma}^\circ_\xi$ sets in ω^ω.

(b) $A \in \text{Sep}(D_\eta(\mathbf{\Sigma}^\circ_\xi), \Gamma) \leftrightarrow A = (A_0 \cap C) \cup (A_1 \backslash C)$ for some $C \in D_\eta(\mathbf{\Sigma}^\circ_\xi)$, $A_0 \in \check{\Gamma}$, $A_1 \in \Gamma$.

(c) $A \in \text{Bisep}(D_\eta(\mathbf{\Sigma}^\circ_\xi), \Gamma, \Gamma') \leftrightarrow A = (A_0 \cap C_0) \cup (A_1 \cap C_1) \cup (B \backslash (C_0 \cup C_1))$ for some disjoint C_0, C_1 in $D_\eta(\mathbf{\Sigma}^\circ_\xi)$, $A_0 \in \Gamma, A_1 \in \check{\Gamma}$ and $B \in \Gamma'$.

(d) $A \in SU(\mathbf{\Sigma}^\circ_\xi, \Gamma)$ with envelope $C \leftrightarrow A = \cup_n (A_n \cap C_n)$ for some sequence of pairwise disjoint $\mathbf{\Sigma}^\circ_\xi$ sets C_n, with $\cup_n C_n = C$, and $A_n \in \Gamma$.

(e) $A \in SD_\eta(SU(\Sigma^\circ_\xi, \Gamma)\Gamma') \leftrightarrow A = \cup_{\theta<\eta}(A_\theta \backslash \cup_{\theta'<\theta} C_{\theta'}) \cup (B \backslash \cup_{\theta<\eta} C_\theta)$, for some increasing sequence $\langle A_\theta,\ \theta < \eta\rangle$ of sets in $SU(\mathbf{\Sigma}^\circ_\xi, \Gamma)$ with respective envelopes C_θ, such that $C_{\theta'} \subseteq A_\theta$ for $\theta' < \theta$, and some $B \in \Gamma'$.

In [1], Louveau selects particular ways of combining the operations of Definition 2, encoded by what he calls "descriptions". In order to simplify later work, we will also use another notion of description. So we refer to the descriptions of [1] as first type descriptions. The encoding is made by elements u in ω_1^ω. Such sequences are sometimes viewed as pairs $\langle u_0, u_1\rangle$ or as sequences $\langle u_n : n \in \omega\rangle$, via some fixed bijections between ω and $\omega.2$, ω^2, respectively. We let $\underline{0}$ be the constant function zero.

DEFINITION 3. ([1], 1.2). The relations "u is a first type description" and "u describes Γ" (written $u \in D_1$ and $\Gamma_u = \Gamma$) are the least relations satisfying:

(a) If $u(0) = 0$, $u \in D_1$ and $\Gamma_u = \{\emptyset\}$.

(b) If $u(0) = \xi \geq 1$, $u(1) = 1$ and $u(2) = \eta \geq 1$, $u \in D_1$ and $\Gamma_u = D_\eta(\mathbf{\Sigma}^\circ_\xi)$.

(c) If $u = \xi^\frown 2^\frown \eta^\frown u^*$, with $\xi \geq 1$, $\eta \geq 1$, $u^* \in D_1$ and $u^*(0) > \xi$, then $u \in D_1$ and $\Gamma_u = \text{Sep}(D_\eta(\mathbf{\Sigma}^\circ_\xi),\ \Gamma_{u^*})$

(d) If $u = \xi^\frown 3^\frown \eta^\frown \langle u_0, u_1\rangle$ with $\xi \geq 1$, $\eta \geq 1$, u_0 and u_1 in D_1, $u_0(0) > \xi$, $u_1(0) \geq \xi$ or $u_1(0) = 0$, and $\Gamma_{u_1} < \Gamma_{u_0}$, then $u \in D_1$ and $\Gamma_u = \text{Bisep}(D_\eta(\mathbf{\Sigma}^\circ_\xi),\ \Gamma_{u_0},\ \Gamma_{u_1})$.

(e) If $u = \xi^\frown 4^\frown \langle u_n : n \in \omega\rangle$, where $\xi \geq 1$, $u_n \in D_1$ for all n, $\Gamma_{u_n} < \Gamma_{u_{n+1}}$, and either for all n $u_n(0) = \xi' > \xi$ or $(u_n(0))_{n\in\omega}$ is strictly increasing with $\sup \xi_n > \xi$, then $u \in D_1$ and $\Gamma_u = SU(\mathbf{\Sigma}^\circ_\xi,\ \cup_n \Gamma_{u_n})$

(f) If $u = \xi^\frown 5^\frown \eta^\frown \langle u_0, u_1\rangle$ with $\xi \geq 1$, $\eta \geq 1$, u_0, u_1 in D_1 with $u_0(0) = \xi$, $u_0(1) = 4$, $u_1(0) \geq \xi$ or $u_1(0) = 0$, and $\Gamma_{u_1} < \Gamma_{u_0}$, then $u \in D_1$ and $\Gamma_u = SD_\eta(\Gamma_{u_0},\ \Gamma_{u_1})$. [Note: As noted by Van Engelen, there is a slight mistake in the original definition [1], 1.2, case e.]

One easily checks that each $u \in D_1$ codes exactly one class Γ_u. And each Γ_u is a non-self dual Borel Wadge class, as can be seen by (inductively) constructing a universal Γ_u set in $\omega^\omega \times \omega^\omega$.

The main result of [1] that we will use, which does not need Borel Wadge determinacy, can be summarized as follows (it corresponds to [1], Lemmas 1.11, 1.14, 1.19, 1.23, 1.24, 1.25 and 1.28).

THEOREM 4. *Let $u \in D_1$. The class $\Delta(\Gamma_u)$ satisfies one of the following three possibilities:*

(i) *There is a description $u^* \in D_1$ with $\Delta(\Gamma_u) = (\Gamma_{u^*})^+$.*

(ii) *There is a sequence (u_n) in D_1 with $\Delta(\Gamma_u) = \langle \Gamma_{u_n} \rangle^+$.*

(iii) *There is a family $(u_\xi)_{\xi<\omega_1}$ in D_1 with $\Delta(\Gamma_u) = \cup_\xi \Gamma_{u_\xi}$.*

In [1], Theorem 4 is just an intermediate step towards proving that any Borel Wadge class is of form Γ_u, $\check{\Gamma}_u$, Γ_u^+ or $\langle \Gamma_{u_n} \rangle^+$ for some u, u_n in D_1, as the case may be. However, this is derived from Theorem 4 by using Borel Wadge determinacy. As the proof is instructive, let us sketch it briefly: First, Borel Wadge Determinacy is used to show that if $A \in \Gamma_u$ but $A \notin \check{\Gamma}_u$, A generates Γ_u – and $\check{A}$ generates $\check{\Gamma}_u$. Similarly if $A \in \Gamma_u^+$ but $A \notin \Gamma_u \cup \check{\Gamma}_u$, A generates Γ_u^+, and if $A \in \langle \Gamma_{u_n} \rangle^+$ but $A \notin \cup_n \Gamma_{u_n}$, A generates $\langle \Gamma_{u_n} \rangle^+$. Finally Borel Wadge Determinacy is used again to prove that $\{\Gamma \cup \check{\Gamma} : \Gamma \text{ a Borel Wadge class}\}$ is well ordered by inclusion (Wadge [7], Martin [3]). One can argue then that if $A \subseteq \omega^\omega$ is Borel, there is a least (for inclusion) class Γ_u with $A \in \Gamma_u \cup \check{\Gamma}_u$. If $A \notin \Delta(\Gamma_u)$, the Wadge class of A is Γ_u or $\check{\Gamma}_u$. If $A \in \Delta(\Gamma_u)$, Theorem 4 applies. Case (iii) is impossible by minimality of u, and Cases (i) and (ii) are solved by the facts above. So in all cases the class of A is described.

In the next sections, we imitate the proof above, except that we will at the same time prove instances of the facts above and the corresponding instances of determinacy; so at the end we will get both that the analysis is exhaustive and that Borel Wadge Determinacy holds.

We now introduce the second type descriptions.

DEFINITION 5. Let $\xi \geq 1$ be a countable ordinal, Γ and Γ' two classes. Then

$$A \in S_\xi(\Gamma, \Gamma') \leftrightarrow A = \cup_n (A_n \cap C_n) \cup (B \backslash \cup_n C_n)$$

for some sequence A_n in Γ, $B \in \Gamma'$, and a sequence (C_n) of pairwise disjoint $\mathbf{\Sigma}^0_\xi$ sets.

Second type descriptions are also elements of ω_1^ω.

DEFINITION 6. The relations "u is a second type description" and "u describes Γ" (written $u \in D_2$ and $\Gamma_u = \Gamma$ – ambiguously) are the least relations satisfying

(a) if $u = \underline{0}$, $u \in D_2$ and $\Gamma_u = \{\emptyset\}$

(b) if $u = \xi^\frown 1^\frown u^*$, with $u^* \in D_2$ and $u^*(0) = \xi$, then $u \in D_2$ and $\Gamma_u = \check{\Gamma}_{u^*}$

(c) if $u = \xi^\frown 2 \cap \langle u_n \rangle$, with $\xi \geq 1$, $u_n \in D_2$, $u_n(0) \geq \xi$ or $u_n(0) = 0$, then $u \in D_2$ and $\Gamma_u = S_\xi(\cup_{n\geq 1}\Gamma_{u_n},\ \Gamma_{u_0})$.

The D_2-encoding is clearly much simpler than the first one. However Theorem 4 would be hard to get using this encoding. Our next step is to show that any class admitting a D_1-description also admits a D_2-description.

PROPOSITION 7. ([1], Lemma 1.4) *Let* $u \in D_1$, *with* $u(0) = \xi \geq 1$. *Then*

(a) $SU(\mathbf{\Sigma}^\circ_\xi,\ \Gamma_u) = \Gamma_u$

(b) Γ_u *is closed under union with a* $\mathbf{\Delta}^\circ_\xi$ *set.*

So in particular both Γ_u and $\check{\Gamma}_u$ are closed under unions or intersections with $\mathbf{\Delta}^\circ_\xi$ sets, and if $A = \cup_n(A_n \cap C_n)$ where $A_n \in \Gamma_u$ (resp. $\check{\Gamma}_u$) and (C_n) is a partition of ω^ω in $\mathbf{\Delta}^\circ_\xi$ sets, then $A \in \Gamma_u$ (resp. $\check{\Gamma}_u$).

THEOREM 8. *Every class admitting a* D_1*-description, and every dual of such a class admit a* D_2*-description.*

Proof. If Γ admits $u \in D_2$ as description, $\check{\Gamma}$ admits $u(0)^\frown 1^\frown u$ as D_2-description. So we prove by induction on $u \in D_1$, that Γ_u admits a D_2-description $v(u)$, with $v(u)(0) = u(0)$.

(a) Clearly if $u(0) = 0$, we can take $v(u) = \underline{0}$.

(b) Let $u(1) = 1$, i.e. $\Gamma_u = D_\eta(\mathbf{\Sigma}^\circ_\xi)$. We use induction on η. For $\eta = 1$, one has $\mathbf{\Sigma}^\circ_\xi = S_\xi(\{\omega^\omega\}, \{\emptyset\})$ so we can take $v(u) = \xi^\frown 2^\frown \langle v_n \rangle$ with $v_0 = 0$ and for $n \geq 1$ $v_n = 0^\frown 1^\frown \underline{0}$. If $\eta = \eta' + 1$, one uses similarly the equality

$$D_\eta(\mathbf{\Sigma}^\circ_\xi) = S_\xi(\widecheck{D_{\eta'}(\mathbf{\Sigma}^\circ_\xi)},\ \mathbf{\Sigma}^\circ_\xi)$$

and if $\lambda = \sup_n(\eta_n + 1)$ is limit

$$D_\lambda(\mathbf{\Sigma}^\circ_\xi) = S_\xi(\cup_n D_{\eta_n}(\mathbf{\Sigma}^\circ_\xi), \{\emptyset\})$$

All these equalities are easy to check.

(c) Suppose now $u = \xi^\frown 2^\frown \eta^\frown u^*$, so that $\Gamma_u = \mathrm{Sep}(D_\eta(\mathbf{\Sigma}^\circ_\xi),\ \Gamma_{u^*})$. We again argue by induction on η, and use the induction hypothesis on Γ_{u^*} and the following equalities

$$\mathrm{Sep}(\mathbf{\Sigma}^\circ_\xi,\ \Gamma_{u^*}) = S_\xi(\check{\Gamma}_{u^*},\ \Gamma_{u^*})$$

$$\mathrm{Sep}(D_{\eta+1}(\mathbf{\Sigma}^\circ_\xi),\ \Gamma_{u^*}) = S_\xi(\widecheck{\mathrm{Sep}(D_\eta(\mathbf{\Sigma}^\circ_\xi),\ \Gamma_{u^*})},\ \mathrm{Sep}(\mathbf{\Sigma}^\circ_\xi,\ \Gamma_{u^*}))$$

and for limit λ

$$\mathrm{Sep}(D_\lambda(\mathbf{\Sigma}^\circ_\xi),\ \Gamma_{u^*}) = S_\xi(\cup_{\eta<\lambda} \mathrm{Sep}(D_\eta(\mathbf{\Sigma}^\circ_\xi),\ \Gamma_{u^*}),\ \{\emptyset\})$$

the proof of which is left to the reader.

(d) $u = \xi^\frown 3^\frown \eta^\frown \langle u_0, u_1\rangle$, so $\Gamma_u =$ $\mathrm{Bisep}(D_\eta(\mathbf{\Sigma}^\circ_\xi),\ \Gamma_{u_0}, \Gamma_{u_1})$. Again by induction on η, one uses the equalities

$$\mathrm{Bisep}\ (\mathbf{\Sigma}^\circ_\xi,\ \Gamma_{u_0},\ \Gamma_{u_1}) = S_\xi(\Gamma_{u_0} \cup \check{\Gamma}_{u_0},\ \Gamma_{u_1})$$

$$\mathrm{Bisep}\ (D_{\eta+1}(\mathbf{\Sigma}^\circ_\xi),\ \Gamma_{u_0},\ \Gamma_{u_1}) = S_\xi(\Gamma \cup \check{\Gamma},\ \Gamma_{u_1})$$

where $\qquad \Gamma = \ \mathrm{Sep}\ (D_\eta(\mathbf{\Sigma}^\circ_\xi),\ \Gamma_{u_0})$

and $\qquad \mathrm{Bisep}\ (D_\lambda(\mathbf{\Sigma}^\circ_\xi),\ \Gamma_{u_0},\ \Gamma_{u_1}) = S_\xi(\cup_{\eta<\lambda}\Gamma_\eta, \Gamma_{u_1})$

where $\qquad \Gamma_\eta = \ \mathrm{Sep}\ (D_\eta(\mathbf{\Sigma}^\circ_\xi),\ \Gamma_{u_0})$.

The equalities do not follow immediately from Proposition 7, so let us sketch one of them, say the successor case (the others are similar, and a bit simpler). Denote by Γ_ℓ and Γ_r the left and right hand side classes.

If $A \in \Gamma_r$, $A = \cup_n(A_n \cap C_n) \cup (B \backslash \cup_n C_n)$ with pairwise disjoint C_n's in $\mathbf{\Sigma}^\circ_\xi$, $B \in \Gamma_{u_1}$ and A_n in $\Gamma = \mathrm{Sep}(D_\eta(\mathbf{\Sigma}^\circ_\xi),\ \Gamma_{u_0})$ or in $\check{\Gamma}$. Write $A_n = (A^\circ_n \cap D_n) \cup (A^1_n \backslash D_n)$ with $D_n \in D_\eta(\mathbf{\Sigma}^\circ_\xi)$, $A^{\varepsilon(n)}_n \in \Gamma_{u_0}$ and $A^{1-\varepsilon(n)}_n \in \check{\Gamma}_{u_0}$, where $\varepsilon(n) = 0$ or 1 depending if A_n is in Γ or $\check{\Gamma}$. Let $D^0_n = D_n \cap C_n$ and $D^1_n = C_n \backslash D_n$. Both D°_n, D^1_n are in $D_{\eta+1}(\mathbf{\Sigma}^\circ_\xi)$, and so are $D^\circ = \cup_n(C_n \cap D^{\varepsilon(n)}_n)$ and $D^1 = \cup_n(C_n \cap D^{1-\varepsilon(n)}_n)$ by an immediate computation. By proposition 7 (as $u_0(0) > \xi$) $A^\circ = \cup_n(C_n \cap A^{\varepsilon(n)}_n) \in \Gamma_{u_0}$ and $A^1 = \cup_n(C_n \cap A^{1-\varepsilon(n)}_n) \in \check{\Gamma}_{u_0}$. And $A = (A^\circ \cap D^\circ) \cup (A^1 \cap D^1) \cup (B \backslash (D^\circ \cup D^1))$, so that $A \in \Gamma_\ell$.

The other inclusion is a bit harder: let $A \in \Gamma_\ell$, i.e. $A = (A_0 \cap C_0) \cup (A_1 \cap C_1) \cup (B \backslash (C_0 \cup C_1))$ with C_0, C_1 disjoint in $D_{\eta+1}(\mathbf{\Sigma}^\circ_\xi)$, $A_0 \in \Gamma_{u_0}$, $A_1 \in \check{\Gamma}_{u_0}$ and $B \in \Gamma_{u_1}$. Let D_0, resp D_1, be the largest $\mathbf{\Sigma}^\circ_\xi$ sets in some constructions of C_0, resp C_1, as $D_{\eta+1}(\mathbf{\Sigma}^\circ_\xi)$ sets, and let D^*_0, D^*_1 reduce D_0, D_1. Clearly $A \backslash (D^*_0 \cup D^*_1) = B \backslash (D^*_0 \cup D^*_1)$, so in order to show that $A \in \Gamma_r$, it is enought to prove that $A \cap D^*_0 \in \Gamma = \mathrm{Sep}(D_\eta(\mathbf{\Sigma}^\circ_\xi), \Gamma_{u_0})$ and similarly $A \cap D^*_1 \in \check{\Gamma}$. Let us prove the first claim, the second one being similar. By the choice of $D_0, D_0 \backslash C_0 \in D_\eta(\mathbf{\Sigma}^\circ_\xi)$, hence $D^*_0 \backslash C_0$ too as $D^*_0 \subseteq D_0$. Now $A \cap D^*_0 \cap C_0 = A_0 \cap D^*_0 \cap C_0$ is in Γ_{u_0}, and $A \cap (D^*_0 \backslash C_0) = A_1 \cap (D^*_0 \cap C_1) \cup (B \backslash (C_0 \cup C_1)) \cap D^*_0$. Both $A_1 \cap D^*_0$ and $B \backslash (C_0 \cup C_1)$ are in $\check{\Gamma}_{u_0}$, and separated by the $\mathbf{\Delta}^\circ_{\xi+1}$ set $C_0 \cup C_1$, hence by Proposition 7 $A \cap (D^*_0 \backslash C_0)$ is in $\check{\Gamma}_{u_0}$ and the equality is proved.

(e) One uses in case $u(1) = 4$ the equality

$$SU(\mathbf{\Sigma}^\circ_\xi,\ \cup_n \Gamma_{u_n}) = S_\xi(\cup_n \Gamma_{u_n}, \{\emptyset\})$$

(f) The final case is when $\Gamma_u = SD_\eta(SU(\mathbf{\Sigma}^\circ_\xi, \cup_n \Gamma_{u_n}), \Gamma_{u^*})$ and is proved as in the $D_\eta(\mathbf{\Sigma}^\circ_\xi)$ case, by using the following (easy) equalities, where $\Gamma = SU(\mathbf{\Sigma}^\circ_\xi, \cup_n \Gamma_{u_n})$

$$SD_{\eta+1}(\Gamma, \Gamma_{u^*}) = S_\xi(SD_\eta(\Gamma, \{\emptyset\}), S_\xi(\cup_n \Gamma_{u_n}, \Gamma_{u^*}))$$

and for limit λ

$$SD_\lambda(\Gamma, \Gamma_{u^*}) = S_\xi(\cup_{\eta<\lambda} SD_\eta(\Gamma, \{\emptyset\}), \Gamma_{u^*})$$

$\dashv$

REMARK: There is a slight defect in the proof above: if one really wants to build a $v(u)$ for $u \in D_1$, one needs at limit steps specific fundamental sequences below limit ordinals. This requires a form of the axiom of choice. The best way to avoid this – which would in any case be necessary for a formalization of the preceding discussion in second order arithmetics – is to replace ordinals by reals coding them, and accordingly descriptions by codes of descriptions in ω^ω. The function $v(u)$ becomes then definable in the codes. However, since working with codes would only create more notational problems to the reader, we will continue with this slight kind of abuse.

Let us denote by

$$W_1 = \{\Gamma_u : u \in D_1\} \cup \{\check{\Gamma}_u : u \in D_1\},$$

$$W_2 = \{\Gamma_u : u \in D_2\}, \quad \text{and}$$

W the set of non self-dual Borel Wadge classes in ω^ω. The preceding theorem says that $W_1 \subseteq W_2$. It is clear that $W_2 \subseteq W$, as can be proved by inductively constructing a universal set for each $\Gamma \in W_2$ in $\omega^\omega \times \omega^\omega$. We will prove later that these inclusions are equalities.

We finish this section with the study of the effect of functions of Baire class η on classes in W_2. For ξ, η countable ordinals, with $\eta \leq \xi$, we denote by $\xi - \eta$ the unique ξ' such that $\eta + \xi' = \xi$. A function $f : \omega^\omega \to \omega^\omega$ is a Baire class η function if for each open $A \subseteq \omega^\omega$ $f^{-1}(A) \in \mathbf{\Sigma}^\circ_{1+\eta}$ (so continuous functions are Baire class 0). One easily checks by induction that if f is of Baire class η and A is $\mathbf{\Sigma}^\circ_\xi$, $f^{-1}(A)$ is in $\mathbf{\Sigma}^\circ_{\eta+'\xi}$, with $\eta +' \xi = 1 + \eta + (\xi - 1)$. Let also $\xi -' \eta$, for $\xi \geq 1$ and $\eta < \xi$, be defined by $\xi -' \eta = 1 + (\xi - (1 + \eta))$, so that $\eta +' (\xi -' \eta) = \xi$.

DEFINITION 9. We define, for each countable η and each $u \in D_2$ a description $u^\eta \in D_2$, and in case $u(0) > \eta$ or $u(0) = 0$, a description ${}^\eta u \in D_2$, by the following clauses:

a) If $u(0) = 0$, $u^\eta = {}^\eta u = u$.

b) If $u = \xi ^\frown 1 ^\frown u^*$, with $\xi \geq 1$

$$u^\eta = (\eta +' \xi) ^\frown 1 \cap (u^*)^\eta$$

and ${}^{\eta}u = (\xi -' \eta)^\frown 1^\frown {}^{\eta}(u^*)$ (for $u(0)$ - hence $u^*(0)$ - bigger than η).

(c) If $u = \xi^\frown 2^\frown \langle u_n \rangle$, with $\xi \geq 1$,

$$u^{\eta} = (\eta +' \xi)^\frown 2^\frown \langle u_n^{\eta} \rangle$$

and ${}^{\eta}u = (\xi -' \eta)^\frown 2^\frown \langle {}^{\eta}u_n \rangle$ (for $\xi > \eta$ - note that ${}^{\eta}u_n$ is defined from some n_0 on, as $\sup u_n(0) > \xi$).

It is clear from the previous definition that $({}^{\eta}u)^{\eta} = u$, when $u(0) = 0$ or $u(0) > \eta$. And one easily gets by induction the following:

PROPOSITION 10.

(i) If $f : \omega^\omega \to \omega^\omega$ is a Baire class η function, and $A \in \Gamma_u$ for some $u \in D_2$, then $f^{-1}(A) \in \Gamma_{u^\eta}$.

(ii) If $u \in D_2$ is such that $u(0) = \xi \geq 1$, there are unique $\underline{u}$ with $\underline{u}(0) = 1$, and $\eta (= \xi - 1)$ such that $u = (\underline{u})^{\eta}$.

In particular, D_2 is the least subset $D \subseteq D_2$ such that $\underline{0} \in D$, $u(0)^\frown 1^\frown u \in D$ if $u \in D$, $1^\frown 2^\frown \langle u_n \rangle \in D$ if for each n $u_n \in D$, and for any η $u^{\eta} \in D$ when $u \in D$.

Let us finally say a few words about relativization: If E is a subset of ω^ω, one can define for any class Γ the relativization $\Gamma(E)$ by using traces on E of sets in Γ. And clearly for u a description of any type, $\Gamma_u(E)$ is the same as the class described by u, starting from Σ^0_ξ subsets of E. And if $f : \omega^\omega \to E$ is continuous and $A \in \Gamma_u(E)$, $f^{-1}(A) \in \Gamma_u$. We will use these remarks in the sequel mainly for $E = 2^\omega$, or $E = 2^\omega \times \omega^\omega$, viewed as a subset of ω^ω.

Section 2. Ramifications of Closed Games. In our first paper [2] on the topic of Borel Wadge determinacy, we proved particular instances of it, namely that if Γ is one of the classes $D_\eta(\Sigma^0_\xi)$, and $A \subseteq 2^\omega$ is a set in $\Gamma \backslash \check{\Gamma}$, the Wadge game $G(A, B)$ is determined, for any Borel B in 2^ω. The main technical tool we introduced to get this result is a specific way of transforming closed games that we called ramifications. We now discuss what will be needed in the sequel about this notion.

Ramifications act on the following kind of games: I plays $\varepsilon \in 2^\omega$, II plays $\beta \in \omega^\omega$, and the game is closed for player II, i.e. specified by a tree J on $2 \times \omega$ – that we will confuse with the game itself. A position $(\varepsilon \restriction_k, \beta \restriction_k)$ is legal in J if $(\varepsilon \restriction_k, \beta \restriction_k) \in J$, and a run (ε, β) is a win for II if for all k $(\varepsilon \restriction_k, \beta \restriction_k)$ is legal in J, i.e. if (ε, β) is a branch through J.

We denote by $J \subseteq \omega^{(2\times\omega)^{<\omega}}$ the (closed) set of all trees on $2 \times \omega$. A strategy for player I in games in J is a function $\sigma : \omega^{<\omega} \to 2$, and we denote by Σ the set $2^{\omega^{<\omega}}$.

DEFINITION 1. A *ramification* of games is a triple (r, ρ, F) of functions, with the following properties:

(a) $r = (r_0, r_1) : (2 \times \omega)^{<\omega} \to (2 \times \omega)^{<\omega}$ satisfies

(i) $r_0(u,v) = r_0(u)$ depends only on $u \in 2^{<\omega}$

(ii) If $n = lh(u,v)$, $t(u,v) = \{r(u \restriction_k, v \restriction_k) : k \leq n\}$ is a subtree of $(2 \times \omega)^{\leq n}$

We let r act on J, by defining a function $R : J \to J$ as follows:

$$(u,v) \in R(J) \leftrightarrow t(u,v) \subseteq J$$

[Intuitively when playing a position (u,v) in $R(J)$, the players are imagining a tree $t(u,v)$ of positions in J, and their position in $R(J)$ is legal if all the imagined positions are legal in J].

(b) $\rho = (\rho_0, \rho_1) : 2^\omega \times \omega^\omega \to 2^\omega \times \omega^\omega$ satisfies

(i) $\rho_0(\varepsilon, \beta) = \rho_0(\varepsilon)$ depends only on $\varepsilon \in 2^\omega$.

(ii) For all ε, β $\rho(\varepsilon, \beta)$ is a branch through the tree $T(\varepsilon, \beta) = \cup_k t(\varepsilon \restriction_k, \beta \restriction_k)$

[Intuitively again, among the positions $T(\varepsilon, \beta)$ associated to a run in some $R(J)$, exists a complete run $\rho(\varepsilon, \beta)$. So in particular if (ε, β) is a win for II in some $R(J), \rho(\varepsilon, \beta)$ is a win for II in J]

(c) $F : J \times \Sigma \to \Sigma$ associates to each game J and each strategy σ (viewed as a strategy for player I in the game $R(J)$) another strategy $\sigma^* = F(J, \sigma)$, which we view as a strategy for I in J. And F satisfies: If σ is winning in $R(J)$, σ^* is winning in J.

Note: In [2], we put some more restrictions on the notion of ramification, in order to be able to inductively construct a nice family of them. But we will have no needs for these refinements.

It is easy to build ramifications – e.g. the identity. But what we need are ramifications for which the function $\rho_0 : 2^\omega \to 2^\omega$ is as complicated a Baire class η function as possible. In order to make this idea precise, let us introduce some more definitions:

DEFINITION 2. Let Γ be a class. A set $H \subseteq 2^\omega$ is Γ-*strategically complete* if

(i) $H \in \Gamma(2^\omega)$

(ii) If $A \subseteq \omega^\omega$ is a Γ set, Player II wins the Wadge game $G(A, H)$ [where I plays $\alpha \in \omega^\omega$, II $\beta \in 2^\omega$ and II wins if $\alpha \in A \leftrightarrow \beta \in H$]

DEFINITION 3. Let $f : 2^\omega \to 2^\omega$, and η a countable ordinal. We say that f is an independent

η-function if

(i) There is a $\pi : \omega \to \omega$ such that for all ε, k, the value of $f(\varepsilon)$ at k depends only on the values of ε on $\pi^{-1}(k)$

(ii) If $\eta = \eta' + 1$ is successor, $\{\varepsilon : f(\varepsilon)(k) = 1\}$ is $\mathbf{\Pi}_{1+\eta'}$ - strategically complete; and if η is limit, then for some increasing sequence (η_n) with sup η, $\{\varepsilon : f(\varepsilon)(k) = 1\}$ is $\mathbf{\Pi}_{1+\eta_k}$ - strategically complete.

The main result of [2] about ramifications ([2], 3.2) can be restated as:

THEOREM 4. *For each countable η, there exists a ramification $(r^\eta, \rho^\eta, F^\eta)$ which satisfies*

(a) *ρ^η and F^η are Baire class η functions.*

(b) *If $\xi < \omega_1$ and $f : 2^\omega \to 2^\omega$ is an independent ξ-function, then $\rho_0^\eta \circ f : 2^\omega \to 2^\omega$ is an independent $\eta + \xi$-function.*

Note: As usual, one cannot pick $(r^\eta, \rho^\eta, F^\eta)$ for each η without some choice, so we should work with a family of ramifications indexed by codes of ordinals. But we will not bother about this in the sequel.

For τ an increasing map: $\omega \to \omega$, let $\tilde{\tau} : 2^\omega \to 2^\omega$ be defined by $\tilde{\tau}(\varepsilon) = \varepsilon \circ \tau$. Clearly $\tilde{\tau}$ is continuous, and in fact an independent 0-function. The next lemma is easy to check.

LEMMA 5. *If $\tilde{\tau} : 2^\omega \to 2^\omega$ is as above, and $f : 2^\omega \to 2^\omega$ is an independent η-function, then $\tilde{\tau} \circ f$ is an independent η-function.*

We let $\mathcal{R}$ be the least set of functions: $2^\omega \to 2^\omega$ which contains the functions ρ_0^η associated with the ramifications of Theorem 4, the functions $\tilde{\tau}$ for τ increasing: $\omega \to \omega$, and is closed under composition. By Theorem 4(b), and Lemma 5 each $f \in \mathcal{R}$ is an independent η-function, for some η we call the order $o(f)$ of f.

Part of our goal now is to define, for each $u \in D_2$ a set $H_u \subset 2^\omega$ which is Γ_u-strategically complete. But the inductive construction needs sets with a slightly stronger property:

DEFINITION 6. Let $u \in D_2$. A set $H \subseteq 2^\omega$ is u-strategically complete if

(i) $H \in \Gamma_u(2^\omega)$

(ii) for each $f \in \mathcal{R}$ of order $o(f) = \eta$, the set $f^{-1}(H)$ is Γ_{u^η}-strategically complete.

THEOREM 7. *Let $u \in D_2$. There exist a u-strategically complete set $H_u \subseteq 2^\omega$, and for each* *pair A_0, A_1 of disjoint Σ^1_1 sets in ω^ω a closed (for II) game $J_u(A_0, A_1)$, where I produces $\varepsilon \in 2^\omega$, II produces $\alpha \in \omega^\omega$ and $\beta \in \omega^\omega$, and a set $C_u(A_0, A_1)$ in $\check{\Gamma}_u(\Sigma \times \omega^\omega)$ such that:*

(i) *If $(\varepsilon, \alpha, \beta)$ is a win for II in $J_u(A_0, A_1)$, then*

$$(\varepsilon \in H_u \to \alpha \in A_0) \quad \textit{and} \quad (\varepsilon \notin H_u \to \alpha \in A_1)$$

(ii) *If for some fixed $\alpha \in \omega^\omega$, σ is a winning strategy for I in the game $J_u(A_0, A_1) \restriction_\alpha$ (where II plays this α), then:*

$$(\alpha \in A_0 \to (\sigma, \alpha) \in C_u(A_0, A_1)) \quad \textit{and} \quad (\alpha \in A_1 \to (\sigma, \alpha) \notin C_u(A_0, A_1))$$

This result is the main result of this section. The proof is by induction on $u \in D_2$. Let us say that u is nice if it satisfies the conclusions of Theorem 7. Using Proposition 1.10, it is enough to prove that $\underline{0}$ is nice, that if u is nice so is $u(0)^\frown 1^\frown u$, and u^η for each $\eta < \omega_1$, and that if $(u_n)_{n\in\omega}$ are nice, so is $1^\frown 2^\frown \langle u_n \rangle$.

LEMMA 8. *The description $\underline{0}$ is nice.*

Proof. We must set $H_{\underline{0}} = \emptyset$ and $C_{\underline{0}}(A_0, A_1) = \Sigma \times \omega^\omega$. For A_0, A_1 disjoint Σ^1_1 sets with associated trees T_0, T_1 respectively on $\omega \times \omega$, let $J_{\underline{0}}(A_0, A_1)$ be the game where I plays ε, II plays α and β and II wins if for all $n(\alpha \restriction_n, \beta \restriction_n) \in T_1$. Clearly for any $f \in \mathcal{R}$ $f^{-1}(H_{\underline{0}}) = \emptyset$ is strategically complete in $\Gamma_{\underline{0}}$. $J_{\underline{0}}(A_0, A_1)$ is closed for II. If $(\varepsilon, \alpha, \beta)$ is a win for II, $\alpha \in A_1$ so (i) is satisfied. And if for some $\alpha \in \omega^\omega$ I has a winning strategy σ in $J_{\underline{0}}(A_0, A_1) \restriction_\alpha$, $\alpha \notin A_1$ hence (ii) is satisfied too. ⊣

LEMMA 9. *Suppose u is nice. Then $\check{u} = u(0)^\frown 1^\frown u$ is nice too.*

Proof. $\Gamma_{\check{u}} = \check{\Gamma}_u$, and one checks that $H_{\check{u}} = \check{H}_u$, $J_{\check{u}}(A_0, A_1) = J_u(A_1, A_0)$ and $C_{\check{u}}(A_0, A_1) = \check{C}_u(A_1, A_0)$ work. ⊣

LEMMA 10. *Suppose u is nice. For each $\eta < \omega_1$, u^η is nice.*

Proof. Let $(r^\eta, \rho^\eta, F^\eta)$ be the ramification of order η, with $\rho^\eta_0 \in \mathcal{R}$. Let $H_u, J_u(A_0, A_1)$ and $C_u(A_0, A_1)$ be associated to u. We define $H_{u^\eta} = (\rho^\eta_0)^{-1}(H_u)$. If now $f \in \mathcal{R}$ is of order ξ, $\rho^\eta_0 \circ f \in \mathcal{R}$, of order $\xi + \eta$, hence $(\rho^\eta_0 \circ f)^{-1}(H_u) = f^{-1}(H_{u^\eta})$ is strategically complete in $\Gamma_{u^{(\xi+\eta)}} = \Gamma_{(u^\eta)^\xi}$, and H_{u^η} is u^η-strategically complete. We now define $J_{u^\eta}(A_0, A_1)$ by

$$J_{u^\eta}(A_0, A_1) \restriction_\alpha = R^\eta(J_u(A_0, A_1) \restriction_\alpha)$$

for all $\alpha \in \omega^\omega$. This is meaningful, for in order to check whether $(\varepsilon \restriction_k, \alpha \restriction_k, \beta \restriction_k)$ is legal in $J_{u^\eta}(A_0, A_1)$, we need to know if $t = \{r^\eta(\varepsilon \restriction_{k'}, \beta \restriction_{k'}) : k' \leq k\}$ is contained in $J_u(A_0, A_1) \restriction_\alpha$. But by the properties of ramifications, $t \subseteq (2 \times \omega)^{\leq k}$ so the knowledge of $\alpha \restriction_k$ is enough for that. Finally we define $C_{u^\eta}(A_0, A_1)$ by

$$(\sigma, \alpha) \in C_{u^\eta}(A_0, A_1) \leftrightarrow (F^\eta(J_u(A_0, A_1) \restriction_\alpha, \sigma), \alpha) \in C_u(A_0, A_1).$$

As $\alpha \mapsto J_u(A_0, A_1) \restriction_\alpha$ is continuous and F^η is Baire class η, the set $C_{u^\eta}(A_0, A_1)$ is in $\check{\Gamma}_{u^\eta}$ by 1.10. It remains to check that they satisfy (i) and (ii) of Theorem 7. For (i), let $(\varepsilon, \alpha, \beta)$ be a win for II in $J_{u^\eta}(A_0, A_1)$. So $(\rho_0^\eta(\varepsilon), \alpha, \rho_1^\eta(\varepsilon, \beta))$ is a win for II in $J_u(A_0, A_1)$. This gives

$$\varepsilon \in H_{u^\eta} \to \rho_0^\eta(\varepsilon) \in H_u \to \alpha \in A_0 \qquad \text{and}$$

$$\varepsilon \notin H_{u^\eta} \to \rho_0^\eta(\varepsilon) \notin H_u \to \alpha \in A_1.$$

For (ii), let σ be winning for I in $J_{u^\eta}(A_0, A_1) \restriction_\alpha$. Then $\sigma^* = F^\eta(J_u(A_0, A_1) \restriction_\alpha, \sigma)$ is winning for I in $J_u(A_0, A_1) \restriction_\alpha$. So $\alpha \in A_0 \to (\sigma^*, \alpha) \in C_u(A_0, A_1) \to (\sigma, \alpha) \in C_{u^\eta}(A_0, A_1)$ and $\alpha \in A_1 \to (\sigma^*, \alpha) \notin C_u(A_0, A_1) \to (\sigma, \alpha) \notin C_{u^\eta}(A_0, A_1)$ ⊣

The preceding proof was trivial – as everything has been embedded in the notion of ramification. The next one is on the other hand long and tedious – but more or less straightforward.

LEMMA 11. *Suppose that for all n u_n is nice. Then so is $u = 1^\frown 2^\frown \langle u_n \rangle$.*

Proof. Let H_n, $J_n = J_n(A_0, A_1)$ and $C_n = C_n(A_0, A_1)$ be associated to u_n. First we choose a bijection $\langle , \rangle$ between $(\omega \cup \{*\}) \times \omega$ and ω such that each $\tau_i = \langle i, \cdot \rangle : \omega \to \omega$ is strictly increasing, for $i \in \omega \cup \{*\}$. We view each $\varepsilon \in 2^\omega$ as a sequence $\varepsilon^*, \langle \varepsilon_i \rangle_{i \in \omega}$, with $\varepsilon^* = \varepsilon \circ \tau^*$, $\varepsilon_i = \varepsilon \circ \tau_i$. Recall that $\Gamma_u = S_1(\cup_{n \geq 1} \Gamma_{u_n}, \Gamma_{u_0})$. Intuitively, we want ε_i to correspond to u_i. As we need repetitions, we choose $\varphi : \omega \setminus \{0\} \to \omega \setminus \{0\}$ such that $\varphi^{-1}(i)$ is infinite for all i. Let also $\psi : \omega \to \omega \setminus \{0\}$ be such that $\psi^{-1}(i)$ is infinite for all i.

We set: $\varepsilon \in H_u \leftrightarrow$ either $\varepsilon^* = \underline{0}$ and $\varepsilon_0 \in H_0$

or for n the least i with $\varepsilon^*(i) = 1$,

$\varepsilon_{\psi(n)} \in H_{\varphi(\psi(n))}$.

We first check that H_u is u-strategically complete. Define $H_0' = \{\varepsilon : \varepsilon_0 \in H_0\}$, and for $n \geq 1$ $H_n' = \{\varepsilon : \varepsilon_n \in H_{\varphi(n)}\}$, and $C_n = \{\varepsilon : \varepsilon^* \neq \underline{0}$ and for m the least i with $\varepsilon^*(i) = 1$, $\psi(m) = n\}$. Clearly the C_n are pairwise disjoint open sets, $H_0' \in \Gamma_{u_0}$ and for $n \geq 1$ $H_n' \in \Gamma_{u_{\varphi(n)}}$, so $H_u = \cup_{n \geq 1}(H_n' \cap C_n) \cup (H_0' \setminus \cup_n C_n)$ is in $\Gamma_u = S_1(\cup_{n \geq 1} \Gamma_{u_n}, \Gamma_{u_0})$. Let now $f \in \mathcal{R}$ be of order η. By 1.10, $H_u^\eta = f^{-1}(H_u)$ is in Γ_{u^η}. Let $\pi : \omega \to \omega$ be associated to f, and let θ_n be the increasing enumeration of $\pi^{-1}(\text{range } \tau_n)$ and θ_n^* the increasing enumeration of $\pi^{-1}(\{i \in \text{range}$

$\tau^* : \psi(\tau^{*-1}(i)) = n\}$. Note that the fact that $\varepsilon \in f^{-1}(H'_n)$ depends only on $\varepsilon \circ \theta_n$. Let then $H^\eta_n = \{\varepsilon \circ \theta_n : f(\varepsilon) \in H'_n\}$, and also $C^\eta_n = \{\varepsilon \circ \theta^*_n : f(\varepsilon)^*$ takes value 1 on some i with $\psi(i) = n\}$. By the hypothesis, H^η_n is strategically complete in $\Gamma_{u^\eta_{\varphi(n)}}$ for $n \geq 1$, in $\Gamma_{u^\eta_0}$ for $n = 0$. Moreover one easily checks that if g is an independent η-function, $\{\varepsilon\ g(\varepsilon) \neq \underline{0}\}$ is $\Sigma^\circ_{1+\eta}$ - strategically complete. So each C^η_n is strategically complete in $\Sigma^\circ_{1+\eta}$. Let then $H^* \subseteq \omega^\omega$ be any set in Γ_{u^η}, say $H^* = \cup_{n\geq 1}(H^*_n \cap C^*_n) \cup (H^*_0 \backslash \cup_n C^*_n)$ with pairwise disjoint C^*_n in $\Sigma^\circ_{1+\eta}$, H^*_0 in $\Gamma_{u^\eta_0}$, and wlog H^*_n in $\Gamma_{u^\eta_{\varphi(n)}}$ (this is where repetitions are used). So player II has for each n a winning strategy σ_n in $G(H^*_n, H^\eta_n)$ and σ^*_n in $G(C^*_n, C^\eta_n)$. Let then II play in $G(H^*, f^{-1}(H))$ against α by playing his strategies σ_n, σ^*_n at the right places – the ranges of θ_n and θ^*_n respectively – against this same α, independently. The result is some ε such that $\varepsilon \circ \theta_n$ wins against α in $G(H^*_n, H^\eta_n)$ and $\varepsilon \circ \theta^*_n$ against α in $G(C^*_n, C^\eta_n)$. This wins, for $\varepsilon \in f^{-1}(H'_n)$ just in case $\alpha \in H^*_n$, and $f(\varepsilon)^*$ takes value 1 on some i with $\psi(i) = n$ just in case $\alpha \in C^*_n$. But as the C^*_n are disjoint, there is at most one n in $\{\psi(i) : f(\varepsilon)^*(i) = 1\}$, and $\varepsilon \in f^{-1}(C_n)$ just in case $\alpha \in C^*_n$. This proves that H_u is u-strategically complete.

We now define $J_u = J_u(A_0, A_1)$. We view ε as $\langle\varepsilon^*, \langle\varepsilon_i\rangle\rangle$ as before, and similarly we decompose β as $\langle\beta^*, \langle\beta_i\rangle\rangle$. In J_u, as long as player I plays 0's on his ε^*-moves, Players I and II must play the game J_0 with ε_0, α and β_0. And once I has played 1 on his ε^*-moves, at say step k of the game, then letting $k_0 = \tau^{*-1}(k), n_0 = \psi(k_0)$, $m_0 = \varphi(n_0)$, the players switch to the game J_{m_0}, played with ε_{n_0} for I, α and the part of β_{n_0} which is played after step k for II (i.e. when switching, I does not revise his previous moves on ε_{n_0}, when II does for β_{n_0}). This defines a closed (for II) game. We now check (i) of Theorem 7: So suppose $(\varepsilon, \alpha, \beta)$ is a win for II in $J_u(A_0, A_1)$. There are two cases:

a) Suppose $\varepsilon^* = \underline{0}$. Then $\varepsilon \in H_u \leftrightarrow \varepsilon_0 \in H_0$. But then $\varepsilon \in H_u \rightarrow \varepsilon_0 \in H_0 \rightarrow \alpha \in A_0$ and $\varepsilon \notin H_u \rightarrow \varepsilon_0 \notin H_0 \rightarrow \alpha \in A_1$ as $(\varepsilon_0, \alpha, \beta_0)$ is a win for II in J_0.

b) for some least k_0 $\varepsilon^*(k_0) = 1$. Let $k = \tau^*(k_0)$, $n_0 = \psi(k_0)$ and $m_0 = \varphi(n_0)$, and $(\beta_{n_0})_{\geq k}$ the sequence of β_{n_0}-moves after k. Then $(\varepsilon_{n_0}, \alpha, (\beta_{n_0})_{\geq k})$ is a win for II in J_{m_0}, and again $\varepsilon \in H_u \leftrightarrow \varepsilon_{n_0} \in H_{m_0}$, so that we get the same conclusion. This proves (i).

We now define $C_u = C_u(A_0, A_1)$.

Let $<$ be an ordering of type ω on $\omega^{<\omega}$, with $u \subseteq v \rightarrow u < v$. Given $\alpha \in \omega^\omega$ and $\sigma \in \Sigma$, let us say that a finite sequence $w \in \omega^{<\omega}$ is (σ, α)-legal if the position in $J_u(A_0, A_1)$ corresponding to the play $\alpha \restriction_{lh(w)}$, w of II and the σ-answer by I is legal. For all (σ, α), let $w(\sigma, \alpha)$ be the least for $<$ sequence which satisfies (i) w is (σ, α)-legal; (ii) the answers by σ on the $*$-moves are 0 up to $lh(w)$; and (iii) $lh(w)$ corresponds to a $*$-move, and $\sigma(w) = 1$ [in other words, we are at a legal switching position in $J_u(A_0, A_1)$ where II follows α and w, and I answers by σ].

The function $(\sigma, \alpha) \mapsto w(\sigma, \alpha)$ is defined on all pairs (σ, α) for which there is a legal beginning w with $\sigma(w)$ a $*$-move with value 1.

We now define, for $\alpha \in \omega^\omega$ and $\sigma \in \Sigma$, a sequence of strategies as follows:

First to each $w_0 \in \omega^{<\omega}$, viewed as a play of II in $J_0 \restriction_\alpha$, associate the play w in $J_u \restriction_\alpha$ consisting in playing w_0 on the 0-moves, and 0 on all other moves, with length such that the next play is the next 0-move. And define $\sigma_0(w_0) = \sigma(w)$. Note that $F_0 : \sigma \mapsto \sigma_0$ is continuous: $\Sigma \to \Sigma$.

Let now $w_0 \in \omega^{<\omega}$. We define $F_{w_0} : \Sigma \to \Sigma$ as follows: We let $F_{w_0}(\sigma) = \underline{0}$ unless the answers by σ to w_0 are 0 on the $*$-moves up to $lh(w_0)$, and $lh(w_0)$ is a $*$-move and $\sigma(w_0) = 1$. And in this case, we associate to each $w \in \omega^{<\omega}$ a position w' as follows: w' is w_0 up to $lh(w_0) = k$, with $\psi(k) = k_0$ say. After that, w' is 0 everywhere except on the k_0-moves, where it is w, and its length is such that the next play will be the next k_0-move. And we define $F_{w_0}(\sigma)(w) = \sigma(w')$. Again each $F_{w_0} : \Sigma \to \Sigma$ is continuous. We can now define $C_u = C_u(A_0, A_1) \subseteq \Sigma \times \omega^\omega$ by

$(\sigma, \alpha) \in C_u \leftrightarrow$ *either* for every (σ, α)-legal sequence $w \in \omega^{<\omega}$
the $*$-answers by σ are 0, and $(F_0(\sigma), \alpha) \in C_0$,
or there is a (σ, α)-legal sequence with $*$-answer
1 by σ, and if $w_0 = w(\sigma, \alpha)$ and $k_0 = \psi(lh(w_0))$,
$(F_{w_0}(\sigma), \alpha) \in C_{\varphi(k_0)}$.

We first check that $C_u \in \check{\Gamma}_u$: let $B_{w_0} = \{(\sigma, \alpha); w_0 = w(\sigma, \alpha)\}$. Clearly, each B_{w_0} is clopen in $\Sigma \times \omega^\omega$, and the B_{w_0}'s are pairwise disjoint. Let $D_0 = \{(\sigma, \alpha) : (F_0(\sigma), \alpha) \in C_0\}$ and for $w_0 \in \omega^{<\omega}$ $D_{w_0} = \{(\sigma, \alpha) : (F_{w_0}(\sigma), \alpha) \in C_{\varphi_0 \psi(lh w_0)}\}$. By continuity of the F_0, F_{w_0}'s, $D_0 \in \Gamma_{u_0}$ and $D_{w_0} \in \Gamma_{u_k}$, $k = \varphi \circ \psi(lh\ w_0)$. And $C_u = \cup_{w_0 \in \omega^{<\omega}} (B_{w_0} \cap D_{w_0}) \cup (D_0 \setminus \cup_{w_0 \in \omega^{<\omega}} B_{w_0})$ hence $C_u \in \check{\Gamma}_u$. It remains to check (ii) of Theorem 7. So we let $\alpha \in \omega^\omega$, and σ winning for I in $J_u(A_0, A_1) \restriction_\alpha$. There are two cases. (a) First, if $w(\sigma, \alpha)$ is undefined, i.e. for any (σ, α) legal sequence w, the $*$-answers are 0. Then $(\sigma, \alpha) \in C_u \leftrightarrow (F_0(\sigma), \alpha) \in C_0$. But we claim $F_0(\sigma)$ is winning for I in $J_0 \restriction_\alpha$, for if $\beta_0 \in \omega^\omega$ defeats $F_0(\sigma)$ in $J_0 \restriction_\alpha$, the play β corresponding to β_0 on the 0-moves and 0 everywhere else is easily seen to defeat σ in $J_u \restriction_\alpha$. So we get $\alpha \in A_0 \to (F_0(\sigma), \alpha) \in C_0 \to (\sigma, \alpha) \in C_u$ and $\alpha \in A_1 \to (F_0(\sigma), \alpha) \notin C_0 \to (\sigma, \alpha) \notin C_u$.

(b) Otherwise, $w_0 = w(\sigma, \alpha)$ is defined, and by definition $(\sigma, \alpha) \in C_u \leftrightarrow (F_{w_0}(\sigma), \alpha) \in C_k$ where $k = \varphi(\psi(lhw_0))$. Again we claim that $F_{w_0}(\sigma)$ is winning for I in $J_k \restriction_\alpha$, which as before will finish the proof. Suppose β is a play of II which defeats $F_{w_0}(\sigma)$ in $J_k \restriction_\alpha$, and let II play in

$J_u \restriction_\alpha$ first w_0, then 0 on all moves except the moves corresponding to $k_0 = \psi(lhw_0)$, where he plays β. One easily checks that all positions are then legal in $J_u \restriction_\alpha$ agains σ, a contradiction which finishes the proof. ⊣

Altogether Lemmas 8, 9, 10 and 11 prove Theorem 7.

Section 3. Proof of Borel Wadge Determinacy. As we said in the introduction, we will prove a slight generalization of Borel Wadge Determinacy. Consider, for $A \subseteq \omega^\omega$ and A_0, A_1 two disjoint subsets of ω^ω the following extended Wadge game $G(A; A_0, A_1)$: I plays $\alpha \in \omega^\omega$, II plays $\beta \in \omega^\omega$, and II wins if $(\alpha \in A \rightarrow \beta \in A_0$ and $\alpha \notin A \rightarrow \beta \in A_1)$. The usual Wadge game $G(A, B)$ corresponds to $B = A_0 = \check{A}_1$. [We will also consider the similar game where $A \subseteq 2^\omega$, and I plays $\alpha \in 2^\omega$, that we will denote ambiguously $G(A; A_0, A_1)$ too]. So Borel Wadge Determinacy is a particular case of

THEOREM 1. *Let $A \subseteq \omega^\omega$ be Borel, and A_0, A_1 two disjoint Σ^1_1 subsets of ω^ω. The extended Wadge game $G(A; A_0, A_1)$ is determined.*

In order to prove Theorem 1, we first prove particular instances of it.

THEOREM 2. *Let $u \in D_2$, and A_0, A_1 two disjoint Σ^1_1 sets in ω^ω*

(i) *If $A \subseteq \omega^\omega$ is in Γ_u, and no set $B \in \check{\Gamma}_u$ separates A_0 from A_1 (i.e. $A_0 \subseteq B \subseteq \check{A_1}$), then player II has a winning strategy in $G(A; A_0, A_1)$.*

(ii) *If $A \subseteq \omega^\omega$ is not in $\check{\Gamma}_u$ and there is a set $B \in \check{\Gamma}_u$ separating A_0 from A_1, then Player I has a winning strategy in $G(A; A_0, A_1)$. In particular, if $A \subseteq \omega^\omega$ is in $\Gamma_u \backslash \check{\Gamma}_u$, $G(A; A_0, A_1)$ is always determined.*

Proof. Let H_u, $J_u(A_0, A_1)$ and $C_u(A_0, A_1)$ be associated to u by Theorem 2.7. Being closed, the game $J_u(A_0, A_1)$ is determined. If I has a winning strategy σ in it, let $\sigma(\alpha)$ be the corresponding winning strategy in $J_u(A_0, A_1) \restriction_\alpha$ obtained by fixing the α-moves. The set $B_\sigma \subseteq \omega^\omega$ defined by $\alpha \in B_\sigma \leftrightarrow (\sigma(\alpha), \alpha) \in C_u(A_0, A_1)$ is then in $\check{\Gamma}_u$, and by (ii) of 2.7, separates A_0 from A_1.

(i) Assume $A \in \Gamma_u$ and no $\check{\Gamma}_u$ set separates A_0 from A_1. Then by the previous discussion, II wins $J_u(A_0, A_1)$. And by forgetting the β-moves in this game, II has a strategy in the game $G(H_u; A_0, A_1)$ which satisfies by (i) of 2.7 $\varepsilon \in H_u \rightarrow \alpha \in A_0$ and $\varepsilon \notin H_u \rightarrow \alpha \in A_1$, i.e. is winning in $G(H_u; A_0, A_1)$. And as $A \in \Gamma_u$ and H_u is Γ_u-strategically complete, II also has a winning strategy in $G(A, H_u)$. Composing his strategies gives a winning strategy for II in $G(A; A_0, A_1)$, and (i) is proved.

(ii) Assume now $A \notin \check{\Gamma}_u$, and let for $n \in \omega$ $A(n) = \{\alpha \in \omega^\omega \; n^\frown\alpha \in A\}$. By Proposition 1.7, one of the $A(n)$'s must satisfy $A(n) \notin \check{\Gamma}_u$. Let n_0 be the least such n. Let also $B \in \check{\Gamma}_u$ separate A_0 from A_1. Applying case (i) to $\check{B}$ and the pair $(A(n_0), \check{A}(n_0))$, we get that player II has a winning strategy τ in $G(\check{B}, A(n_0))$. Let then I play first n_0, and then follow the strategy τ against II's play. At the end, one gets $n_0^\frown\alpha$ and β, and $n_0^\frown\alpha \in A \leftrightarrow \alpha \in A(n_0) \leftrightarrow \beta \notin B$, so that $\beta \in A_0 \to n_0^\frown\alpha \notin A$ and $\beta \in A_1 \to n_0^\frown\alpha \in A$ and this strategy is winning for I in $G(A; A_0, A_1)$, and (ii) is proved. And the final statement immediately follows from (i) and (ii). ⊣

Recall that we associated to each class Γ a class Γ^+ by $A \in \Gamma^+ \leftrightarrow A = (B \cap D) \cup (C \backslash D)$ for some $D \in \mathbf{\Delta}^0_1$, $B \in \Gamma$ and C in $\check{\Gamma}$.

THEOREM 3. *Let $u \in D_2$, A_0, A_1 two disjoint $\mathbf{\Sigma}^1_1$ sets in ω^ω and A a set in $\Gamma^+_u \backslash (\Gamma_u \cup \check{\Gamma}_u)$. The game $G(A; A_0, A_1)$ is determined.*

Proof. For each $s \in \omega^{<\omega}$, let $A(s) = \{\alpha : s^\frown\alpha \in A\}$. Let $T_A = \{s \in \omega^{<\omega} : A(s) \in \Gamma^+_u - (\Gamma_u \cup \check{\Gamma}_u)\}$. Clearly T_A is a tree, $\emptyset \in T_A$, and as $A \in \Gamma^+_u, T_A$ is well founded. Let $T_{A_0,A_1} = \{t \in \omega^{<\omega}$: no $\Gamma_u \cup \check{\Gamma}_u$ set separates $A_0(t)$ from $A_1(t)\}$. Again T_{A_0,A_1} is a tree, which now may be empty or not well-founded. Let G^* be the game where I and II play integers, and I loses if he gets off T_A before II gets off T_{A_0,A_1}. (So if in particular T_{A_0,A_1} is empty, I wins before the game starts). As T_A is well-founded, G^* is clopen. We claim that whoever wins G^* also wins $G(A; A_0, A_1)$:

Case (a): I has a winning strategy in G^*. Let him play it in $G(A; A_0, A_1)$. Then a position (s, t) must be reached such that $s \in T_A$ but $t \notin T_{A_0,A_1}$ (we use $\emptyset \in T_A$ here). This means that $A_0(t)$ is separable from $A_1(t)$ by some set in $\Gamma_u \cup \check{\Gamma}_u$, say $\check{\Gamma}_u$ to be specific. As $A(s) \notin \check{\Gamma}_u$, Player I has a winning strategy in $G(A(s);\ A_0(t),\ A_1(t))$ by Theorem 2, and switching to it is clearly winning in $G(A; A_0, A_1)$.

Case (b) is similar: By playing his winning strategy in G^*, II reaches a position (s, t) with $t \in T_{A_0,A_1}$, but any extension of s gets off T_A. Let then I play n_0. The set $B = \{\alpha \in A(s) \; \alpha(0) = n_0\}$ is in $\Gamma_u \cup \check{\Gamma}_u$, say in Γ_u to be specific, and $A_0(t)$ cannot be separated from $A_1(t)$ by a set in $\check{\Gamma}_u$, hence II has a winning strategy in $G(B; A_0(t), A_1(t))$ by Theorem 2, and switching to it is clearly winning in $G(A; A_0, A_1)$. ⊣

For u_n a sequence of type 2 descriptions with $\Gamma_{u_n} < \Gamma_{u_{n+1}}$, we defined $\langle\Gamma_{u_n}\rangle^+ = (\cup_n \Gamma_{u_n})^+$. The next result is entirely similar to the preceding one, and we omit the proof.

THEOREM 4. *If (u_n) is a sequence in D_2 with $\Gamma_{u_n} < \Gamma_{u_{n+1}}$ for all n, A_0, A_1 are two disjoint $\mathbf{\Sigma}^1_1$ sets in ω^ω and $A \subseteq \omega^\omega$ is a set in $\langle\Gamma_{u_n}\rangle^+ \backslash \cup_n \Gamma_{u_n}$, the game $G(A; A_0, A_1)$ is determined.*

The last step in the proof of Theorem 1 is to show that Theorems 2, 3, 4 cover all possible cases. And to do this, the last ingredient is the following precise version of Martin's result on the well-foundedness of Wadge's ordering.

THEOREM 5. (Martin). *Let $(A_n)_{n\in\omega}$ be a sequence of Borel sets in ω^ω. Then Player I cannot at the same time have a winning strategy in all games $G(A_n, A_{n+1})$ and $G(A_n, \check{A}_{n+1})$.*

Proof. By contradiction. Let σ_n^0, resp σ_n^1 be winning for I in $G(A_n, A_{n+1}^0)$, resp $G(A_n, A_{n+1}^1)$ where by definition $A_{n+1}^0 = A_{n+1}$, $A_{n+1}^1 = \check{A}_{n+1}$. Associate to each $\varepsilon \in 2^\omega$ a sequence α_n^ε by induction by $\alpha_n^\varepsilon(0) = \sigma_n^{\varepsilon(n)}(\emptyset)$, and $\alpha_n^\varepsilon(k) = \sigma_n^{\varepsilon(n)}(\alpha_{n+1}^\varepsilon \restriction k)$. This sequence is clearly obtained continuously in ε, as $\alpha_n^\varepsilon \restriction k$ depends on $\varepsilon \restriction n+k$. Moreover α_n^ε depends only on the values of ε for $m \geq n$, and for all n the pair $(\alpha_n^\varepsilon, \alpha_{n+1}^\varepsilon)$ is a run in $G(A_n, A_{n+1}^{\varepsilon(n)})$ where I follows $\sigma_n^{\varepsilon(n)}$, so $\alpha_n^\varepsilon \in A_n \leftrightarrow \alpha_{n+1}^\varepsilon \notin A_{n+1}^{\varepsilon(n)}$. Let $B = \{\varepsilon : \alpha_0^\varepsilon \in A_0\}$. The set B is Borel. On the other hand if $s \in 2^{<\omega}$, neither B nor $\check{B}$ is comeager on $N_s = \{\varepsilon : s \subseteq \varepsilon\}$, for if $\varepsilon \in N_s$ and $\bar{\varepsilon}$ is defined by $\bar{\varepsilon}(p) = \varepsilon(p)$ for $p \neq p_0 = lh(s)$, and $\bar{\varepsilon}(p_0) = 1 - \varepsilon(p_0)$, then $\bar{\varepsilon} \in N_s$, and $\alpha_p^{\bar{\varepsilon}} = \alpha_p^\varepsilon$ for $p > p_0$, and for $p \leq p_0$ $\alpha_p^{\bar{\varepsilon}} \in A_p \leftrightarrow \alpha_p^\varepsilon \notin A_p$, so that in particular $\alpha_0^{\bar{\varepsilon}} \in B \leftrightarrow \alpha_0^\varepsilon \notin B$. This shows B does not possess the Baire property, a contradiction which finishes the proof. ⊣

Recall that we defined, for $u, u' \in D_2$, $\Gamma_u < \Gamma_{u'}$ if $\Gamma_u \subseteq \Delta(\Gamma_{u'})$.

COROLLARY 6. *The relation $<$ is well-founded on the set $W_2 = \{\Gamma_u : u \in D_2\}$.*

Proof. If not, let $(u_n) \in D_2$ be such that (Γ_{u_n}) is a $<$-decreasing sequence, and A_n any set in $\Gamma_{u_n} \setminus \check{\Gamma}_{u_n}$. By Theorem 2, player I wins all games $G(A_n, A_{n+1})$ and $G(A_n, \check{A}_{n+1})$, contradicting Martin's Theorem 5. ⊣

Proof of Theorem 1. We argue by contradiction. The set $W_1 = \{\Gamma_u, \check{\Gamma}_u : u \in D_1\}$ is a subset of W_2 (Theorem 1.8), hence is well-founded for $<$ by Corollary 6 above, and is cofinal in $\mathbf{\Delta}_1^1$. So if Theorem 1 fails, we can find a $<$-minimal class Γ in W_1 and sets $A \in \Gamma$, A_1 and A_1 disjoint $\mathbf{\Sigma}_1^1$ sets in ω^ω such that $G(A; A_0, A_1)$ is not determined. Now $\Gamma = \Gamma_u$ or $\check{\Gamma}_u$ for some $u \in D_1$, hence $\Gamma = \Gamma_v$ for some $v \in D_2$.

Case (a): $A \in \Gamma \setminus \check{\Gamma}$. Then $A \in \Gamma_v \setminus \check{\Gamma}_v$, and by Theorem 2 $G(A; A_0, A_1)$ is determined, a contradiction. So $A \in \Delta(\Gamma)$, and we can apply Theorem 1.4. Note that Case 1.4 (iii) (i.e. $\Delta(\Gamma) = \cup_{\xi<\omega_1} \Gamma_{u_\xi}$ for some $u_\xi \in D_1$ is impossible by $<$-minimality of Γ.

Case (b): For some $u^* \in D_1$, $\Delta(\Gamma) = (\Gamma_{u^*})^+$. Then by $<$-minimality of Γ, $A \in \Gamma_{u^*}^+ \setminus (\Gamma_{u^*} \cup \check{\Gamma}_{u^*})$, hence for some $v \in D_2$ $A \in \Gamma_v^+ \setminus (\Gamma_v \cup \check{\Gamma}_u)$ and $G(A; A_0, A_1)$ is determined by Theorem 3, a contradiction.

The last case where $\Delta(\Gamma) = \langle \Gamma_{u_n} \rangle^+$ for some (u_n) in D_1 is handled similarly, using Theorem 4. ⊣

Let us finish this section with a brief discussion of some other results which are by-products of the preceding proof.

1. We get from the preceding proof that $W = W_1 = W_2$.

2. The sets H_u, $u \in D_2$ we constructed are in $\Gamma_u(2^\omega)$. It follows that (a) If Γ is a non-self dual Wadge class, so is $\Gamma(2^\omega)$; and (b) $\Gamma = \{f^{-1}(A) : A \subseteq 2^\omega,\ A \in \Gamma(2^\omega)\}$.

3. For Γ a class and η a countable ordinal, we can always define its η-expansion Γ^η as the class of $f^{-1}(A)$, $A \in \Gamma$ and $f : \omega^\omega \to \omega^\omega$ of Baire class η. Then one gets $(\Gamma_u)^\eta = \Gamma_{u^\eta}$ for all $u \in D_2$: inclusion $\subseteq$ is easy (Proposition 1.10), and in the other direction, the set H_{u^η} which generates Γ_{u^η} is in $(\Gamma_u)^\eta$ by its very definition. Using this and the equality $W = W_2$, one gets that the class W of non-self dual Borel Wadge classes is the least family of classes containing $\{\emptyset\}$ and closed under complementations, S_1, and η-expansions for all $\eta < \omega_1$.

4. The arguments we gave for Borel Wadge classes and Borel Wadge games on ω^ω easily translate to the case of 2^ω. In particular one gets that the non-self dual Borel Wadge classes on 2^ω are exactly the $\Gamma_u(2^\omega)$ for $u \in D_2$. One also gets the determinacy of the games $G(A; A_0, A_1)$ played on 2^ω, by noticing that the H_u's are strategically complete in $\Gamma_u(2^\omega)$, and that Theorem 1.4 holds for 2^ω too. There is however a slight difference between the two cases: For 2^ω, case (ii) in 1.4. trivializes, as $\langle \Gamma_n \rangle^+(2^\omega) = \cup_n \Gamma_n(2^\omega)$ by compactness. This also shows that fact 2 above fails for self-dual classes.

Let us make now some comments on how further information can be obtained from the existence of the "unfolded" games Ju:

5. It is clear, by looking at their definition, that the games $J_u(A_0, A_1)$ are defined uniformly in A_0 and A_1. In fact if one codes the pairs of $\mathbf{\Sigma}^1_1$ sets reasonably – e.g. by coding pairs of associated trees on $\omega \times \omega$, one easily checks that for each $u \in D_2$ the tree of the game $J_u(A_0, A_1)$ is continuous in the codes. Using this uniformity, one easily gets the following result, where for $A \subseteq \omega^\omega \times \omega^\omega$ and $x \in \omega^\omega$, A_x denotes the section of A at x:

THEOREM. *Let Γ be some Borel Wadge class.*

(i) *If A_0, A_1 are $\mathbf{\Sigma}^1_1$ sets in $\omega^\omega \times \omega^\omega$*

$$B_\Gamma = \{x : (A_0)_x \text{ is separable from } (A_1)_x \text{ by a } \Gamma - \text{set}\}$$

is $\mathbf{\Pi}^1_1$, hence in particular for Borel B, $\{x : B_x \in \Gamma\}$ is $\mathbf{\Pi}^1_1$

(b) *If* A_0, A_1 *are* $\mathbf{\Sigma}_1^1$ *in* $\omega^\omega \times \omega^\omega$ *and for all* x $(A_0)_x$ *can be separated from* $(A_1)_x$ *by some* Γ *set, there is a Borel set* B *with sections in* Γ *which separates* A_0 *from* A_1.

To prove this, one can easily reduce the case where Γ is self dual to the non-self dual case, say $\Gamma = \Gamma_u$ for some $u \in D_2$. One gets then a continuous map: $x \mapsto J_{\check{u}}((A_0)_x, (A_1)_x) = J_x$, and by Theorem 2, $x \in B_\Gamma \leftrightarrow$ I wins $J_x \leftrightarrow$ II does not win J_x and this last statement is $\mathbf{\Pi}_1^1$. This gives (a). By general standard facts, (b) is a consequence of (a). One can also use that if $B_\Gamma = \omega^\omega$, one can find by the selection theorem for strategies in open games a Borel function $x \mapsto \sigma_x$, with σ_x winning for I in J_x. And by Theorem 2 again $B = \{(x, \alpha) : (\sigma_x(\alpha), \alpha) \in C_{\check{u}}((A_0)_x, (A_1)_x)\}$ is a set which separates A_0 from A_1, and is Borel with Γ_u sections.

6. We now discuss the uniformity of our constructions in u. This leads to lightface versions of our results.

First by encoding countable ordinals by reals, one gets a coding of descriptions. Let us say that a description u (in D_1 or D_2) is HYP if some code of it is Δ_1^1, and that Γ is a HYP_1 (a HYP_2) class if $\Gamma = \Gamma_u$ or $\check{\Gamma}_u$ for some HYP u in D_1 ($\Gamma = \Gamma_u$ for some HYP u in D_2). The proof of 1.8 easily gives $\text{HYP}_1 \subseteq \text{HYP}_2$. Let also W_{HYP} be the family of non-self dual Wadge classes which are generated by a Δ_1^1 set. One also checks easily that $\text{HYP}_2 \subseteq W_{\text{HYP}}$.

Fix now some HYP description u (in D_1 or D_2). One can then define the notion of a HYP-in-Γ_u subset of ω^ω: Intuitively, these are the sets in Γ_u which admit a HYP construction as a Γ_u set. Formally, one has to go through codes. This is done precisely for first type descriptions in the second part of Louveau's paper [1], and it is proved there that for u HYP in D_1, $\Gamma_u \cap \Delta_1^1 =$ HYP-in-Γ_u.

Our games allow to prove a similar result for $u \in D_2$: The point is that for recursive ordinal η, one can choose the ramification $(r^\eta, \rho^\eta, F^\eta)$ so that all functions are Δ_1^1-recursive, and then it is not hard to check that for HYP $u \in D_2$,

(a) $H_u \in \Delta_1^1$;

(b) for A_0, A_1 Σ_1^1 sets, $J_u(A_0, A_1)$ is Δ_1^1, and $C_u(A_0, A_1)$ is HYP-in-$\check{\Gamma}_u$. Using this, one gets $\Gamma_u \cap \Delta_1^1 =$ HYP-in-Γ_u for u HYP in D_2: For if A is Δ_1^1 in Γ_u, Player I wins J_u $(A, \check{A})$ by Theorem 2, and this game is open Δ_1^1 for I, hence I has a Δ_1^1 winning strategy σ. But then one gets $A = \{\alpha : (\sigma(\alpha), \alpha) \in C_{\check{u}}(A, \check{A})\}$, which proves that A is HYP-in-Γ_u.

The second step is to get a "lightface" analog of Theorem 1.4. First one easily defines the notions of HYP-in-Γ_u^+ and HYP-in $\langle \Gamma_{u_n} \rangle^+$ for u HYP in D_1, and $\langle u_n \rangle$ a Δ_1^1 sequence of HYP descriptions in D_1. Then one can prove the analog of 1.4, analyzing the family $\Delta_1^1 \cap \Delta(\Gamma_u) =$

(HYP-in-Γ_u) $\cap$ (HYP-in-$\check{\Gamma}_u$) in terms of HYP-in-$\Gamma_{u'}$ classes, for u''s HYP in D_1 with $\Gamma_{u'} < \Gamma_u$. The proof essentially follows the proof of 1.4 in [1], and although a bit tedious, necessitates no new ideas.

By combining the two previous steps, one finally gets

THEOREM (i) *every* W_{HYP} *class* Γ *admits a* D_1 *(and hence a* D_2*)* HYP *description - as* $\Gamma_u, \check{\Gamma}_u, (\Gamma_u)^+$ *or* $\langle \Gamma_{u_n} \rangle^+$*, as the case may be*

(ii) *if* A *is* Δ^1_1 *and in a* HYP *class* Γ*, it is* HYP*-in-*Γ. *Hence in particular the Wadge class* $\Gamma(A)$ *of* A *is a* HYP *class, and* A *is HYP-in-*$\Gamma(A)$.

Finally notice that when considering a Borel Wadge game $G(A, B)$ with Δ^1_1 sets A, B, the winning strategies in this game are recursively obtained, using Theorems 2, 3, 4, from winning strategies for player II in associated closed games depending on the Wadge class $\Gamma(A)$, so by the previous result which can be chosen Δ^1_1. One then finally gets

THEOREM. *If* A, B *are* Δ^1_1 *in* ω^ω*, one of the two players in* $G(A, B)$ *has a winning strategy which is recursive in Kleene's* $\mathcal{O}$.

Note that this result is the best possible along these lines, as by considering $A = \omega^\omega$ and B some non empty Π^0_1 set with no Δ^1_1 members, we see that there may be no HYP winning strategy.

7. A final word on the games J_u: The fact that the β-part of II's play in J_u are in ω^ω is not an essential feature, and one can define similar games with $\beta \in \kappa^\omega$ for κ some infinite cardinal. Such extensions were used in [2] to study separation of κ-Suslin sets by $\mathbf{\Sigma}^0_\xi$-sets, and separation of lightface projective sets by $\mathbf{\Sigma}^0_\xi$ sets, using strong set-theoretic hypotheses. Similar results could be obtained, along the same lines, for all Borel Wadge classes.

Section 4. Wadge Classes in Metric Separable Spaces. Unless the underlying space is 2^ω or ω^ω, there is no clear notion of Wadge game available, and to extend Theorem 3.1 to more general situations, we must first rephrase (and weaken) it a bit. Let Γ be a Borel Wadge class in ω^ω, and A_0, A_1 a pair of disjoint sets in some metric separable space E. We say that (A_0, A_1) *reduces* Γ if for any $B \subseteq \omega^\omega$ in Γ, there is a continuous $f : \omega^\omega \to E$ with $f(\omega^\omega) \subseteq A_0 \cup A_1$ and $f^{-1}(A_0) = B$. Note that this property is intrinsic, i.e. depends only on $A_0 \cup A_1$. We say A reduces Γ if $(A, \check{A})$ does.

Let us say, for A_0, A_1 in ω^ω, that Γ separates (A_0, A_1) if A_0 is separable from A_1 by some B in Γ. With this terminology, Theorem 3.1. gives:

THEOREM 0.

(a) *If* Γ *is a Borel Wadge class in* ω^ω *and* $A \subseteq \omega^\omega$ *is Borel and does not reduce* Γ, *then* $A \in \check{\Gamma}$.

(b) *If* Γ *is a Borel Wadge class in* ω^ω, A_0, A_1 *are pairwise disjoint* $\mathbf{\Sigma}^1_1$ *sets in* ω^ω *and* (A_0, A_1) *does not reduce* Γ, *then* $\check{\Gamma}$ *separates* (A_0, A_1).

Part (a) is usually called "Wadge's lemma". It is of course a consequence of part (b), but we stated the two versions, for their fate will be slightly different in the sequel.

The main point in order to extend Theorem 0 to other spaces is to define, for each Wadge class Γ on ω^ω, a corresponding class $\Gamma(E)$ of sets in E. This can be done in various ways, and most of the work will consist in showing that the various possible definitions lead to the same classes.

From now on, we will consider only classes Γ in W, i.e. non-self-dual Borel Wadge Classes. In the first section, we briefly looked at the case of subsets E of ω^ω, where one can define $\Gamma(E)$ using traces. As any dim 0 metric separable space is homeomorphic to such a subset of ω^ω, this gives one way of defining $\Gamma(E)$ for dim 0 space E. Another way is to consider continuous preimages, still another way to consider Hausdorff operations performed on open sets in E. And there is a more subtle possibility, coming from Wadge's lemma: to define the class $\Gamma(E)$ by the properties of its continuous preimages. The next result shows the equivalence of all these possible definitions, at least when E is a dim 0 *Suslin* space, i.e. is absolute $\mathbf{\Sigma}^1_1$ metrizable separable.

THEOREM 1. *Let* E *be a dim* 0 *Suslin space,* $\Gamma \in W$, *and* A *a subset of* E. *The following are equivalent*

1. $\exists j$ *1-1 embedding:* $E \to \omega^\omega$ *and* $B \in \Gamma$ $A = j^{-1}(B)$
2. $\forall j$ *1-1 embedding* $E \to \omega^\omega$ $\exists B \in \Gamma$ $A = j^{-1}(B)$
3. $\exists f$ *continuous* $E \to \omega^\omega$ $\exists B \in \Gamma$ $A = f^{-1}(B)$
4. $\exists D$ *Hausdorff operation* $\exists (U_n)$ *open in* E $(\Gamma = D(\mathbf{\Sigma}^0_1)$ *and* $A = D((U_n))$
5. $\forall D$ *Hausdorff operation with* $\Gamma = D(\mathbf{\Sigma}^0_1)$ $\exists (U_n)$ *open in* E *with* $A = D((U_n))$
6. $\forall g$ *continuous:* $\omega^\omega \to E$, $g^{-1}(A) \in \Gamma$.

We then define $\Gamma(E)$ as the class of sets in E which satisfy one of these equivalent properties.

Proof. $1 \Rightarrow 3$ is trivial, and $3 \Rightarrow 5$ and $4 \Rightarrow 6$ come from preservation of Hausdorff operations on $\mathbf{\Sigma}^0_1$ sets by continuous preimages. $2 \Rightarrow 1$ comes from the existence of an embedding from E

into ω^ω, i.e. the fact that E is dim 0. $5 \Rightarrow 4$ comes from the existence of a Hausdorff operation generating Γ, i.e. the fact that Γ is non-self dual. It remains to prove $6 \Rightarrow 2$. So let $j : E \to \omega^\omega$ be an embedding, and let A be such that for all continuous $g : \omega^\omega \to E$ $g^{-1}(A) \in \Gamma$. As E is Suslin, let $\pi : \omega^\omega \twoheadrightarrow E$ be a continuous surjection. Then $\pi^{-1}(A)$ is in Γ, so is Borel, and $A_0 = j(A)$ and $A_1 = j(E \setminus A)$ are both Σ^1_1 in ω^ω. Applying Theorem 0 to $\check{\Gamma}$ and (A_0, A_1), one gets that Γ separates (A_0, A_1), else A_0, A_1 would reduce $\check{\Gamma}$ and for some $f : \omega^\omega \to E$ $f^{-1}(A)$ would be the complete $\check{\Gamma}$-set in ω^ω. But if $B \in \Gamma$ separates A_0 from A_1, $A = j^{-1}(B)$ as desired. ⊣

With the previous definition, it is clear that if $f : E \to F$ is continuous, for E, F dim 0 Suslin spaces, and $A \in \Gamma(E)$, then $f^{-1}(A) \in \Gamma(F)$. And one immediately gets, by applying Theorem 1 to $A_0 \cup A_1$, the following extension of Theorem 0.

THEOREM 2. *Let E be a dim 0 Suslin space, $\Gamma \in W$ and A_0, A_1 two disjoint Σ^1_1 subsets of E. If the pair (A_0, A_1) does not reduce Γ, the class $\check{\Gamma}(E)$ separates (A_0, A_1).*

We now study the general case, where E is still Suslin but not necessarily 0-dimensional. There are clearly difficulties: There are no more embeddings in ω^ω, and the only continuous functions into ω^ω may be the constants, so that definitions by (1) (2) (3) of Theorem 1 cannot be used.

The first remaining possibility is to use descriptions and the corresponding operations, but even this approach has to be changed a bit for descriptions u with $u(0) = 1$, as there might not be enough families of pairwise disjoint open sets in E. So we adopt the following definition.

DEFINITION 3. Let E be metric separable. For each $u \in D_1$ the class Γ^E_u is defined (together with $\check{\Gamma}^E_u$, where ˇ refers to complementation inside E) by the following:

(a) If $u(0) = 0$ $\Gamma^E_u = \{\emptyset\}$

(b) If $u = \xi^\frown 1^\frown \eta^\frown u^*$ with $u^*(0) = 0$ $\Gamma^E_u = D_\eta(\Sigma^0_\xi(E))$

(c) If $u = \xi^\frown 2^\frown \eta^\frown u^*$ $\Gamma^E_u = \mathrm{Sep}(D_\eta(\Sigma^0_\xi), \Gamma^E_{u^*})$

(d) If $\xi^\frown 3^\frown \eta^\frown \langle u_0, u_1 \rangle$ and $\xi \geq 2$, $\Gamma^E_u = \mathrm{Bisep}(D_\eta(\Sigma^0_\xi), \Gamma^E_{u_0}, \Gamma^E_{u_1})$

(d′) If $u = 1^\frown 3^\frown \eta^\frown \langle v_0, v_1 \rangle$, $A \subseteq E$ is in Γ^E_u if there are $D_\eta(\Sigma^0_\xi)$ sets C_0 and C_1 in E, and $B \in \Gamma^E_{u_1}$, such that $A \cap C_0 \in \Gamma^E_{u_0}$, $A \cap C_1 \in \check{\Gamma}^E_{u_0}$, and $A \setminus (C_0 \cup C_1) = B \setminus (C_0 \cup C_1)$

(e) If $u = \xi^\frown 4^\frown \langle u_n : n \in \omega \rangle$ and $\xi \geq 2$ $\Gamma^E_u = SU(\Sigma^0_\xi, (\Gamma^E_{u_n}))$

(e′) If $u = 1^\frown 4^\frown \langle u_n,\ n \in \omega \rangle$, $A \subseteq E$ is in Γ^E_u if every point $x \in A$ admits a nbhd V with $A \cap V \in \cup_n \Gamma^E_{u_n}$

(f) If $u = \xi^\frown 5^\frown \eta^\frown \langle u_0, u_1\rangle$, $\Gamma_u^E = SD_\eta(\Sigma_\xi^0, \Gamma_{u_0}^E, \Gamma_{u_1}^E)$.

Note that in case E has dimension 0, one has the reduction property for open sets (it is actually equivalent), so that (d′) and (e′) above correspond then to the usual definition, i.e. $\Gamma_u^E = \Gamma_u(E)$ when E is a dim 0 Suslin space.

Note also that although our definitions d′ and e′ look natural, it is now unclear that they correspond to some Hausdorff operation on open sets – and in fact they do not – and it is also unclear that if u and v describe the same class Γ in ω^ω, they also describe the same class in any E. This will be true for Suslin E by Theorem 5 below.

Another possible way for defining Γ in a space E comes from (6) of Theorem 1.

DEFINITION 4. Let E be metric separable, and $\Gamma \in W$. We define $\Gamma(E)$ as the family of those sets $A \subseteq E$ such that for every continuous $f : \omega^\omega \to E$, $f^{-1}(A) \in \Gamma$.

Note that in Definition 4, one could replace "ω^ω" by "all dim 0 Suslin spaces", by using Theorem 1.

THEOREM 5. *Let E be a Suslin space, and u a first type description. Then $\Gamma_u^E = \Gamma_u(E)$. Hence part (a) of Theorem 0 extends to arbitrary Suslin spaces, i.e. if $A \subseteq E$ is Borel and A does not reduce Γ_u, $A \in \check{\Gamma}_u^E$.*

The second statement follows from the first, as before. But the proof of the first statement is more involved, and uses a transfer method via open maps.

Recall that $f : E \to F$ is *open* if $f(U)$ is open in F for all U in E. It follows from a result of Saint-Raymond [4] that if $f : E \twoheadrightarrow F$ is a continuous open surjection and E is Polish, then F is Polish too. Conversely, one has the following classical fact:

PROPOSITION 6. *Let E be a (non-empty) Polish space. Then there exists a continuous open surjection $\pi : \omega^\omega \twoheadrightarrow E$.*

Proof. Let D be a metric for which E is complete, and construct by induction on $lh(s)$, $s \in \omega^{<\omega}$, non-empty open sets U_s in E with diam$U_s < 2^{-lhs}$, $\bar{U}_{s^\frown n} \subseteq U_s$, and $U_\emptyset = E, U_s = \cup_n U_{s^\frown n}$. This is easily done. For each $\alpha \in \omega^\omega$ $\cap_{s \subseteq \alpha} U_s$ reduces to a singleton $\{\pi(\alpha)\}$, and $\pi : \omega^\omega \to E$ defined this way clearly works. ⊣

The next result is a variant of a selection theorem of Saint-Raymond ([5], Section 4).

THEOREM 7. *Let E, F be Polish spaces, π a continuous open surjection $E \twoheadrightarrow F$, and f a Baire class 1 function: $E \to X$ for some Polish X. Then there exists a* *selection $s : F \to E$ of π (i.e. $\pi \circ s = \mathrm{id}_F$) such that $f \circ s : F \to X$ is a Baire class 1 function.*

Proof. Let d_E, d_X be complete metrics on E, X respectively. Say that a function S from F into the non-empty closed subsets of E is lsc if for each open U in E $\{x \in F : S(x) \cap U \neq \emptyset\}$ is open in F. For example $x \mapsto \pi^{-1}(x)$ is lsc, as π is open. We construct by induction a sequence S_n of applications from F into the non-empty closed subsets of E such that

(i) $S_0(x) = \pi^{-1}(x)$, $x \in F$.

(ii) $S_{n+1}(x) \subseteq S_n(x)$

(iii) For each n, there is a countable partition $(F^n_k)_{k\in\omega}$ of F in $\mathbf{\Delta}^0_2$ sets such that $S_n \restriction_{F^n_k}$ is lsc for all k, and if x, y are in F^n_k, $\alpha \in S_n(x)$, $\beta \in S_n(y)$, then $d_E(\alpha, \beta) < \frac{1}{n}$ and $d_x(f(\alpha), f(\beta)) < \frac{1}{n}$.

Case $n = 0$ is immediate, by taking $F^0_k = F$. Suppose $(F^n_k)_{k\in\omega}$ and S_n have been defined. Fix k, and construct by transfinite induction a sequence (T_ξ) of closed subsets of F^n_k by letting $T_0 = F^n_k$, $T_\lambda = \cap_{\xi<\lambda} T_\xi$ for limit λ. And if $T_\xi \neq \emptyset$, let $H_\xi = \overline{\cup_{x\in T_\xi} S_n(x)}$, and as f is Baire class 1, let U_ξ be open, of diameter $< \frac{1}{n+1}$, with $U_\xi \cap H_\xi \neq \emptyset$, and such that $\mathrm{diam}_X(f(U_\xi \cap H_\xi)) < \frac{1}{n+1}$. Let then $T_{\xi+1} = T_\xi \backslash \{x\ S_n(x) \cap U_\xi \neq \emptyset\}$. As F is separable, there is a countable η such that $T_\eta = \emptyset$, and the $(T_\xi \backslash T_{\xi+1})_{\xi<\eta}$ form a partition of F^n_k into $\mathbf{\Delta}^0_2$ sets. And if one defines $S_{n+1}(x)$, for $x \in T_\xi \backslash T_{\xi+1}$, by $\overline{S_n(x) \cap U_\xi}$, S_{n+1} is lsc on $T_\xi \backslash T_{\xi+1}$; so by rearranging the $T_\xi \backslash T_{\xi+1}$ for all F^n_k's, we get S_{n+1} and $(F^{n+1}_k)_{k\in\omega}$ as desired. Now $\cap_n S_n(x)$ is a singleton $\{s(x)\}$ for each $x \in F$, and this clearly defines a selection $s : F \to E$ of π. It remains to show that $f \circ s$ is a first class function. Choose for all n, k a point $\alpha_{n,k}$ in the set $\cup\{S_n(x) : x \in F^n_k\}$, and define $g_n : F \to X$ by $g_n(x) = f(\alpha_{n,k})$ if $x \in F^n_k$. As the partition of F is in $\mathbf{\Delta}^0_2$ sets, g_n is a Baire class 1 function. And by condition (iii), $f \circ s$ is the uniform limit of the g_n's, hence is Baire class 1 too. ⊣

We now state the transfer result from which Theorem 5 will follow.

THEOREM 8. *Let E be a dim 0 Polish space, F a Polish space, f a continuous open surjection: $E \twoheadrightarrow F$, $u \in D_1$, and A_0, A_1 two $\mathbf{\Sigma}^1_1$ sets in F*

(a) *If $\Gamma_u(E)$ separates $(f^{-1}(A_0), f^{-1}(A_1))$, then the set A_0 is in $\Gamma_u^{A_0\cup A_1}$*

(b) *If either (i) $u(0) \geq 2$, <u>or</u> (ii) F is dim 0, <u>or</u> (iii) $A_1 = \check{A}_0$, then if $\Gamma_u(E)$ separates $(f^{-1}(A_0), f^{-1}(A_1))$, Γ^F_u separates (A_0, A_1).*

Proof. Part b(iii) is of course a particular case of part (a). Assume first that $u(0) \geq 2$. One easily defines, by induction on u, a Hausdorff operation D such that $\Gamma_u, \Gamma_u(E), \Gamma^F_u$ are respectively $D(\mathbf{\Sigma}^0_2)$, $D(\mathbf{\Sigma}^0_2(E))$ and $D(\mathbf{\Sigma}^0_2(F))$: the only point is that for $\xi \geq 2$ the $D_\eta(\mathbf{\Sigma}^0_\xi)$ sets do have reduction in any Polish space F. Let then C_n be a sequence of $\mathbf{\Sigma}^0_2(E)$ sets such that

$C = D((C_n))$ separates $f^{-1}(A_0)$ from $f^{-1}(A_1)$, and let $g : E \to [0,1]^\omega$ be a Baire class 1 function with $C_n = \{x \in E : g_n(x) > 0\}$. Applying Theorem 7 to $f : E \to F$ and $g : E \to [0,1]^\omega$ gives a selection $s : F \to E$ of f with $g \circ s$ of Baire class 1. Let $B_n = \{x \in F : g_n \circ s(x) > 0\}$ and $B = D((B_n))$. The set $B \in \Gamma_u^F$, and as $B = s^{-1}(C)$, it separates A_0 from A_1. This proves both (a) and b(i). So it remains to study the case $u(0) = 1$ (We forget about $\{\emptyset\}$!). It is done by inspection of the various cases.

1. $\Gamma_u = D_\eta(\Sigma_1^0)$. If $C \in \Gamma_u$ separates $f^{-1}(A_0)$ from $f^{-1}(A_1)$, let $(C_\theta)_{\theta<\eta}$ be an increasing sequence of open sets in E building C. The sequence $f(C_\theta)$ is an increasing sequence of open sets in F which build some set $B \in D_\eta(\Sigma_1^0(F))$, and clearly separates A_0 from A_1.

2. $\Gamma_u = \mathrm{Sep}(D_\eta(\Sigma_1^0), \Gamma_{u^*})$ with $u^*(0) > 1$. Let $C \in D_\eta(\Sigma_1^0)$, built say from $(C_\theta)_{\theta<\eta}$, be such that $\check{\Gamma}_{u^*}(E)$ separates the pair $(f^{-1}(A_0) \cap C,\ f^{-1}(A_1) \cap C)$, and $\Gamma_{u^*}(E)$ separates the pair $(f^{-1}(A_0)\backslash C,\ f^{-1}(A_1)\backslash C)$. Let $D_\theta = f(C_\theta)$, and $f_\theta = f\restriction_{C_\theta}\colon C_\theta \twoheadrightarrow D_\theta$. Clearly as C_θ is open, f_θ is a continuous open surjection. And depending on the parity of θ, $\Gamma_{u^*}(C_\theta)$ or $\check{\Gamma}_{u^*}(C_\theta)$ separates the pair $(f_\theta^{-1}(A_0\backslash \cup_{\theta'<\theta} D_{\theta'}),\ f_\theta^{-1}(A_1\backslash \cup_{\theta'<\theta} D_{\theta'}))$. And as $u^*(0) \geq 2$, we can apply the first case, and the same class separates in D_θ $(A_0\backslash \cup_{\theta'<\theta} D_{\theta'},\ A_1\backslash \cup_{\theta'<\theta} D_{\theta'})$. And sticking the pieces together gives, for D built from the D_θ's, that $\check{\Gamma}_{u^*}^F$ separates $A_0 \cap D$ from $A_1 \cap D$ and $\Gamma_{u^*}^F$ separates $(A_0\backslash D, A_1\backslash D)$, as desired.

3. $\Gamma_u = \mathrm{Bisep}(D_\eta(\Sigma_1^0), \Gamma_{u_0}, \Gamma_{u_1})$, with $u_0(0) \geq 2$. At first sight, this case looks very similar to the preceding case. But in fact it is where the problem arises: The same argument as before does give two sets D_0 and D_1 in $D_\eta(\Sigma_1^0(F))$ such that $\Gamma_{u_0}^F$ separates $(A_0 \cap D_0, A_1 \cap D_0)$, $\check{\Gamma}_{u_0}^F$ separates $(A_0 \cap D_1, A_1 \cap D_1)$ and $\Gamma_{u_1}^F$ separates $A_0\backslash(D_0 \cup D_1,\ A_1\backslash(D_0 \cup D_1)$. But we cannot conclude that Γ_u^F separates A_0 from A_1 without sticking the pieces together, and this time D_0 and D_1 are not necessarily disjoint. So we can conclude only if there is no sticking to be done, for we work in $A_0 \cup A_1$ and the only possibility is A_0 – this gives part (a) in this case, or if we can reduce D_0 and D_1, which is possible if F is 0-dimensional, and this give b(ii). [We will see later that this obstruction cannot be avoided].

Case 4. $\Gamma_u = SU(\Sigma_1^0, \langle \Gamma_{u_n} \rangle)$ with $u_n(0) \geq 2$. The proof is similar to case 3: Again one gets open sets D_n in F such that $\Gamma_{u_n}^F$ separates $A_0 \cap D_n$ from $A_1 \cap D_n$, and $A_0 \subseteq \cup_n D_n$. This is enough to conclude for part (a) and for (bii). Note that in this case one can always stick the pieces together, by using locally finite refinements of the D_n's, and the closure properties of the Γ_{u_n}'s, but we won't need this.

Case 5. $\Gamma_u = SD_\eta(\Sigma_1^0, \langle \Gamma_{u_n} \rangle, \Gamma_{u^*})$ creates no difficulty, the argument being as in Case 2, and is left to the reader. ⊣

Proof of Theorem 5. We want to show that $\Gamma_u^E = \Gamma_u(E)$ for any Suslin space E. One easily checks, using Theorem 1, that $\Gamma_u^E \subseteq \Gamma_u(E)$. If now $A \in \Gamma_u(E)$, let F be Polish with $E \subseteq F$, and by Proposition 6 let $\pi : \omega^\omega \twoheadrightarrow F$ be continuous open surjection. By the definition of $\Gamma_u(E)$, $\pi^{-1}(A) \in \Gamma_u(\pi^{-1}(E))$, and we can apply Theorem 8(a) to ω^ω, F, π, and $A_0 = A, A_1 = E - A$, which gives $A \in \Gamma_u^E$. ⊣

Concerning separation of analytic sets, Theorem 8 does not give the full strength of Theorem 0, part (b), even for Polish spaces. One still gets

COROLLARY 9. *Let E be Polish, A_0, A_1 two disjoint $\mathbf{\Sigma}_1^1$ sets in E and Γ a class in W. The following are equivalent:*

(i) $A_0 \in \Gamma(A_0 \cup A_1)$.

(ii) *For any Polish* dim 0 *space F and continuous $f : F \to E$ $\Gamma(F)$ separates $(f^{-1}(A_0), f^{-1}(A_1))$.*

(iii) *There is a Polish* dim 0 *space F and an open continuous surjection $\pi : F \to E$ such that $\Gamma(F)$ separates $(\pi^{-1}(A_0),\ \pi^{-1}(A_1))$.*

In most cases, (i) above can be replaced by the stronger "$\Gamma(E)$ separates (A_0, A_1)", e.g. if dim $E = 0$, or if Γ is Γ_u for some $u \in D_1$ with $u(0) \geq 2$, or some specific Γ_u's with $u(0) = 1$, like the $D_\eta(\mathbf{\Sigma}_1^0)$'s. But the next example shows it does not work in general (and hence that in general $\Gamma_u(E)$ is not always obtained by some Hausdorff operation performed on open sets).

EXAMPLE 10. Let $E = [0,1] \times 2^\omega$. E is a dim 1 compact metrizable space. Let D_0, D_1 be two disjoint countable dense sets in 2^ω, and let $\Gamma = \text{Bisep}(\mathbf{\Sigma}_1^0, \mathbf{\Sigma}_2^0)$ and $A_0 = \{(x,\alpha) \in E : (x < 1$ and $\alpha \in D_0) \vee (x = 1$ and $\alpha \notin D_1)\}$, $A_1 = \{(x,\alpha) \in E : (x = 0$ and $\alpha \notin D_0) \vee (x > 0$ and $\alpha \in D_1)\}$. Then $A_0 \in \Gamma(A_0 \cup A_1)$, but $\Gamma(E)$ does not separate A_0 from A_1.

Proof. On $[0,1[\times 2^\omega$, $\mathbf{\Sigma}_2^0$ separates A_0, A_1, and on $]0,1] \times 2^\omega$, $\mathbf{\Pi}_2^0$ separates (A_0, A_1), so $A_0 \in \Gamma(A_0 \cup A_1)$. Suppose, towards a contradiction, that $C \in$ Bisep $(\mathbf{\Sigma}_1^0, \mathbf{\Sigma}_2^0)(E)$ separates (A_0, A_1). Let $U = \{(x,\alpha) : C$ is locally $\mathbf{\Sigma}_2^0$ at $(x,\alpha)\}$ and $V = \{(x,\alpha) : C$ is locally $\mathbf{\Pi}_2^0$ at $(x,\alpha)\}$. By assumption, $C \subseteq U \cup V$. Now for $x = 0$, $(A_0)_x = (\check{A_1})_x = D_0$ is not $\mathbf{\Pi}_2^0$, hence $\{0\} \times 2^\omega \cap V = \emptyset$, and similarly $\{1\} \times 2^\omega \cap U = \emptyset$. Fix $\alpha \in D_0$, and let $U_\alpha = \{y : (y,\alpha) \in U\}$ and $V_\alpha = \{y : (y,\alpha) \in V\}$. U_α and V_α are open non-empty by the preceding facts, and cover $[0,1]$ as $[0,1] \times \{\alpha\} \subseteq A_0$. By connectedness, $U_\alpha \cap V_\alpha \neq \emptyset$, so $\exists x \in]0,1[$ $(x,\alpha) \in U \cap V$, hence a neighborhood W of (x,α) with $W \cap C$ and $W \cap \check{C}$ $\mathbf{\Pi}_2^0$. But both $W \cap C$ and $W \cap \check{C}$ must be dense in W by choice of D_0 and D_1, contradicting the Baire Category theorem. ⊣

Section 5. Hurewicz Tests and Hurewicz-Type Results.

Let Γ be a Wadge class in W. A *Hurewicz-test* for Γ is a pair (K, H) consisting of a dim 0 metric compact space K, and a $\Gamma(K)$ subset H of K which satisfies: For every Borel set B in some (non-empty) Suslin space E, $B \notin \check{\Gamma}(E) \leftrightarrow \exists$ 1-1 continuous map $\varphi : K \to E$ with $\varphi^{-1}(B) = H$ (in other words, E contains a homeomorphic copy of the space K on which B is the corresponding copy of H).

Our terminology comes from the well known theorem of Hurewicz on the characterization of Polish spaces, which states, with our terminology, that if D is dense countable in 2^ω, $(2^\omega, D)$ is a Hurewicz test for the class Σ^0_2.

It follows from work of Steel [6] that every class Γ_u with $u(0) \geq 2$ and $D_\omega(\Sigma^0_2) \subseteq \Gamma_u$ admits a Hurewicz test - at least for Borel subsets of ω^ω. In [2], we independently reproved Steel's result by a method which yields

THEOREM 1 [2]. *Let u be a description with $u(0) \geq 2$. The class Γ_u admits a Hurewicz test (K, H), with $K = 2^\omega$.*

We extend here this result to all Borel Wadge classes $\Gamma \in W$:

THEOREM 2. *Every class Γ in W admits a Hurewicz test. Hence in particular, the membership of a Borel set $B \subseteq E$ Suslin in $\Gamma(E)$ depends only on the dim 0 compact subsets of E.*

We will prove the result by induction on descriptions. For doing this, it is easier to work with type 2 descriptions, except that one has to reduce a bit the allowed combinations:

DEFINITION 3. (i) Let (Γ_n) be a sequence of classes in W. The sequence (Γ_n) is admissible if whenever $A = (A_0 \cap C) \cup (A_1 \backslash C)$ with $C \in \Sigma^0_1$, $A_0 \in \bigcup_{n \geq 1} \Gamma_n$ and $A_1 \in \check{\Gamma}_0$, $A \in \bigcup_{n \geq 1} \Gamma_n$.

(ii) We let D'_2 be the least set of type 2 descriptions satisfying $\underline{0} \in D'_2$, $u(0)^\frown 1^\frown u \in D'_2$ if $u \in D'_2$, and $\xi^\frown 2^\frown \langle u_n \rangle \in D'_2$ if $u_n \in D'_2$ for all n and the sequence Γ_{u_n} is admissible.

PROPOSITION 4. *Every class Γ in W admits a D'_2-description.*

Proof. It is enough to check that any Γ_u, $u \in D_1$ admits such a description. In fact, one can check that the function $u \mapsto v(u)$ defined in 1.8 takes values in D'_2, by induction on $u \in D_1$. $\dashv$

Note that the only change between D'_2 and D_2 descriptions occurs for $u(0) = 1$, because by the closure properties of the classes Γ_u, $u(0) \geq 2$, any sequence Γ_{u_n} with $u_n(0) \geq 2$ is automatically admissible.

It is immediate to see that if K is a singleton and $H \subseteq K$ is $\emptyset$, $\langle K, H\rangle$ is a Hurewicz-test for $\Gamma = \{\emptyset\}$, and that if $\langle K, H\rangle$ is a Hurewicz-test for Γ, $\langle K, \check{H}\rangle$ is a Hurewicz test for $\check{\Gamma}$. So the proof of Theorem 2 follows from Theorem 1 and the following:

THEOREM 5. *Let* (Γ_n) *be an admissible sequence of classes in* W *which admit Hurewicz-tests. The class* $\Gamma = S_1(\langle \Gamma_n \rangle)$ *admits a Hurewicz-test too.*

LEMMA 6. *Let* E *be a Suslin space, and* $\Gamma = S_1(\langle \Gamma_n \rangle)$. *A Borel set* $A \subseteq E$ *is in* $\Gamma(E)$ *iff*

(*) There is a sequence $(C_n)_{n\geq 1}$ *of* $\mathbf{\Sigma}^0_1$ *sets in* E *such that for* $n \geq 1$ $A \cap C_n \in \Gamma_n(C_n)$, *and* $A \backslash \cup_{n\geq 1} C_n \in \Gamma_0(E \backslash \cup_n C_n)$.

Proof. If A satisfies (*) and $f : \omega^\omega \to E$ is continuous, $f^{-1}(A)$ satisfies (*) in ω^ω, and by reducing the $f^{-1}(C_n)$, one gets that $f^{-1}(A) \in \Gamma$ in ω^ω. As this is true for any such f, $A \in \Gamma(E)$. If now $A \in \Gamma(E)$, let $f : \omega^\omega \to F$ be a continuous open map onto some Polish space F containing E. Applying 4.8, we know that some set D in Γ separates $(f^{-1}(A),\ f^{-1}(E\backslash A))$ in ω^ω, and if $(B_n)_{n\geq 1}$ are $\mathbf{\Sigma}^0_1$ sets with $D \cap B_n \in \Gamma_n$ for $n \geq 1$, and $D \backslash \cup_{n\geq 1} B_n = D_0 \backslash \cup_{n \geq 1} B_n$ for some $D_0 \in \Gamma_{u_0}$, then by 4.8 again, the sequence $C_n = f(B_n) \cap E$ witnesses (*) for A. ⊣

Proof of Theorem 5. Let (K_n, H_n) be Hurewicz-tests for the classes Γ_n, and $\Gamma = S_1(\langle \Gamma_n \rangle)$. We define a Hurewicz-test (K, H) for Γ as follows:

First we fix a bijection $n \mapsto (n)_0, (n)_1$ between $\omega\backslash\{0\}$ and $(\omega\backslash\{0\}) \times \omega$, and a dense sequence $(x^0_p)_{p\in\omega}$ in K_0. Let K be the set defined by

$$(n, x) \in K \leftrightarrow n \in \omega \wedge [(n = 0 \text{ and } x \in K_0) \text{ or } (n \neq 0 \text{ and } x \in K_{(n)_0})]$$

that we topologize as follows: the topology is generated by the sets $\{n\} \times V$ for $n \neq 0$ and V open in $K_{(n)_0}$, and by the sets $U_{q,v} = (\{0\} \times V) \cup (\cup\{\{n\} \times K_{(n)_0} : n \geq q,\ x^0_{(n)_1} \in V\})$, where $q \in \omega$ and V is open in K_0. One easily checks that K is compact metrizable of dimension 0. Note that $K^{(n)} = \{n\} \times K_{(n)_0}$, for $n \geq 1$, is clopen in K, and the sequence $K^{(\langle n,p\rangle)}$ converges, for fixed p, to the point $(0, x^0_p)$ of $K^0 = \{0\} \times K_0$.

We now define $H \subseteq K$ by $(n, x) \in H \leftrightarrow (n = 0 \text{ and } x \in H_0) \wedge (n \neq 0 \text{ and } x \in H_{(n)_0})$. Clearly $H \in S_1((\Gamma_n))(K)$, as witnessed by the sequence of open sets $U_n = \bigcup_{(p)_0 = n} K^{(p)}$, $n \geq 1$.

As (K_n, H_n) is a Hurewicz-test for Γ_n, $H_n \notin \check{\Gamma}_n(K_n)$. We claim that $H \notin \check{\Gamma}(K)$: For let $V \subseteq K$ be the open set of all points $z \in K$ admitting a neighborhood V_z with $\check{H} \cap V_z \in \cup_{n\geq 1} \Gamma_n$. By the density of the sequence (x^0_p) in K_0 and the convergence property of the $K^{(\langle n,p\rangle)}$, any neighborhood of a $z \in K^{(0)}$ contains sets $K^{(n)}$ for arbitrary large n, hence V is disjoint from

$K^{(0)}$. So if $\check{H}$ were in $\Gamma(K)$, one would get $\check{H} \cap K^{(0)}$ in $\Gamma_0(K^{(0)})$, hence H_0 in $\check{\Gamma}_0(K_0)$ a contradiction.

This proves that if E is Suslin and $B \subseteq E$ satisfies $\exists f$ 1-1 continuous: $K \to E$ with $f^{-1}(B) = H$, the set B is not in $\check{\Gamma}(E)$ – else H would be in $\check{\Gamma}(K)$. It remains to prove the converse, for B Borel in E. So we assume $B \notin \check{\Gamma}(E)$. Let for $n \geq 1$ $V_n = \{x \in E \; \exists V_x$ open $x \in V_x$ and $V_x \cap \check{B} \in \Gamma_n(V_x)\}$ and let $F = E \backslash \cup_{n \geq 1} V_n$. By Lemma 6, if $\check{B} \cap F \in \Gamma_0(F)$, one gets $\check{B} \in \Gamma(E)$, a contradiction. So we can find a 1-1 continuous $f_0 : K_0 \to F$ with $f_0^{-1}(B) = H_0$. We now inductively construct a sequence $f_n, n \geq 1$ of 1-1 continuous functions, with $f_n : K_{(n)_0} \to E$, such that

(i) the images $f_0(K_0)$, $f_n(K_{(n)_0})$, are all pairwise disjoint

(ii) $f_n^{-1}(B) = H_{(n)_0}$ for $n \geq 1$

(iii) If d is a distance on E, $d(f_0(x^0_{(n)_0}), \; f_n(K_{(n)_0})) \leq \frac{1}{(n)_1+1}$.

If we can do this , we are clearly done, for $f : K \to E$ defined by $f(n, x) = f_n(x)$ is 1-1 by (i), continuous by (iii) and satisfies $f^{-1}(B) = H$ by (ii).

Suppose the f_n have been constructed for $n < p$. Let $y = f_0(x_{(p)_0}) \in f_0(K_0)$. As for $n < p$ $f_0(K_0)$ is disjoint from $f_n(K_{(n)_0})$, we can find an open U in E with $y \in U$, $U \cap f_n(K_{(n)_0}) = \emptyset$ for $n < p$ and $U \subseteq \{z \; d(y, z) \leq \frac{1}{(p)_1+1}\}$. We claim that $\check{B} \cap (U \backslash f_0(K_0)) \notin \Gamma_{(p)_0}(E)$: If not, by admissibility of $\langle \Gamma_n \rangle$, the set $\check{B} \cap U$ would be in $\cup_{n \geq 1} \Gamma_n$, contradicting the fact that $U \not\subset \cup_{n \geq 1} V_n$ (as $y \in U \backslash \cup_{n \geq 1} V_n$). But then we can find a 1-1 continuous $f_p : K_{(p)_0} \to U \backslash f_0(K_0)$ with $f_p^{-1}(B) = H_{(p)_0}$, and this f_p clearly works. ⊣

References

[1] Louveau, A. *Some Results in the Wadge hierarchy of Borel Sets.* Cabal Sem. 79-81, Lecture Notes 1019, (1983) 28-55.

[2] Louveau, A. and Saint-Raymond, J. *Borel Classes and Closed Games: Wadge-Type and Hurewicz-Type Results.* To appear in Trans. Amer. Math. Soc.

[3] Martin, D.A. *Wadge Degrees are Well-Founded.* Unpublished.

[4] Saint-Raymond, J. *Caractérisation d'espaces polonais d'aprēs des travaux récents de J.P.R Christensen et D.Preiss, Sém. Choquet*, 11-12 années, 1971-73, (1973) I.H.P, Paris

[5] Saint-Raymond, J. *Fonctions Boréliennes sur un Quotient.* Bull. Sc. Math 100(1976),

141-147

[6] Steel, J.R. *Analytic Sets and Borel Isomorphisms.* Fund. Math 108(1980), pp. 83-88

[7] Wadge, W.W. Thesis, Berkeley 1984.

MORE CLOSURE PROPERTIES OF POINTCLASSES

Howard S. Becker[1]
Department of Mathematics
California Institute of Technology
Pasadena, California 91125

and

Department of Mathematics
University of South Carolina
Columbia, South Carolina 29208

We work in ZF + DC + AD. The theory of arbitrary boldface pointclasses, also known as the theory of Wadge degrees, is one of the topics in descriptive set theory (under AD) which has been a major area of study in recent years. This paper is a contribution to that topic. Specifically, it is about closure properties of arbitrary pointclasses. It will be shown that pointclasses closed under countable union and intersection are also closed under quantification by various types of measure and category quantifiers. The canonical reference for descriptive set theory and AD is Moschovakis [6], whose notation and terminology we will generally follow. Oxtoby [7] is a good reference for the subject of measure and category. Van Wesep [9] is an introduction to Wadge degrees, Kechris [2] is about measure and category in descriptive set theory, and Steel [8] is a paper which is also about closure properties of arbitrary pointclasses.

In this paper we work with the Cantor space, $\mathcal{C} = 2^\omega$; all of our results are also valid for ω^ω. Let μ denote the product measure on 2^ω, where the measure on each factor space is the usual probability measure on $2 : m(\{0\}) = m(\{1\}) = 1/2$. To begin with, we work solely with the measure μ; at the end of the paper we will discuss more general measures. For any $B \subset \mathcal{C}^2$ and any $x \in \mathcal{C}$, let $B_x = \{y \in \mathcal{C} : B(x,y)\}$. To say that a pointclass $\mathbf{\Gamma}$ is *closed under quantification of the form "for a comeager set of y's"* means that if $B \subset \mathcal{C}^2$ is in $\mathbf{\Gamma}$ then the set $\{x : B_x$ is comeager$\}$ is also in $\mathbf{\Gamma}$. A similar interpretation holds for other quantifiers. Call a pointclass $\mathbf{\Gamma}$ *nice* if it is $\mathcal{C}$-parametrized, contains all open sets, and is closed under continuous preimages, countable unions and countable intersections.

Theorem 1. *Let $\mathbf{\Gamma}$ be a nice pointclass. Then $\mathbf{\Gamma}$ is closed under quantification of the forms:*

(a) *For a comeager set of y's.*

(b) *For μ-a.e. y.*

[1]Research partially supported by NSF Grant MCS 82-11328.

This is the main theorem of this paper. Before proving it we will point out some of its corollaries, which follow very easily.

Corollary 2. *Let* Γ *be a nice pointclass and let* N_i *be the* ith *basic open set in* C. *Then* Γ *is closed under quantification of the following forms:*

(a) *For a non-meager set of y's.*

(b) *For a comeager-in-*N_i *set of y's.*

(c) *For a non-meager-in-*N_i *set of y's.*

(d) *For a set of y's of positive* μ*-measure.*

Corollary 3. *Let* Γ *be a nice pointclass and let* r *be a real number. If* Γ *is closed under either* $\exists^{\mathbf{R}}$ *or* $\forall^{\mathbf{R}}$, *then* Γ *is closed under quantification of the following forms (uniformly in* r*):*

(a) *For a set of y's of* μ*-measure* $> r$.

(b) *For a set of y's of* μ*-measure* $\geq r$.

Proof. We prove (a) where Γ is closed under $\forall$.

$$\mu(A) > r \Leftrightarrow \forall z \text{ [if } (z \text{ encodes a } G_\delta \text{ set}$$
$$S_z \ \&\ \mu(S_z) \leq r) \text{ then } \mu(A \backslash S_z) > 0].$$

By Kechris [2], $\{z : \mu(S_z) \leq r\}$ is Borel. ⊣

The theorem also has a different sort of closure property as a corollary.

Corollary 4. *Let* Γ *be a nice pointclass. If* Γ *is closed under* $\exists^{\mathbf{R}}$, *then* Γ *is closed under quantification of the form:*

For uncountably many y.

Proof. (For uncountably many y) $P(y)$ $\Leftrightarrow$ $\exists z$ [z encodes a perfect tree $T_z \subset 2^{<\omega}$ & (for a comeager set of branches $y \in [T_z]$) $P(y)$]. ⊣

In Corollary 4, closure under $\exists^{\mathbf{R}}$ is necessary. If Γ is closed under "for uncountably many", then it is also closed under $\exists^{\mathbf{R}}$, because if $Q(x,y) \Leftrightarrow P(x)$ then $\exists x\, P(x)$ is equivalent to

$$\text{(For uncountably many } \langle x,y \rangle) Q(x,y).$$

I do not know whether the hypothesis of closure under $\exists^{\mathbf{R}}$ or $\forall^{\mathbf{R}}$ in Corollary 3 is necessary.

Theorem 1 (and its corollaries) were already known for many specific examples of nice pointclasses. Call Γ a *Kleene-type* pointclass if Γ is nice, Γ is closed under $\forall^{\mathbf{R}}$ but not under $\exists^{\mathbf{R}}$

and $\mathbf{\Delta}$ is closed under $\forall^{\mathbf{R}}$. The smallest Kleene-type pointclass is $\mathbf{Env}(^3E)$, the class of sets Kleene semirecursive in ^3E and a real (see Moschovakis [5]). To use the terminology of Kechris-Solovay-Steel [4], the theorem was known to hold for all inductive-like pointclasses, and for all projective-like pointclasses *except* Kleene-type classes and their duals (that is, *except* the first class in a projective-like hierarchy of type III). For these classes, the theorem was essentially proved in Kechris [2] (and Corollary 4 was proved in Kechris [3]). It was the open question for Kleene-type classes which motivated this work, but it turns out that Theorem 1 holds in much greater generality than that, as there are many nice pointclasses which are not projective-like, e.g. $\mathbf{Env}(A,\ ^2E)$, the smallest boldface Spector pointclass containing A and $\neg A$, where $A \subset \mathcal{C}$ is not Borel (see Moschovakis [5]).

Many of the properties of measure and category that were proved in Kechris [2] for the projective pointclasses can be generalized to other classes using Theorem 1 (together with the techniques of Kechris [2]). For example if $\mathbf{\Gamma} = \mathbf{Env}(^3E)$, then every $\mathbf{\Gamma}$ relation with non-meager sections has a $\mathbf{\Delta}$ uniformization.

As usual in the theory of Wadge degrees, the results here are all boldface. I do not know whether Theorem 1 holds for lightface pointclasses, or whether it holds for the particular pointclass $\mathrm{Env}(^3E)$. But it does follow from the S-m-n Theorem that for nice $\mathbf{\Gamma}$, Theorem 1 and its corollaries hold for the lightface classes $\Gamma(x)$, for a cone of x's.

We now prove Theorem 1.

Definition. For any $A \subset \mathcal{C}$, let $\tilde{A} =$

$$\{y : \exists y' \ (y \text{ and } y' \text{ agree on all but finitely many coordinates and } y' \in A)\}.$$

For any $B \subset \mathcal{C}^2$, let $\hat{B} = \{(x, y) : y \in \widetilde{(B_x)}\}$.

Lemma 5. (a) *For any $A \subset \mathcal{C}$, $\tilde{A}$ is either meager or comeager. Moreover, $\tilde{A}$ is meager iff A is meager.*

(b) *For any $A \subset \mathcal{C}$, $\mu(\tilde{A})$ is either 0 or 1. Moreover, $\mu(\tilde{A}) = 0$ iff $\mu(A) = 0$.*

Proof. For the measure case, use Kolmogorov's zero-one law (Oxtoby [7]). $\dashv$

Proof of Theorem 1. (a) Let Q be the quantifier "for a comeager set of y's". (Thus QB denotes the pointset $\{x : B_x \text{ is comeager}\}$.) Consider the pointclass $Q\mathbf{\Gamma} = \{QB : B \in \mathbf{\Gamma}\}$. To prove (a) we must show that $Q\mathbf{\Gamma} \subset \mathbf{\Gamma}$. Since $\mathbf{\Gamma}$ is parametrized, $Q\mathbf{\Gamma}$ is also parametrized, hence not self-dual. Let $C \subset \mathcal{C}$ be a set in $Q\mathbf{\Gamma}$ such that $\neg C$ is not in $Q\mathbf{\Gamma}$. By definition of $Q\mathbf{\Gamma}$, there is a $D \subset \mathcal{C}^2$, D in $\mathbf{\Gamma}$, such that $C(x) \Leftrightarrow D_x$ is comeager. Let B be the complement of D. Then

$B \in \check{\Gamma}$.

$$\begin{aligned} \neg C(x) &\iff \neg(D_x \text{ is comeager}) \\ &\iff \neg(B_x \text{ is meager}) \\ &\iff B_x \text{ is non-meager} \\ &\iff \widetilde{(B_x)} \text{ is comeager} \qquad \text{(by Lemma 5)} \\ &\iff (\hat{B})_x \text{ is comeager} \\ &\iff x \in Q\hat{B}. \end{aligned}$$

The pointclass $Q\Gamma$ is closed under Q – this is the Kuratowski-Ulam Theorem (Oxtoby [7]):

$$Qx_1Qx_2\ P(x_1, x_2) \iff Q\langle x_1, x_2\rangle\ P(x_1, x_2).$$

Since $\neg C$ is not in $Q\Gamma$ and $\neg C$ is $Q\hat{B}$, clearly $\hat{B}$ is not in $Q\Gamma$. Since B is in $\check{\Gamma}$, by definition of $\hat{B}$ and the closure properties of Γ, $\hat{B}$ is also in $\check{\Gamma}$. Therefore $\check{\Gamma}$ is not a subclass of $Q\Gamma$. So by Wadge's Lemma, $Q\Gamma$ is a subclass of Γ, which completes the proof.

(b) The proof of (b) is just like that of (a), using Fubini's Theorem rather than the Kuratowski-Ulam Theorem. ⊣

We now consider more general measures. For any σ-finite Borel measure on $\mathcal{C}$, Corollaries 2 and 3 can be deduced from Theorem 1 by the same proof as for μ. So the only question remaining is whether arbitrary measures satisfy Theorem 1. (Note that Lemma 5 is false for arbitrary measures.) Theorem 6, below, gives an affirmative answer to this question for all nice pointclasses which are closed under preimages by Borel-measurable functions.

In the original version of this paper, I asked the following question: Is *every* nice pointclass closed under preimages by Borel-measurable functions? About a year and a half later, John Steel proved that this is indeed the case. The proof of Steel's theorem (which will not be given here) is similar to the proof of Theorem 2.1 of Steel [8]. Hence the extra hypothesis on Γ can be removed from Theorem 6.

Theorem 6. *Let ν be an arbitrary σ-finite Borel measure on $\mathcal{C}$, and let Γ be a nice pointclass which is closed under preimages by Borel-measurable functions. Then Γ is closed under quantification of the form:*

$$\text{For } \nu\text{-a.e. } y.$$

By a *Borel measure* we mean that every Borel set is measurable, hence by AD, every set is measurable (in the completed measure). Call a measure ν *regular* if $\nu(\mathcal{C}) = 1$ and for any $x \in \mathcal{C}$,

$\nu(\{x\}) = 0$. Call ν ***principal*** if there is an $x \in C$ such that for any $A \subset C$,

$$\nu(A) = \begin{cases} 1, & \text{if } x \in A \\ 0, & \text{if } x \notin A. \end{cases}$$

Call a measure *good* if it satisfies Theorem 6; we must show that all measures are good. To prove Theorem 6, we need a result from measure theory (which is a theorem of ZF).

Theorem 7. *Let ν_1 and ν_2 be arbitrary regular Borel measures on C. There is a bijection $I : C \to C$ such that:*

(a) *I is a Borel isomorphism, that is, I and I^{-1} are both Borel-measurable functions.*

(b) *For any (measurable) set $A \subset C$,*

$$\nu_1(A) = \nu_2(I[A]).$$

Theorem 7 is a combination of Lemma 6.2 of Aumann-Shapley [1] and Theorem 1G.4 of Moschovakis [6].

Proof of Theorem 6. Since $\mathbf{\Gamma}$ is closed under preimages by Borel-measurable functions, it follows from Theorem 7 that if one regular measure is good then all regular measures are. So by Theorem 1 (b), all regular measures are good. It is trivial that all principal measures are good. Now let ν be an arbitrary σ-finite Borel measure on C such that $\nu(C) > 0$. There is a sequence $\nu_0, \nu_1, \nu_2, \ldots$ of measures on C and a sequence $r_0, r_1, r_2, \ldots$ of positive real numbers such that for any $A \subset C$,

$$\nu(A) = \sum_{i=0}^{\infty} (r_i \cdot \nu_i(A)),$$

and such that each measure ν_i is either regular or principal. Then

$$(\text{For } \nu\text{-a.e. } y) P(y) \iff (\forall i \in \omega)(\text{For } \nu_i\text{-a.e. } y) P(y),$$

and since each measure ν_i is good, ν is also good. ⊣

References

[1] R. J. Aumann and L. S. Shapley, *Values of Non-Atomic Games,* Princeton University Press, 1974.

[2] A. S. Kechris, Measure and category in effective descriptive set theory, *Ann. Math. Logic* 5 (1973), 337-384.

[3] A. S. Kechris, The theory of countable analytical sets, *Trans. Amer. Math. Soc.* 202 (1975), 259-297.

[4] A. S. Kechris, R. M. Solovay, and J. R. Steel, The axiom of determinacy and the prewellordering property, *Cabal Seminar 77-79,* A. S. Kechris, D. A. Martin, and Y. N. Moschovakis (Eds.), *Lecture Notes in Mathematics,* Vol. 839, Springer-Verlag 1981, 101-125.

[5] Y. N. Moschovakis, Hyperanalytic predicates, *Trans. Amer. Math. Soc.* 129 (1967) 249-282.

[6] Y. N. Moschovakis, *Descriptive Set Theory,* North- Holland, 1980.

[7] J. C. Oxtoby, *Measure and Category,* Springer, 1971.

[8] J. R. Steel, Closure properties of pointclasses, *Cabal Seminar 77-79,* A. S. Kechris, D. A. Martin, and Y. N. Moschovakis (Eds.), *Lecture Notes in Mathematics,* Vol. 839, Springer-Verlag 1981, 147-163.

[9] R. Van Wesep, Wadge degrees and descriptive set theory, *Cabal Seminar 76-77,* A. S. Kechris and Y. N. Moschovakis (Eds.), *Lecture Notes in Mathematics,* Vol. 689, Springer-Verlag 1978, 151-170.

DEFINABLE FUNCTIONS ON DEGREES

Theodore A. Slaman*
Department of Mathematics
University of Chicago
Chicago, IL 60637

and

John R. Steel*
Department of Mathematics
University of California
Los Angeles, CA 90024-1555

The work we shall describe was motivated by some attractive conjectures of D. A. Martin. Let $\leq_T$ and $\equiv_T$ be Turing reducibility and equivalence. Unless otherwise specified, a degree is a Turing degree, that is, an $\equiv_T$-equivalence class; $\deg(x)$ is the degree of x. A real is an element of ${}^\omega 2$, the Cantor space. A cone of reals is a set of the form $\{y|x \leq_T y\}$ for some real x; a cone of degrees is a set of the form $\{\deg(y)|x \leq_T y\}$ for some real x. A property P of reals (degrees) holds on a cone iff $\{x|P(x)\}$ contains a cone of reals (degrees). A function $f : {}^\omega 2 \to {}^\omega 2$ is degree invariant iff $\forall x, y(x \equiv_T y \Rightarrow f(x) \equiv_T f(y))$; f is increasing on a cone iff $x \leq_T f(x)$ on a cone, order-preserving on a cone iff $\exists z \; \forall x, y(z \leq_T x \leq_T y \Rightarrow f(x) \leq_T f(y))$, and constant on a cone iff $\exists y(f(x) \equiv_T y$ on a cone). This last term is an abuse of language since not f, but the induced function on degrees, is literally constant on a cone. For $f, g : {}^\omega 2 \to {}^\omega 2$, let $f \leq_m g$ iff $f(x) \leq_T g(x)$ on a cone. The Turing jump of x is x'.

Conjectures. (Martin). Assume $ZF + AD + DC$. Then

I. If f is degree invariant and not increasing on a cone, then f is constant on a cone.

II. $\leq_m$ prewellorders the set of degree invariant functions which are increasing on a cone. If f has $\leq_m$-rank α, then f' has $\leq_m$-rank $\alpha + 1$, where $f'(x) = f(x)'$ for all x.

A special case of conjecture II was raised as a question by Sacks in [7].

AD is the axiom of determinacy. The AD hypothesis in the conjectures amounts to a definability restriction on the functions to be considered. If the conjectures are true and all games in $L(\mathbf{R})$ are determined, then I and II are true for all functions in $L(\mathbf{R})$. Moreover, any proof of the conjectures will almost certainly use determinacy "locally", and so show e.g. that Borel

*Partially supported by NSF grant.

determinacy implies I and II for Borel functions. Our partial results use determinacy locally.

One of the main reasons determinacy is useful in this area is a well known theorem of Martin. Let $X^{<\omega}$ be the set of finite sequences from X; a tree on X is a subset of $X^{<\omega}$ closed under initial segment. If T if a tree on X, then $[T] = \{f \in {}^{\omega}X | \forall n \in \omega (f \restriction n \in T)\}$ is the set of infinite branches of T. For $s, t \in X^{<\omega}$, s is compatible with t iff $s \subseteq t$ or $t \subseteq s$; T is perfect iff $T \neq \emptyset$ and every sequence in T has incompatible extensions in T. We shall consider mainly trees on $2 = \{0,1\}$; a tree T on 2 is pointed iff T is perfect and $\forall\, x \in [T](T \leq_T x)$.

THEOREM (Martin [4]). Assume AD. *Suppose* $P \subseteq {}^{\omega}2$ *and* $\forall x \exists y \geq_T x P(y)$. *Then there is a pointed* T *such that* $[T] \subseteq P$.

If T is pointed then $\forall x (T \leq_T x \Rightarrow \exists y \in [T](y \equiv_T x))$. Thus if $P \subseteq {}^{\omega}2$ is degree invariant, that is, $(P(x) \wedge y \equiv_T x) \Rightarrow P(y)$, then assuming AD either P or ${}^{\omega}2 - P$ contains a cone. A simple corollary is that, under AD, there is no choice function on degrees. For if $f : D \rightarrow {}^{\omega}2$, where D is the set of degrees, then $\forall n \in \omega$ ($f(d)(n)$ is constant on a cone), so $f(d)$ is constant on a cone, so $f(d) \notin d$ on a cone. (One can show there is no choice function on degrees assuming only all sets have the Baire property; cf. §3.) Notice that one can easily construct counterexamples to I and II from a choice function on degrees, so the fact that AD rules out such a function is reassuring. We elaborate a bit on this theme in §3.

The only definable, degree invariant functions known are the constant functions and the various jump operators (Turing jump, Δ^1_n jump, sharp, etc.). The true intent of I and II is to assert there are no others. This assertion can be made more explicit as follows. Let $\{i\}^x$ be the ith real recursive in x in a standard enumeration. We say $x \equiv_T y$ via (i,j) iff $\{i\}^x = y$ and $\{j\}^y = x$. A degree invariant $f : {}^{\omega}2 \rightarrow {}^{\omega}2$ is uniformly degree invariant iff there is a pointed tree T and a function $t : \omega \times \omega \rightarrow \omega \times \omega$ such that $\forall x, y \in [T](x \equiv_T y$ via $(i,j) \Rightarrow f(x) \equiv_T f(y)$ via $t(i,j))$. We shall show in §2 that any uniformly invariant function which is not increasing on a cone is constant on a cone (assuming AD). Becker [1] shows that the uniformly invariant functions which increase on a cone are precisely the jump operators, in the following sense: for any such f there is a poinclass Γ such that $f(x) \equiv_T$ the universal $\Gamma(x)$ subset of ω on a cone. (It follows that any uniformly invariant function is uniformly order preserving. Becker's results require AD.) Steel [11] shows that conjecture II holds when restricted to uniformly invariant functions (Lachlan proved a special case of this result earlier in [3].) Thus the following conjecture implies Martin's conjectures, and seems to capture their true intent.

Conjecture III. (Steel [11].) Assume AD. Then every degree invariant function is uniformly degree invariant.

Incidentally, conjectures I and II imply the restriction of III to arithmetic functions. For if f is invariant, then I and II imply f is constant on a cone or $f(x) \equiv_T x^n$ (the nth jump of x) on a cone, so in either case $f \equiv_m g$ for some uniformly invariant g. But then by Martin's theorem we can fix a pointed T and an (i,j) such that $\forall x \in [T](f(x) \equiv_T g(x)$ via $(i,j))$, so f is uniformly invariant.

In this paper we shall prove some theorems related to the foregoing conjectures. In §2 we prove I for uniformly invariant functions. We also prove I, and hence III, for functions f such that $f(x) <_T x$ on a cone. Finally, we prove III, and hence II, for Borel functions which are increasing and order-preserving on a cone. In §3 we prove that, under AD, there is no function assigning to each degree d a linear order of d of order type $\omega * + \omega$. This strengthens the theorem that there is no choice function on degrees, and is important for the conjectures since such a function would yield counterexamples to them. In a sequel to this paper, [10], we shall prove one further technical result related to conjectures: we shall construct an r.e. operator $x \mapsto W_e^x$ which acts like a counterexample to II on the degrees r.e. in $0'$.

§2. Our first theorem completes the work of [11] on the uniformly invariant case of conjectures I and II.

THEOREM 1. *Assume* AD. *Suppose f is uniformly invariant and not increasing on a cone. Then f is constant on a cone.*

Proof. Martin's theorem implies $x \not\leq_T f(x)$ on a cone. We show first that $f(x) <_T x$ on a cone. Let T be a pointed tree and $t : \omega \times \omega \to \omega \times \omega$ witness the uniform invariance of f on $[T]$. Consider the game from [11] comparing f to the identity function: I plays $x \in {}^{\omega}\omega$ and II plays $y \in {}^{\omega}\omega$, the players alternating moves as usual. Let $x(0)$ code $(e,n) \in \omega \times \omega$ and let $x^-(n) = x(n+1)$. Then I loses unless $x^- \in [T]$ and $\{e\}^x = y$; if I succeeds in this then II loses unless $x \leq_T y$ and $(f(x^-)(n) = 0 \Leftrightarrow y(y(0)) = 0)$.

Suppose s were a winning strategy for I. We shall show $z \leq_T f(z)$ on the cone above (T,s), a contradiction. So fix $z \in {}^{\omega}2$ such that $T, s \leq_T z$. Let (e_0, n_0) be the first move dictated by s. For $m \in \omega$, let $y_m = \langle m \rangle ^\frown z$, and let $x_m \in [T]$ be such that $s(y_m) = (e_0, n_0)^\frown x_m$. Now there is a fixed (a,b) such that for all $m, x_m \equiv_T x_{m+1}$ via (a,b). [For example, to compute x_{m+1} from x_m compute $\{e_0\}^{x_m} = y_m = \langle m \rangle ^\frown z$. Then compute s from z. Then compute $x_{m+1} = s(\langle m+1 \rangle ^\frown z)^-$. This procedure is independent of m (since e_0 computes m from x_m).] But then $f(x_m) \equiv_T f(x_{m+1})$ via $t(a,b)$, for all m. So $\langle f(x_m) | m < \omega \rangle$ is recursive in $f(z)$. Since s is winning for I we have $z(m-1) = 0$ iff $y_m(y_m(0)) = 0$ iff $f(x_m)(n_0) \neq 0$. Thus $z \leq_T f(z)$.

So by AD we have a winning strategy s for II. We shall show $f(z) \leq_T z$ on the cone above (T,s). So let $T, s \leq_T z$. To compute $f(z)$ from z, proceed as follows: given n, find effectively, using the recursion theorem relative to z, an e such that $s(\langle(e,n)\rangle^\frown z) = \{e\}^z$. Let $y = s(\langle(e,n)\rangle^\frown z)$. Then $f(z)(n) = 0$ iff $y(y(0)) = 0$; otherwise $f(z)(n) = 1$.

Thus $f(x) <_T x$ on a cone. Theorem 2 to follow implies f is constant on a cone, and with no further need for the uniformity hypothesis on f. Its proof, however, works only for Turing degrees. We give here a different proof which does make use of the uniformity of f, but works for other notions of degree.

For each $e \in \omega$ let $P_e = \{x \in [T] | \{e\}^x = f(x)\}$; then by Martin's theorem we can fix an e and a pointed $T' \subseteq T$ such that $[T'] \subseteq P_e$. Thus f is recursive on $[T']$. The usual splitting argument gives a pointed $U \subseteq T'$ such that either f is constant on U or f is one-one on U. If f is constant on U we are done, so assume toward a contradiction that f is one-one on U.

By passing to a subtree we may assume U in uniformly pointed, that is, there is a fixed i such that $\forall x \in [U](U = \{i\}^x)$. We may also assume $f(x) <_T x$ for $x \in [U]$. Finally, let $U^* = \{s \in U | s^\frown\langle 0\rangle \in U \wedge s^\frown\langle 1\rangle \in U\}$. We may assume there is a map $\phi \leq_T U, \phi : U \to 2^{<\omega}$, such that $\forall x \in [U](f(x) = \cup_{n<\omega}\phi(x \restriction n))$ and $\forall s,t \in U^*$ (s is incompatible with $t \Rightarrow \phi(s)$ is incompatible with $\phi(t)$).

Let $V = \{v \in 2^{<\omega} | \exists u \in U(v \subseteq \phi(u))\}$, so that V is a tree and $[V] = f''[U]$. Clearly we can find an $x_0 \in [U]$ such that $x_0 \leq_T U$ and $U \leq_T (V, f(x_0))$. It will be enough to show $V \leq_T f(x_0)$, since then $x_0 \leq_T f(x_0)$ and $x_0 \in [U]$, a contradiction.

Notice that there is a recursive, Lipschitz homeomorphism π of ${}^\omega 2$ so that for any x, $\{\pi^n(x) | n \in \omega\}$ is dense in ${}^\omega 2$. [Define $\pi : 2^{<\omega} \to 2^{<\omega}$ by induction on length. Set $\pi(\emptyset) = \emptyset$. Let $\pi(u^\frown\langle i\rangle) = \pi(u)^\frown\langle i\rangle$ unless $u = \emptyset$ or $\forall m \in \operatorname{dom} u(u(m) = 0)$; in that case set $\pi(u^\frown\langle i\rangle) = \pi(u)^\frown\langle 1-i\rangle$. Finally, let $\pi(x) = \cup_{n<\omega} \pi(x \restriction n)$.] Fix a recursive in U isomorphism $\sigma : (2^{<\omega}, \subseteq) \simeq (U^*, \subseteq)$; we also use σ for the induced homeomorphism from ${}^\omega 2$ to $[U]$. Let $\sigma(y_0) = x_0$. Notice that there is a fixed a such that for all y, $\sigma(y) \equiv_T \sigma(\pi(y))$ via a (this makes use of the uniform pointedness of U.) Thus $f(\sigma(y)) \equiv_T f(\sigma(\pi(y)))$ via $t(a)$ for all y, so that $\langle f(\sigma(\pi^n(y_0))) | n < \omega\rangle$ is recursive in $f(\sigma(y_0)) = f(x_0)$. But $\{\pi^n(y_0) | n < \omega\}$ is dense in ${}^\omega 2$, so its image under σ is dense in $[U]$ and its image under $f \circ \sigma$ is dense in $[V]$ thus

$$v \in V \iff \exists n(v \subseteq f(\sigma(\pi^n(y_0)))),$$

so that V is r.e. in $f(x_0)$.

Say that u n-splits on V iff $u \in V$ and $\{k \in \operatorname{dom} u | u \restriction k^\frown\langle 0\rangle \in V \wedge u \restriction k^\frown\langle 1\rangle \in V\}$ has cardinality at least n. Since V is r.e. in $f(x_0)$, so is $\{(u,n)|u\ n\text{-splits on } V\}$. But

$$v \notin V \iff \exists n \exists u_0 \cdots \exists u_n (u_0 = u_n \wedge \forall i \leq n[u_i \subseteq f(\sigma(\pi^i(y_0))) \wedge u_i$$
$$\text{is incompatible with } v \wedge u_i \text{ dom } v + 2 - \text{splits on } V]),$$

so that V is co-r.e. in $f(x_0)$, finishing the proof. The left-to-right direction of the displayed equivalence is obvious. Now suppose $u_0 \cdots u_n$ are as on the right. Let $r_i = \pi^i(y_0) \upharpoonright \text{dom } v$. Since $\phi(\sigma(r_i))$ is compatible with u_i and does not dom $v+2$-split on V, $\phi(\sigma(r_i)) \subseteq u_i$. Thus $r_0 = r_n$, and then since π is Lipschitz, $\pi^k(r_0) = \pi^j(r_0)$ if $k \equiv j \pmod n$. Thus $\forall k \exists i \leq n(\phi(\sigma(r_i)) \subseteq f(\sigma(\pi^k(y_0))))$. But $\phi(\sigma(r_i))$ is incompatible with v since $\phi(\sigma(r_i)) \subseteq u_i, u_i$ is incompatible with v, and $\text{dom}(\phi(\sigma(r_i))) \geq \text{dom } v$. Thus $\forall k(v \not\subseteq f(\sigma(\pi^k(y_0))))$, so that $v \notin V$. □

The arguments of [11] adapt easily to most notions of degree coarser than $\leq_T$, and our proof of Theorem 1 adapts similarly. For example, let $x \equiv_{\Delta^1_n} y$ via (ϕ, ψ) iff ϕ and ψ are Σ^1_n formulae such that $(x(k) = j \iff \phi(y,k,j))$ and $(y(k) = j \iff \psi(x,k,j))$. Suppose $f : {}^\omega 2 \to {}^\omega 2$ and $t : HF \to HF$ are such that $x \equiv_T y$ via $a \Rightarrow f(x) \equiv_{\Delta^1_n} f(y)$ via $t(a)$. Suppose $x \not\leq_{\Delta^1_n} f(x)$ on a (Turing) cone. Then, granting AD, $\exists y(f(x) \equiv_{\Delta^1_n} y$ on a cone). Notice that it only strengthens the hypothesis on f to require that $x \equiv_{\Delta^1_n} y$ via $a \Rightarrow f(x) \equiv_{\Delta^1_n} f(y)$ via $t(a)$. A similar result holds for constructibility degrees if we let "$x \equiv_L y$ via (ϕ, ψ)" mean that $L[x] = L[y]$ and $(x(k) = j \iff L[y] \models \phi[y,k,j,c^y_0 \cdots c^y_n])$ and $(y(k) = j \iff L[x] \models \psi[x,k,j,c^x_0 \cdots c^x_n])$, where $c^x_i = c^y_i$ is the ith canonical indiscernible of $L[x] = L[y]$. Arithmetic equivalence is the one anomaly we know of in this area; the uniformly invariant cases of both conjectures I and II for arithmetic degrees are open. Let $x \equiv_a y$ via $(\phi, \psi) \iff \phi$ and ψ are Σ^0_n formulae, for some n, and $(x(k) = j \iff \phi(y,k,j))$ and $(y(k) = j \iff \psi(x,k,j))$. The trouble with generalizing our proof of Theorem 1, or the proofs of [11], is that there are b and $\langle x_m | m < \omega \rangle$ such that $\forall m(x_m \equiv_a x_{m+1}$ via $b)$ but $\langle x_m | m < \omega \rangle \not\leq_a x_0$; take, for example, $x_m = 0^m$.

Q1. Do the analogues for uniformly arithmetically invariant functions of conjectures I and II hold?

We pass now to the non-uniform case of Martin's conjectures.

THEOREM 2. *Assume* AD. *Let f be degree invariant and $f(x) <_T x$ on a cone. Then f is constant on a cone.*

Proof. For $e \in \omega$ let $P_e = \{x | f(x) = \{e\}^x\}$. by Martin's theorem we have a fixed e and a pointed tree U such that $[U] \subseteq P_e$. Thus f is recursive on $[U]$. This is the only use of AD in the proof we are giving.

We can thin U to a pointed T such that either f is constant on $[T]$ or f is one-one on $[T]$. If f is constant on T we are done, so assume toward a contradiction that f is one-one on T. By further thinning we may assume that $f(x) <_T x$ for $x \in [T]$, and that there is an $f^* : T \to 2^{<\omega}$ recursive in T such that

$$\forall u, v \in [T][(u \subseteq v \Rightarrow f^*(u) \subseteq f^*(v)) \wedge$$

$$(u \text{ incompatible with } v \Rightarrow f^*(u) \text{ incompatible with } f^*(v))]$$

$$\text{and } \forall x \in [T](f(x) = \cup_{n<\omega} f^*(x \restriction n)).$$

The crux of our proof is embodied in the next lemma, which implies that any $h : \omega \to \omega$ is dominated by a function recursive in some $f(x)$.

DELAY LEMMA. *Let T, f, and f^* be as described above. Let $h : \omega \to \omega$ and $h \leq_T x$, where $x \in [T]$. Then there is an $e \in \omega$ such that, letting*

$$\ell(0) = 0$$

and

$$\ell(n+1) = \sum_{i=0}^{\ell(n)} (\text{number of steps to compute } \{e\}^{f(x)}(i)),$$

ℓ is total and for all sufficiently large n, $h(\ell(n)) < \ell(n+1)$.

Proof. We shall define by initial segments a $y \in [T]$ such that $y \equiv_T x$. We shall be "h-slow" about committing ourselves to $f(y)$, and thereby guarantee that if the conclusion of the lemma fails for e, then $\{e\}^{f(x)} \neq f(y)$. Since $\{e\}^{f(x)} = f(y)$ for some e, this is enough.

By thinning T we may assume $x \equiv_T T$. For any $s \in T$ let $u(s)$ and $v(s)$ be incompatible extensions of s on T of minimal length, and let $m(s)$ be the least i such that

$$f^*(u(s))(i) \neq f^*(v(s))(i).$$

We now define $s_n \in T$ by induction on n; we then set $y = \cup_{n<\omega} s_n$. As we define the s_n's we declare various $e \in \omega$ dead. We begin by setting $s_0 = \emptyset$. Now suppose s_n is given. Let $u = u(s_n), v = v(s_n)$, and $m = m(s_n)$.

Case 1. There is an $e \leq n$ such that e is not yet dead and $\{e\}^{f(x)}(m)$ converges in $\max\{h(i) | i \leq m(u) + m(v)\}$ steps.

In this case, let e_0 be the least such e. Let s_{n+1} be the unique $r \in \{u, v\}$ such that $f^*(r)(m) \neq \{e_0\}^{f(x)}(m)$, and declare e_0 dead.

Case 2. Otherwise.

Then set $s_{n+1} = u$.

Now clearly $y = \cup_n s_n \in [T]$. Since T is pointed, $x \le_T y$. On the other hand, $\langle s_n | n < \omega \rangle$ is recursive in x, T, and h, hence in x. Thus $y \equiv_T x$. So pick e such that $\{e\}^{f(x)} = f(y)$. We claim that e satisfies the lemma.

Let ℓ be defined from e as in the statement of the lemma. Let $j \ge e$ be large enough that no $e' \le e$ is declared dead at some step $n \ge j$. We claim that $h(\ell(n)) < \ell(n+1)$ for $n \ge m(s_j)$. So fix $n > m(s_j)$. Pick k such that

$$m(s_k) \le \ell(n) < m(s_{k+1}).$$

Now since $m(s_j) \le n \le \ell(n), j \le k$. At step k no $e' \le e$ is declared dead. Thus the hypothesis of case 1 does not apply to e at step k. Since s_{k+1} is one of $u(s_k)$ and $v(s_k)$, we have $\ell(n) < m(u(s_k)) + m(v(s_k))$. The failure of the case 1 hypothesis for e means that $\{e\}^{f(x)}(m(s_k))$ takes at least $h(\ell(n))$ steps to compute. By the definition of ℓ then, $h(\ell(n)) < \ell(n+1)$. □

Notice that the Delay lemma immediately gives $0' \le_T f(x)$ on a cone (and more). For let $h(n)$ be the least m such that $0' \cap n$ is enumerated by step m. Let $x \in [T]$ and $h \le_T x$. Let ℓ be as in the claim. Then $\ell \le_T f(x)$, and since from ℓ we can compute a function dominating $h, 0' \le_T \ell$.

For $s \in \omega^{<\omega}$ we define $u_s \in T$ as follows: $u_\emptyset = \emptyset$. Given u_s, let $u_{s^\frown\langle i \rangle}$ be the $v \supseteq u_s, v \in T$, of minimal length such that $v(\text{dom } v - 1) = 0$ and $v \restriction (\text{dom } v - 1)^\frown\langle 1 \rangle \in T$ and

$$\{k \in \text{ dom } v - \text{ dom } u_s | v(k) = 1 \wedge v \restriction k^\frown\langle 0 \rangle \in T\}$$

has cardinality i. For $h : \omega \to \omega$, let

$$x_h = \bigcup_{s \subseteq h} u_s.$$

So x_h turns right on T $h(0)$ times, then left, then right $h(1)$ times, then left, etc. Since T is pointed, $h \le_T x_h$.

Our plan now is to construct $h_1, h_2 \le_T T$ so that if ℓ_1 and ℓ_2 come from applying the Delay lemma to (h_1, x_{h_1}) and (h_2, x_{h_2}), then $T \le_T \ell_1 \oplus \ell_2$. Since $x_{h_1} \equiv_T x_{h_2} \equiv_T T$, this will be enough. The only trouble comes in guessing the e_1 and e_2 appropriate for the h_1 and h_2 we are constructing. But since e_i is a Borel function of h_i, we can fix e_i on a nonmeager set. This turns out to be good enough.

So for any $h : \omega \to \omega$ let e_h be the least e satisfying the conclusion of the Delay Lemma for h, x_h. Let

$$\ell_h(0) = 0,$$

$$\ell_h(n+1) = \sum_{i=0}^{\ell_h(n)} (\text{number of steps to compute } \{e_h\}^{f(x_h)}(i)).$$

Finally, let m_h be the least m such that $h(\ell_h(n)) < \ell_h(n+1)$ for all $n \geq m$. Since the function $h \mapsto (e_h, m_h)$ is Borel, it is constant on a nonmeager set. Thus we fix an $r \in \omega^{<\omega}$ and $e, m \in \omega$ so that for any $s \supseteq r, s \in \omega^{<\omega}$, there is an $h \supseteq s$ such that $e = e_h$ and $m = m_h$.

For $s \in \omega^{<\omega}$, let

$$\ell_s(0) = 0$$

and

$$\ell_s(n+1) = \sum_{i=0}^{\ell_s(n)} (\text{numbers of steps to compute } \{e\}^{f^*(u_s)}(i)),$$

where $\ell_s(n+1)$, and hence $\ell_s(k)$ for $k \geq n+1$, are considered undefined if for some $i \leq \ell_s(n) \{e\}^{f^*(u_s)}(i)$ does not converge within $\mathrm{dom}(f^*(u_s))$ steps. Notice that $s \subseteq t \Rightarrow \ell_s \subseteq \ell_t$, and if $r \subseteq h$ and $e = e_h$, then $\ell_h = \cup_{n<\omega} \ell_{h \restriction n}$. Also, the function $s \mapsto \ell_s$ is recursive in T. Also

$$(*) \qquad \ell_s(i) = k \Rightarrow \ell_{s \restriction k}(i) = k.$$

This is true because if $\ell_s(i) = k$, then the e-computations which verify this only use $f^*(u_s) \restriction k$, which is determined by $u_s \restriction k$ and thus by $s \restriction k$. Notice finally that

$$(**) \qquad \text{if } r \subseteq t,\ i \geq m, \text{ and } \ell_t(i) \notin \text{ dom } t, \text{ then } i+1 \notin \text{ dom } \ell_t.$$

For suppose $\ell_t(i+1) = k$. Pick $h \supseteq t$ such that $e = e_h$ and $m = m_h$ and $h(\ell_t(i)) > k$. Then $\ell_t \subseteq \ell_h$, and as $i \geq m = m_h, h(\ell_h(i)) < \ell_h(i+1)$, a contradiction.

Fix now a $p \supseteq r$ and an $m_0 \geq m$ such that $m_0 \in \text{dom } \ell_p$ and $\ell_p(m_0) = \text{dom } p$. [Take any $h \supseteq r$ such that $e = e_h$. Let $m_0 \geq m$ be such that $\ell_h(m_0) \geq \text{dom } r$. Let $p = h \restriction \ell_h(m_0)$ and apply (*).] Fix also an $x \in {}^{\omega}2$ such that $x \equiv_T T$. We now define $s_i, t_i \in \omega^{<\omega}$ by induction on i (and recursively in T). Let

$$s_0 = t_0 = p.$$

Suppose now we are given s_i and t_i such that $m_0 + i \in \text{dom } s_i \cap \text{ dom } t_i$ and $\text{dom } s_i = \ell_{s_i}(m_0 + i)$ and $\text{dom } t_i = \ell_{t_i}(m_0 + i)$. It follows by (**) then that $m_0 + i + 1 \notin (\text{dom } \ell_{s_i} \cup \text{ dom } \ell_{t_i})$.

Case 1. $x(i) = 0$.

Find recursively in T an $s \supseteq s_i$ such that $\ell_s(m_0 + i + 1)$ is defined. Let

$$s_{i+1} = s \restriction \ell_s(m_0 + i + 1).$$

Say $\ell_s(m_0+i+1)=k$. Find recursively in T a $t \supseteq t_i^\frown\langle k\rangle$ such that $\ell_t(m_0+i+1)$ is defined. Set

$$t_{i+1} = t \upharpoonright \ell_t(m_0+i+1).$$

Notice that by (*) our inductive hypotheses still hold of s_{i+1} and t_{i+1}. Also

$$\ell_{s_{i+1}}(m_0+i+1) = k < \ell_{t_{i+1}}(m_0+i+1)$$

since there is an $h \supseteq t_{i+1}$ with $e = e_h$ and $m = m_h$.

Case 2. $x(i) = 1$.

Reverse the roles of s_i and t_i.

Now let $h = \cup_{i<\omega} s_i$ and $g = \cup_{i<\omega} t_i$. By construction, $g, h \leq_T T$. So $x_h \equiv_T x_g \equiv_T T$. Let $\ell_1 = \cup_{i<\omega}\ell_{s_i}$ and $\ell_2 = \cup_{i<\omega}\ell_{t_i}$.

By construction ℓ_1 and ℓ_2 are total; by their definitions $\ell_1 \leq_T f(x_h)$ and $\ell_2 \leq_T f(x_g)$. by construction,

$$x(i) = 0 \iff \ell_1(m_0+i) > \ell_2(m_0+i).$$

So $T \leq_T x \leq_T f(x_h) \oplus f(x_g)$. Since $f(x_h) \equiv_T f(x_g)$, $x_h \leq_T f(x_h)$, contradicting the fact that $x_h \in [T]$. □

Our proof of Theorem 2 seem to give no information about the analogous question for other notions of degree.

Q2. Let $f : {}^\omega 2 \to {}^\omega 2$ be such that $\forall x \forall y (x \equiv_{\Delta^1_1} y \Rightarrow f(x) \equiv_{\Delta^1_1} f(y))$ and $f(x) <_{\Delta^1_1} x$ on a cone. Granting AD, must there be a y such that $f(x) \equiv_{\Delta^1_1} y$ on a cone?

We turn now to conjecture II and order-preserving functions. For $x \in {}^\omega 2$, let ω_1^x be the least ordinal not recursive in x. For $\alpha < \omega_1^x$, let x^α be the αth Turing jump of x; that is, $x^\alpha = H_e^x$ where e is the least $i \in \mathcal{O}^x$ such that i is a notation for α. We shall need the following slight generalization of a theorem of Posner and Robinson ([6]; cf. also [2] and [9]).

THEOREM (Posner, Robinson). *Let* $1 \leq \alpha < \omega_1^x$, *and suppose* $x^\beta <_T y$ *for all* $\beta < \alpha$. *Then there is a* $z \geq_T x$ *such that* $z^\alpha \equiv_T (y, z)$.

THEOREM 3. *Assume* AD. *Suppose* $f : {}^\omega 2 \to {}^\omega 2$ *is order preserving and increasing on a cone. Then either*

(a) $\exists \alpha < \omega_1\ (f(x) \equiv_T x^\alpha$ *on a cone)*

or

(b) $\forall\alpha < \omega_1^x\ (x^\alpha <_T f(x))$ *on a cone.*

Proof. Suppose that (b) is false. Martin has shown that if $h : {}^\omega 2 \to \omega_1$ and $x \equiv_T y \Rightarrow h(x) = h(y)$ and $h(x) < \omega_1^x$ on a cone, then $\exists\alpha(h(x) = \alpha$ on a cone). (By Martin's theorem we can fix an e and a pointed T such that $\{e\}^x$ codes a wellorder of type $h(x)$ for all $x \in [T]$. By boundedness, $h(x) < \omega_1^T$ for all $x \in [T]$. But ω_1^T is countable, so applying Martin's theorem again we have a pointed $S \subseteq T$ such that h is constant on $[S]$.) So we can fix an $\alpha < \omega_1$ such that on a cone, $x^\alpha \not<_T f(x)$ but $x^\beta <_T f(x)$ for all $\beta < \alpha$. Pick any x in this cone, and let $y = f(x)$. The Posner-Robinson theorem yields a $z \geq_T x$ such that $z^\alpha \equiv_T (y, z)$. Now $z \leq_T f(z)$, and $y = f(x) \leq_T f(z)$ since f is order preserving. Thus $z^\alpha \leq_T f(z)$, so that $z^\alpha \equiv_T f(z)$ by our choice of α and x. We have shown then that $\forall x \exists z \geq_T x(f(z) \equiv_T z^\alpha)$, so that $f(z) \equiv_T z^\alpha$ on a cone by Martin's theorem. □

COROLLARY. *Assume* AD. *Let f be increasing and order preserving, and suppose $f(x) \in \Delta_1^1(x)$ on a cone. Then f is uniformly invariant.*

Theorem 3 and its corollary beg for improvement. Woodin (unpublished) has extended the Posner-Robinson theorem by showing that if $x^\alpha <_T y$ for all $\alpha < \omega_1^x$, then $\exists z \geq_T x(\mathcal{O}^z \equiv_T (y, z))$. It follows that we can strengthen alternative (b) of Theorem 3 to read "$\mathcal{O}^x \leq_T f(x)$ on a cone". It is plausible that these ideas will lead to a proof of the strengthening of the corollary obtained by replacing "$\Delta_1^1(x)$" by "$L[x]$". The following proposition is relevant to the attempt to do this.

PROPOSITION. *Assume* AD. *Let $g : {}^\omega 2 \to \mathrm{OR}$ be such that $x \equiv_T y \Rightarrow g(x) = g(y)$. Then for a cone of $x, \{y | g(x) = g((x, y))\}$ is comeager.*

Proof. For a cone of $x, z \geq_T x \Rightarrow g(z) \geq_T g(x)$, as otherwise we get a descending chain of ordinals. If for a cone of x, $\{y | g(x) < g((x, y))\}$ is nonmeager, then the Kuratowski-Ulam theorem gives a sequence $\langle y_i | i < \omega \rangle$ such that for some x, $g((x, \langle y_i | n \leq i < \omega \rangle)) > g((x, \langle y_i | n{+}1 \leq i < \omega \rangle))$. This again is a descending chain of ordinals. □

The proposition says that, under AD, Cohen forcing preserves all ordinal assignments. (If Cohen had known enough about AD, this might have motivated him!) It might be interesting to find analogues of the proposition for other notions of forcing. The generalization of the Posner-Robinson theorem needed to push Theorem 3 and its corollary thorough $L[x]$ seems to require this.

Of course, something more general, to which the Posner-Robinson theorem seems irrelevant, must be true.

Q3. Assume AD. Let $F : {}^{\omega}2 \rightarrow {}^{\omega}2$ be increasing and order preserving on a cone. Must f be uniformly invariant?

In view of Theorem 3 and its corollary, the answer is almost certainly "yes".

We shall now attempt to clarify the relationships between Theorems 1-3 and the conjectures. To construct a counterexample to any of the conjectures is to construct a degree invariant function f. We can regard such a construction as a family of constructions, one for each $x \in {}^{\omega}2$, done simultaneously; the x-construction has "coding requirements", designed to guarantee $y \equiv_T x \Rightarrow f(y) \equiv_T f(x)$, and additional requirements designed to guarantee whatever else we have in mind.

In this light, Theorem 1 and the results of [11] show that the x-coding requirements must be satisfied in a way that is not uniform over $[T]$ for any pointed T. It is at least superficially plausible that the injuries to these requirements during the x-construction could produce this non-uniformity.

Theorem 2 shows that if the x-construction is to be recursive in x, then the coding requirements become so difficult to guess that we can only meet them uniformly. The reason is that the sets to be coded by $f(x)$, that is, the $f(y)$ for $y \equiv_T x$, can appear very slowly. This fact, embodied in the Delay lemma, is the key to our proof. Unfortunately, there is no such limitation on nonrecursive constructions. The following open question illustrates the shortcomings of Theorem 2 in this regard. Let $x \equiv_m y \iff \{n|x(n) = 0\}$ has the same many-one degree as $\{n|y(n) = 0\}$.

Q4. Is there a continuous, one-one $f : {}^{\omega}2 \rightarrow {}^{\omega}2$ such that for a cone of x, $\forall y(x \equiv_m y \Rightarrow f(x) \equiv_T f(y))$ and $0' \not\leq_T f(x)$?

Notice that, given such an f, the function $g(x) = f(x')$ is a counterexample to conjecture I.

Theorem 3 shows that if we require $f(y) \leq_T f(x)$ for $y \leq_T x$, then again the coding requirements overwhelm everything else. The reason is that this imposes a burden on the y-construction as well as the x-construction; the y-construction must guarantee that $f(y)$ is tame enough that $f(x)$ can afford to code it. Thus the y-construction must cooperate with $2^{\aleph_0}$ other constructions, a heavy burden indeed. Of course, if we want only $f(x) \equiv_T f(y)$ for $y \equiv_T x$, then every construction need only cooperate with countably many others, and we have some hope. As a diluted form of cooperation within a degree we offer the following proposition. (We also mention this proposition because it rules out one natural approach to proving conjecture II.) Let W_e^x be the eth subset of ω r.e. in x.

PROPOSITION. *There is an e such that whenever $x_1 \equiv_T x_2 \equiv_T \cdots \equiv_T x_n$, then*

$$x_1 <_T W_e^{x_1} \oplus \cdots \oplus W_e^{x_n} <_T x_1'.$$

One can obtain such an e by recursively associating to each x an infinite injury construction of W_e^x, and arranging for the x and y constructions to cooperate when $x \equiv_T y$. (One can also simply cite the relativisation of a theorem of Cooper and Yates; cf. [5]: there is a nonrecursive r.e. A such that for all r.e. $B <_T 0', A \oplus B <_T 0'$.) In contrast, notice that the Posner-Robinson theorem implies that if $x <_T f(x)$ for all x, then there are y and x such that $y \leq_T x$ and $x' \leq_T f(y) \oplus f(x)$.

We close this section with a few remarks on arbitrary, not necessarily definable, functions on degrees. So we assume AC for the duration of this and the next paragraph. The proof of Theorem 3 gives: if f is order preserving and $x <_T f(x)$ on a cone, then $\forall y \exists x \geq_T y (x' \leq_T f(x))$. Thus, even with AC, one cannot construct an order preserving $g : D \to D$ such that $d < g(d) < d'$ for all d. The proof of Theorem 2 gives no information about undefinable "pressdown" functions (functions f such that $f(d) < d$ on a cone). Of course the Friedberg jump inversion theorem implies there is a function f such that $f(d)' = d$ for $d \geq 0'$, and thus a pressdown function one-one on the cone above $0'$. We know of one limitation on possible undefinable pressdown functions.

PROPOSITION. *Suppose $f(d) < d$ on a cone. Then it is not the case that $\forall c(c \leq f(d)$ on a cone).*

Proof. Suppose $\forall c(c \leq f(d)$ on a cone). Let d_0 be arbitrary, and let d_{i+1} be such that $d_i \leq f(d)$ whenever $d_{i+1} \leq d$. Now let u be a minimal upper bound for $\{d_i | i \in \omega\}$ (cf. [8]). By the definition of the d_i's $f(u)$ is an upper bound for $\{d_i | i \in \omega\}$. Thus $f(u) \not<_T u$. Since d_0 was arbitrary, f is not a pressdown function. □

We don't know whether there are nontrivial order preserving pressdown functions.

Q5. (a) Is there a pressdown function which is order preserving and one-one on a cone?

(b) Let $I \subseteq D$ be a countable jump ideal (i.e. closed under join, Turing jump, and downward under Turing reducibility). Let $A \subseteq I$ be such that $\forall b \in I(A \not\subseteq \{c | c \leq b\})$. Must there be an upper bound u for I such that no $v < u$ is an upper bound for A?

A positive answer to (b) would strengthen Sacks' theorem on the existence of minimal upper bounds (cf. [8]). One could use this strengthening as in the proposition above to obtain a negative answer to (a).

§3. In order to construct a degree invariant function we must assign to each degree d a structure on d telling the x-constructions for $x \in d$ how to cooperate with one another. The structure easiest to use would be a distinguished element of d, or equivalently, a wellorder of d; however, AD implies there is no function assigning to all d such a structure. A somewhat weaker structure, which still suffices to construct counterexamples to the conjectures, is a linear order of d of order type $\omega * + \omega$. We now show that, assuming AD, there is no function putting this weaker structure on an arbitrary degree.

The following reformulation will be useful. A resolution of an equivalence relation E is a sequence $\langle E_n | n < \omega \rangle$ of equivalence relations such that $E = \cup_{n<\omega} E_n$. We call a resolution $\langle E_n | n < \omega \rangle$ finite iff $\forall x \forall n([x]_{E_n}$ is finite). Since we are not assuming AC, the next lemma has content.

LEMMA 1. *Let E be an equivalence relation on ${}^{\omega}2$. Then the following are equivalent:*

(a) *There is a finite resolution of E,*

(b) *There is a function assigning to each E-equivalence class d a linear order of d of order type $\omega * + \omega$.*

Proof. (a) $\Rightarrow$ (b): Let $\langle E_n | n < \omega \rangle$ be a finite resolution of E. Let $<$ be the usual lexicographic linear order of ${}^{\omega}2$. Let d be a given E-equivalence class. For $x, y \in d$ let n be least such that $x E_n y$. If $n = 0$, put $x <_d y \iff x < y$. If $n = m + 1$, let w and z be the $<$-least elements of $[x]_{E_m}$ and $[y]_{E_m}$ respectively,and put $x <_d y \iff w < z$.

It is easy to check that $<_d$ has order type $\omega * + \omega, \omega$, or $\omega *$. In the latter cases we can pass canonically to a linear order of d of order type $\omega * + \omega$.

(b) $\Rightarrow$ (a): Suppose that for all E-equivalence classes $d, <_d$ is a linear order of d of order type $\omega * + \omega$. For $x E y$, let

$$D(x, y) = \text{card}(\{z | x <_d z <_d y \text{ or } y <_d z <_d x\}).$$

For d an E-equivalence class, let

$$T_d = \{s \in 2^{<\omega} | \exists x \in d(s \subseteq x)\}$$

and

$$x_d = \text{leftmost branch of } T_d.$$

We define $E_n \cap (d \times d)$ by cases on the relation of x_d to $<_d$.

Case 1. $x_d \in d$. Then for $y, z \in d$

$$yE_nz \iff y = z \vee (D(y, x_d) < n \wedge D(z, x_d) < n).$$

Case 2. $x_d \notin d$ and there is a $k \in \omega$ and $y \in d$ such that $z <_d y \Rightarrow z \restriction k \neq x_d \restriction k$. Then let k be least such that some such y exists, and let y_d be the $<_d$-largest such y for k. Set

$$yE_nz \iff y = z \vee (D(y, y_d) < n \wedge D(z, y_d) < n).$$

Case 3. Cases 1 and 2 fail to hold, and there is a $k \in \omega$ and $y \in d$ such that $z >_d y \Rightarrow z \restriction k \neq x_d \restriction k$. Proceed as in case 2.

Case 4. Otherwise. Then for $y, z \in d$, let

$$yE_nz \iff y = z \vee \forall w[(y <_d w \leq_d z \vee z <_d w \leq_d y) \Rightarrow w \restriction n \neq x_d \restriction n].$$

It is not difficult to check that $\langle E_n | n < \omega \rangle$ is a finite resolution of E. □

The relation $xEy \iff \exists m \forall n \geq m(x(n) = y(n))$ is a simple example of an equivalence relation admitting a finite resolution. It is fairly easy to show that if E admits a finite resolution, then there are pathological E-invariant functions; for example, functions f such that $xEy \Rightarrow f(x) \equiv_T f(y)$ and $x <_T f(x) <_T x'$ for all x and y.

We shall show AD implies $\equiv_T$ admits no finite resolution. This phrasing is misleading, however; it seems one must prove a stronger conclusion from a weaker hypothesis. If G is a group of homeomorphisms of ${}^{\omega}2$, then let $x \sim_G y \iff \exists \pi \in G(\pi(x) = y)$. If the homeomorphisms in G are recursive, then $\sim_G$ refines $\equiv_T$. We shall show that if G is modestly complicated, then $\sim_G$ admits no finite resolution. The relevant hypothesis is that all sets of reals are Lebesgue measurable.

Consider an alphabet with symbols π, π^{-1}, σ, and σ^{-1}. A reduced word is a sequence $\langle a_0 \cdots a_k \rangle$ of symbols such that for no $i < k$ do we have $a_i = \pi$ and $a_{i+1} = \pi^{-1}$, or $a_i = \pi^{-1}$ and $a_{i+1} = \pi$, or $a_i = \sigma$ and $a_{i+1} = \sigma^{-1}$, or $a_i = \sigma^{-1}$ and $a_{i+1} = \sigma$. Given homeomorphisms $\bar{\pi}$ and $\bar{\sigma}$ of ${}^{\omega}2$, let $(\overline{\pi^{-1}}) = \bar{\pi}^{-1}$ and $(\overline{\sigma^{-1}}) = \bar{\sigma}^{-1}$ and for $w = \langle a_0 \cdot a_k \rangle$ a reduced word, let $\bar{w} = \bar{a}_k \circ \cdots \circ \bar{a}_0$. Notice that if $\bar{\pi}$ and $\bar{\sigma}$ are Lipschitz, so that $\bar{\pi}(x) = \cup_{n<\omega} \bar{\pi}(x \restriction n)$ where $\bar{\pi} \restriction {}^{n}2$ is a permutation of ${}^{n}2$ and similarly for $\bar{\sigma}$, then $\bar{w}$ is Lipschitz and $\bar{w} \restriction {}^{n}2$ is determined by $\bar{\pi} \restriction {}^{n}2$ and $\bar{\sigma} \restriction {}^{n}2$. We call a pair $\{\bar{\pi}, \bar{\sigma}\}$ of homeomorphisms independent iff whenever w is a reduced word then $\bar{w}(x) \neq x$ for all $x \in {}^{\omega}2$; equivalently, iff whenever s and v are distinct reduced words, than $\bar{w}(x) \neq \bar{v}(x)$ for all $x \in {}^{\omega}2$.

LEMMA 2. *There is an independent pair of recursive, Lipschitz homeomorphisms of* ${}^{\omega}2$.

Proof. We define $\bar{\pi} \restriction {}^{n}2$ and $\bar{\sigma} \restriction {}^{n}2$ by induction on n. Let R be the set of reduced words, and $<$ a recursive wellorder of $R \times 2^{<\omega}$ in order type ω.

Step 1. Set

$$\bar{\pi}(\langle 0\rangle) = \bar{\sigma}(\langle 0\rangle) = \langle 1\rangle \text{ and } \bar{\pi}(\langle 1\rangle) = \bar{\sigma}(\langle 1\rangle) = \langle 0\rangle.$$

Step $n+1$. Let $\langle w, s\rangle$ be the $<$-least pair such that $\mathrm{dom}(s) \leq n, \bar{w}(t) = t$ for some $t \in {}^{n}2$ such that $s \subseteq t$, and $\bar{u}(s) \neq \bar{v}(s)$ for all distinct initial segments u and v of w. We call $\langle w, s\rangle$ the critical pair at $n+1$. Let

$$F = \{t \in {}^{n}2 | s \subseteq t \wedge \bar{w}(t) = t\},$$

and for $t \in F$,

$$C_t = \{\bar{u}(t) | u \text{ is an initial segment of } w\}.$$

Thus $t \in C_t$ and C_t has dom w distinct elements, all in ${}^{n}2$. Notice that if $r \neq t$ then $C_r \cap C_t = \emptyset$. For otherwise $\bar{u}(r) = \bar{v}(t)$ for u, v initial segments of w, and since $s \subseteq r$ and $s \subseteq t$, and $\bar{u}, \bar{v}$ preserve $\subseteq$, $\bar{u}(s) = \bar{v}(s)$. Since $\langle w, s\rangle$ is critical, $u = v$. Since $\bar{u}(r) = \bar{u}(t)$ and $\bar{u}$ is one-one, $r = t$.

Let $w = \langle a_0, \ldots, a_k\rangle$. It is now reasonably easy to see that we can extend $\bar{\pi}$ and $\bar{\sigma}$ to $\{u \in {}^{n+1}2 | u \restriction n \in \cup_{t \in F} C_t\}$ so that for $t \in F$ and $i \in \{0, 1\}$,

$$\bar{a}_0(t^\frown\langle i\rangle) = \bar{a}_0(t)^\frown\langle 1 - i\rangle$$

and

$$\bar{a}_j(\overline{w \restriction j}(t)^\frown\langle i\rangle) = \overline{w \restriction (j+1)}(t)^\frown\langle i\rangle \text{ for } 1 \leq j \leq k.$$

(A problem might arise if $a_0 = a_k^{-1}$; but then for $t \in F$, $\bar{a}_k(\bar{a}_0(t)) = t = \bar{a}_k(\overline{w \restriction k}(t))$, so $\bar{a}_0(t) = \overline{w \restriction k}(t)$, so $\bar{a}_0(s) = \overline{w \restriction k}(s)$, so $a_0 = w \restriction k$, and $w = a_0 a_0^{-1}$ is not reduced.)

We complete step $n+1$ by extending $\bar{\pi}$ and $\bar{\sigma}$ arbitrarily to all of ${}^{n+1}2$. Notice $\bar{w}(t) \neq t$ for all $t \in {}^{n+1}2$ such that $s \subseteq t$, so that $\langle w, s\rangle$ is never again critical.

This completes the definition of $\bar{\pi}$ and $\bar{\sigma}$. Suppose $\bar{w}(x) = x$ for some reduced w and $x \in {}^{\omega}2$. We may assume $\bar{u}(x) \neq \bar{v}(x)$ for all distinct initial segments u and v of w by passing to a shorter word if necessary. Then there is an n such that $(w, x \restriction n)$ is eligible at all steps $\geq n+1$ to be critical. Since every pair is critical at most once. $(w, x \restriction n)$ is critical at some step, hence never eligible to be critical again, a contradiction. □

It is just a matter of more notation to construct a family of $2^{\aleph_0}$ Lipschitz homeomorphisms which is independent in the obvious sense.

THEOREM 4. *Assume all sets of reals are Lebesgue measurable. Let G be a group of homeomorphisms of ${}^{\omega}2$ containing an independent pair of Lipschitz homeomorphisms. Then $\sim_G$ admits no finite resolution.*

Proof. Suppose $\langle E_n | n < \omega \rangle$ is a resolution of $\sim_G$. Let $\{\bar{\pi}, \bar{\sigma}\}$ be an independent pair of Lipschitz homeomorphisms from G. We assume $\bar{\pi}$ and $\bar{\sigma}$ have been extended so as to act on $2^{<\omega}$.

For any $x \in {}^{\omega}2$, let n_x be the least m such that $\{\bar{\pi}(x), \bar{\sigma}(x), \bar{\pi}^{-1}(x), \bar{\sigma}^{-1}(x)\} \subseteq [x]_{E_m}$. Then we can fix an n and a $C \subseteq {}^{\omega}2$ closed such that C has measure $\geq 3/4$ and $n_x < n$ for $x \in C$. Let $T = \{s \in 2^{<\omega} | \exists x \in C(s \subseteq x)\}$, so that T is a tree on $\{0,1\}$ and $[T] = C$. Then for any symbol $a \in \{\pi, \pi^{-1}, \sigma, \sigma^{-1}\}$ we have

$$xE_n\bar{a}(x), \qquad \forall x \in [T].$$

We now define a tree U on $\{0,1\} \times \{\pi, \sigma, \pi^{-1}, \sigma^{-1}\}$ as follows:

$$(s,w) \in U \iff \exists k(s \in {}^{k}2 \cap T \wedge \text{ dom } w = k \wedge w \text{ is a reduced word}$$
$$\wedge\, \bar{v}(s) \in T \text{ for every initial segment } v \text{ of } w.)$$

Clearly U is finitely branching. If U is infinite, then we have an $(x,f) \in [U]$. Let $w_i = f \restriction i+1$. Then $x \in [T]$ and $\bar{w}_i(x) \in [T]$ for all i, so

$$xE_n\bar{w}_0(x)\ E_n\bar{w}_1(x)E_n \cdots$$

On the other hand, since $\{\bar{\sigma}, \bar{\pi}\}$ is independent and each w_i is a reduced word, $\bar{w}_j(x) \neq \bar{w}_i(x)$ for $j \neq i$. So $[x]_{E_n}$ is infinite, and we're done.

So it is enough to show U is infinite. Let i be given; we seek a node of U on level i. We may assume i is large enough that

$$\forall a \in \{\pi, \pi^{-1}, \sigma, \sigma^{-1}\} \forall s \in {}^{i}2 \quad (\bar{a}(s) \neq s)$$

and

$$\forall a, b \in \{\pi, \pi^{-1}, \sigma, \sigma^{-1}\} \forall s \in {}^{i}2(a \neq b \Rightarrow \bar{a}(s) \neq \bar{b}(s)).$$

Consider the graph (that is, symmetric irreflexive binary relation) E on ${}^{i}2$ defined by

$$(s,t) \in E \iff \exists a \in \{\pi, \pi^{-1}, \sigma, \sigma^{-1}\} \quad (\bar{a}(s) = t).$$

For each s there are exactly 4 t's such that $(s,t) \in E$, so there are $4 \cdot 2^i$ pairs in E. So the number of edges of E (that is, unordered pairs $\{s,t\}$ such that (s,t) and (t,s) are in E) is $2 \cdot 2^i$. Now let F be the subgraph of E defined by

$$(s,t) \in F \iff (s,t) \in E \wedge s \in T \wedge t \in T.$$

Since $[T]$ has measure $\geq 3/4$, ${}^i2 - T$ has no more than $1/4\ 2^i$ members. So in passing from E to F we remove no more than $4(1/4\ 2^i)$ edges from E. thus F has at least 2^i edges.

A well known easy induction shows that any graph on k vertices with at least k edges contains a circuit. (That is, a sequence $\langle v_0 \cdots v_m \rangle$ of vertices such that $v_0 = v_m, \{v_j, v_{j+1}\}$ is an edge for $j < m$, and $\{v_j, v_{j+1}\} \neq \{v_\ell, v_{\ell+1}\}$ for $j, \ell < m$ such that $j \neq \ell$.) Let $\langle s_0 \cdots s_m \rangle$ be a circuit in F. For $j < m$ there is a unique $a_j \in \{\pi, \pi^{-1}, \sigma, \sigma^{-1}\}$ such that $\bar{a}_j(s_j) = s_{j+1}$. Let $v = \langle a_0 \cdots a_{m-1} \rangle$. Then v is reduced, as otherwise $\{s_j, s_{j+1}\} = \{s_{j+1}, s_{j+2}\}$ for some j, and $v^2 = v^\frown v, v^3 = v^\frown v^\frown v$, etc., are reduced since otherwise $\{s_0, s_1\} = \{s_{m-1}, s_m\}$. Let $w = v^i \restriction i$. Then (s_0, w) is on level i of U, and we are done. □

We say $x, y \in {}^\omega 2$ are recursively isomorphic if there is a recursive permutation π of ω such that $\forall n (x(n) = 0 \iff y(\pi(n)) = 0)$. If E and F are equivalence relations, then F is coarser than $E \iff E \subseteq F$.

COROLLARY. *Assume all sets are Lebesgue measurable. Then no equivalence relation on* ${}^\omega 2$ *coarser than recursive isomorphism admits a finite resolution.*

Proof. If F is coarser than E and F admits a finite resolution, so does E. Let G be the group generated by an independent pair of recursive Lipschitz homeomorphisms. Then $\equiv_T$ is coarser than $\sim_G$. Since $\sim_G$ admits no finite resolution, neither does $\equiv_T$.

Now suppose $\langle E_n | n < \omega \rangle$ is a finite resolution of recursive isomorphism. Let $x F_n y \iff x' E_n y'$. Then $\langle F_n | n < \omega \rangle$ is a finite resolution of $\equiv_T$, a contradiction. The corollary follows. □

By the way, one can show that, assuming all sets are Lebesgue measurable (or all sets have the Baire property) that if G is a group of homeomorphisms of ${}^\omega 2$ such that every $\sim_G$ class is dense, then there is no function picking a member from each $\sim_G$ class. [If such a function exists, then there is a linear order $<^*$ of the $\sim_G$ classes. Let $(s, t) \in 2^{<\omega} \times 2^{<\omega}$ be such that e.g. $\{(x, y) | [x]_{\sim_G} <^* [y]_{\sim_G}\}$ is comeager in the neighborhood determined by (s, t). A symmetry argument gives a contradiction. This proof is due perhaps to Sierpinski and Mycielski.] Letting G be generated by a single homeomorphism, we get an example where $\sim_G$ admits a finite resolution (by Lemma 1), but no choice function on equivalence classes. We would guess that there is a model of $ZF + DC$ in which $\equiv_T$ admits a finite resolution, but no choice function on equivalence classes.

The following two questions occur naturally if one attempts to push Theorem 1 closer to a proof of the conjectures. Call a resolution $\langle E_n | n < \omega \rangle$ recursively finite iff $\forall x \forall n$ (there is no sequence $\langle y_i | i < \omega \rangle \leq_T x$ of distinct elements of $[x]_{E_n}$).

Q6. Assume AD. Does $\equiv_T$ admit a recursively finite resolution?

Q7. Let G be a group generated by an independent pair of recursive, Lipschitz homeomorphisms of ${}^{\omega}2$. Is there a continuous, one-one $f: {}^{\omega}2 \to {}^{\omega}2$ such that for all x, y

$$x \sim_G y \Rightarrow f(x) \equiv_T f(y)$$

and

$$0' \not\leq_T f(x).$$

(If G is generated by a single homeomorphism, then there is an f as called for in Q7, essentially because $\sim_G$ admits a finite resolution.)

The next question seems unrelated to the conjectures, but perhaps of some interest.

Q8. Assume AD. Suppose we have a function assigning to each Turing degree d a linear order $<_d$ of d. Then

(a) must the rationals embed order preservingly in $<_d$, for a cone of d's?

(b) must there be a linear order $<$ of ${}^{\omega}2$ such that $<_d = < \cap (d \times d)$ on a cone?

(A positive answer to (b) implies a positive answer to (a); if $<$ is a linear order of ${}^{\omega}2$ then, assuming all sets have the Baire property, the rationals embed into $< \cap (d \times d)$ on a cone.) We cannot at the moment rule out a function assigning to each d a linear order of d of order type $(\omega^* + \omega)(\omega^* + \omega)$.

References

[1] H. Becker, *Pointclass jumps*, to appear.

[2] C. Jockusch and R. Shore, *REA operators, r.e. degrees, and minimal covers*, Proc. of Symposia in Pure Math. of the AMS, v. 42, 1985, pp. 33-11.

[3] A. H. Lachlan, *Uniform enumeration operations*, JSL v. 40, 1975, pp. 401-409.

[4] D. A. Martin, *The axiom of determinacy and reduction principles in the analytic hierarchy*, Bull. Amer. Math. Soc. v. 74, 1968, pp. 687-689.

[5] D. P. Miller, *High recursively enumerable degrees and the anti-cupping property*, Logic year 1979-80, Springer Lecture notes in Math., v. 859.

[6] D. Posner and R. W. Robinson, *Degrees joining to $\underset{\sim}{0}'$*, JSL v. 46, 1981, pp. 714-722.

[7] G. E. Sacks, *On a theorem of Lachlan and Martin*, Proc. Amer. Math. Soc. v.18, 1967, pp. 140-141.

[8] G. E. Sacks, *Degrees of unsolvability*, Ann. Math. Studies, v.55 1966, Princeton Univ. Princeton, N.J.

[9] T. A. Slaman and J. R. Steel, *Complementation in the Turing degrees*, to appear.

[10] T. A. Slaman and J. R. Steel, *A construction degree invariant below* $0'$, to appear.

[11] J. R. Steel, *A classification of jump operators*, JSL v.47, 1982, pp. 347-358.

LONG GAMES

John R. Steel[1]
Department of Mathematics
University of California
Los Angeles, CA 90024-1555

The hypothesis that definable games are determined has proven very powerful in its realm, the realm of reals and definable sets of reals. For example, ZFC + $AD^{L(\mathbf{R})}$ seems to yield a "complete" theory of $L(\mathbf{R})$, in the same way that ZFC alone yields a "complete" theory of L. This success makes it natural to investigate stronger forms of definable determinacy, and the universes larger than $L(\mathbf{R})$ which these might civilize. One might hope to ultimately bring determinacy techniques to bear on questions involving quantification over arbitrary sets of reals, for example, the question of the prewellordering property for Π^2_n.

Various papers, in particular Becker [1], Blass [2], Martin [4], Martin [5], Solovay [7], and Woodin [8], have contributed to this investigation. These papers have been concerned with games of length strictly less than ω_1, on ω or on $\mathbf{R}$. In this paper we shall go a bit further and consider certain clopen (i.e., decided after countably many moves) games of length ω_1.

In §1 we show that the game quantifiers associated to these clopen games propagate scales, and in §2 we show that the games have canonical winning strategies. Of course, both results require the determinacy of the games in question. Our methods here extend those of Moschovakis ([6], Chapter 6), who proved these "third periodicity" theorems for game of length ω on ω, and of Martin ([4]) who extended Moschovakis' proof to games of length less than ω_1 on ω or $\mathbf{R}$. In §3 we show that the determinacy of clopen games of length ω_1 with payoff $\underset{\sim}{\Pi}^1_1$ in the codes implies the existence of a natural inner model in which all games of length less than ω_1 on $\mathbf{R}$ (with arbitrary payoff) are determined. This section makes use of the results of §1 and §2, but not of their proofs. It also makes heavy use of Woodin [8], which shows how to construct such a model if one exists; our contribution is just to show that some form of definable determinacy implies its existence. We also indicate in §3 how to construct variants of Woodin's model satisfying stronger forms of determinacy for arbitrary payoff. In §4 we consider the problem of proving the determinacy hypotheses we have been using, and obtain a partial result using the methods of Blass [2] and Martin [3]. Along the way we show that the inner model of §3 satisfies the determinacy of certain games on $P(\mathbf{R})$. Finally, §5 is devoted to remarks and questions.

[1]The author was partially supported by National Science Foundation grant DMS-3802555.

As an expository device we work in ZF + DC throughout, and state our additional hypotheses as we need them. By $\mathbf{R}$ we mean ${}^{\omega}\omega$, the Baire space. What concepts we take for granted, in particular that of a scale, are explained in [6].

§1. Some game quantifiers which propagate scales.

Let us call $T \subseteq \cup_{\alpha<\omega_1} {}^{\alpha}\omega$ a tree if $\forall p \in T \forall \beta \in \mathrm{dom}(p)(p \restriction \beta \in T)$. Associated to such a T is

$$E = \{p \in \bigcup_{\alpha<\omega_1} {}^{\alpha}\omega | p \notin T \wedge \forall \beta \in \mathrm{dom}(p) \quad (p \restriction \beta \in T)\}.$$

Suppose we are given some such T with associated E, and suppose that some $A \subseteq E$ is given. We then have a game $\mathcal{G}(A,T)$: I and II alternate playing natural numbers, I moving first at limit ordinals. The game is over when they reach a position $p \in E$, in which case I wins iff $p \in A$. (It is convenient to agree that if p is a position, then for all $\alpha \in \mathrm{dom}(p)$, $p(\alpha) = \langle m, n\rangle$ where m is I's αth move and n is II's αth move.) Let

$$\circlearrowleft^{T} A = \{p \in {}^{\omega}\omega | p \in T \wedge \text{I has a winning strategy in } \mathcal{G}(A,T) \text{ starting from } p\},$$

Of course, if $p \in E \Rightarrow \mathrm{dom}(p) < \omega_1$, so that T has no ω_1-branches, then $\mathcal{G}(A,T)$ is just the general form for a clopen (that is, decided after countably many moves) game of length ω_1. We want to restrict this notion a bit. Let WO be the set of (codes of) wellorders of ω. Let

$$F : (T \cup E) \cap \{p | \ \mathrm{dom}(p) \geq \omega\} \to \mathrm{WO},$$

where

$$\mathrm{dom}(p) = |F(p)| = \text{order type of } F(p)$$

for all $p \in \mathrm{dom}(F)$. For $p \in T \cup E$ such that $\mathrm{dom}(p) \geq \omega$, let

$$p^* = \langle F(p), x\rangle$$

where

$$x(n) = p(|n|_{F(p)}) = p \text{ (ordinal rank of } n \text{ in } F(p)).$$

(Here $\langle F(p), x\rangle \in {}^{\omega}\omega \times {}^{\omega}\omega \approx {}^{\omega}\omega$). For any $S \subseteq \mathrm{dom}(F)$, we let $S^* = \{p^* | p \in S\}$.

Now let

$$C(i,j,k,y) \text{ iff } \exists p \in \mathrm{dom}(F) \quad (y = p^* \wedge |i|_{F(p)} = |j|_{F(p \restriction \alpha)}, \text{ where } \alpha = |k|_{F(p)}).$$

We call F a *scaled coding of* T if both T^* and C admit HOD($\mathbf{R}$) scales.

Example Fix $n \geq 1$. Let

$$T = \left\{ p \in \bigcup_{\alpha < \omega_1} {}^{\alpha}\omega | \forall \beta \leq \text{dom}(p) \quad ((L_\beta[p \restriction \beta], \epsilon, p \restriction \beta) \text{ is not } \Sigma_n \text{ admissible}) \right\}.$$

For $q \in T \cup E$, $\text{dom}(q) \geq \omega$, let $F(q) =$ first wellorder of ω of order type $\text{dom}(q)$ constructed in $L[q]$. Then T^* and C are Π^1_1, so that F is a scaled coding of T.

We shall now prove our general scale propagation theorem. We remark afterward on refinements of the theorem involving weaker determinacy hypotheses and better definability estimates on the scale produced.

Theorem 1. Assume that all clopen games of length ω_1 with HOD ($\mathbf{R}$) payoff are determined. Let T admit a scaled coding, and suppose that $A \subseteq E$ is such that A^* admits a HOD($\mathbf{R}$) scale. Then $\mathfrak{D}^T A$ admits a HOD($\mathbf{R}$) scale.

Proof. We assume that if $p \in T$ and $\alpha \in \text{dom}(p)$, then $p(\alpha)_1 \in \{0, 1\}$; that is, II can only play 0's and 1's in $\mathcal{G}(A, T)$. This is no loss of generality.

Let F be a scaled coding of T, and $\vec{\rho}$ a HOD($\mathbf{R}$) scale on the associated C. This gives us some useful norms on A^* relating "global and local codes" of ordinals. For $p^* \in A^*$, let

$$\theta^0_i(p^*) = (n, \rho_{(i)0}(n, (i)_1, (i)_2, p^*))$$

where n is unique s.t. $C(n, (i)_1, (i)_2, p^*)$.

Remarks

(a) We code elements of $\omega^{<\omega}$ by prime powers, so that $\langle n_0 \dots n_k \rangle = 2^{n_0+1} \cdot 3^{n_1+1} \cdot \dots$, and $(i)_k =$ exponent of p_k in $i) - 1$. Let $(o)_k = 0$.

(b) $\theta^0_i(p^*)$ is an ordinal gotten by using the lexicographic order of $\omega \times$ range $\vec{\rho}$.

Let also, for $p^* \in A^*$,

$$\theta^1_i(p^*) = (n, \rho_{(i)_0}((i)_1, n, (i)_2, p^*))$$

where n is unique s.t. $C((i)_1, n, (i)_2, p^*)$ if any such n exists (i.e., if $|(i)_1|_{F(p)} < |(i)_2|_{F(p)}$. Let $\theta^1_i(p^*) = 0$ if no such n exists.

Let $\vec{\sigma}$ be a HOD($\mathbf{R}$) scale on T^*, and for $p^* \in A^*$ let

$$\theta^2_i(p^*) = \sigma_{(i)_0}((p \restriction \alpha)^*)$$

where $\alpha = |(i)_1|_{F(p)}$, and

$$\theta_i^3(p^*) = |i|_{F(p)}$$

and

$$\theta_i^4(p*) = n, \text{ where } |n|_{F(p)} = i.$$

Finally, let $\vec{\theta}_i^5$ be a very good scale on A^*, and set, for $p^* \in A^*$,

$$\psi_i(p^*) = \langle \mathrm{dom}(p), \theta_0^0(p^*), \ldots, \theta_0^5(p^*), \theta_1^0(p^*) \ldots \theta_1^5(p^*), \ldots \theta_i^0(p^*) \ldots \theta_i^5(p^*) \rangle.$$

We now describe some games which lead to a scale on $\Game^T A$. As in Moschovakis' proof that the ordinary game quantifier propagates scales, we compare positions $p, q \in \Game^T A$ by playing out $\mathcal{G}(A,T)$ from each position simultaneously on two boards. This assigns an ordinal value to each such position. The new ingredient is that in these comparison games the players must now make additional moves. These moves reflect the ordinal value they assign to one-move variants of intermediate positions they reach during the game.

So let $p, q \in \Game^T A$, and let $k \in \omega$. We shall define a game $G_k(p,q)$. The players in $G_k(p,q)$ are F and S. They play on two boards, the p board and the q board, and make additional moves lying on neither board. On the p board S plays $\mathcal{G}(A,T)$ from p as I while F plays as II. On the q board F plays $\mathcal{G}(A,T)$ from q as I while S plays as II. Play is divided into rounds; we now describe the typical round.

Round α

(a) F makes I's αth move on the q board, then S makes I's αth move on the p board.

(b) F now "proposes" some i such that $0 \leq i \leq k$ and $(i)_0 \in \{0,1\}$.

(c) S either accepts i or proposes some $i', 0 \leq i' < i$ and $(i')_0 \in \{0,1\}$.

(d) Let $t \leq k$ be the least proposal made during (b) and (c):

Case 1. $t \neq 0$. Then F and S must play $(t)_0$ as II's αth move on the p and q boards respectively.

Case 2. $t = 0$. Then F plays any $m \in \{0,1\}$ as II's αth move on the p board, after which S plays any $m' \in \{0,1\}$ as II's αth move on the q board.

This completes round α.

Remarks.

(a) We call the 0-proposal "freedom."

(b) Our description of round α is valid only if neither board has reached a position in E. As soon as one board reaches a position in E, F and S start simply playing $\mathcal{G}(A,T)$ in their proper roles on the other board, until it too reaches a position in E. $G_k(p,q)$ ends when both boards have reached a position in E.

(c) A position in $G_k(p,q)$ is a function $u : \alpha \to \omega$, where $u(\beta)$ codes the (up to) 6 moves during round β.

Suppose now that u is a run of $G_k(p,q)$ and that $r \supseteq p$ and $s \supseteq q$ are the runs of $\mathcal{G}(A,T)$ produced in u on the two boards. Let $e \in \omega$ be the least n such that, letting $\alpha = |(n)_0|_{F(r)}$, either $\alpha \geq \text{dom}(s)$ or $(n)_1$ was the least proposal made during round α. Let $\alpha = |(e)_0|_{F(r)}$. If either $\alpha \geq \text{dom}(s)$ or F proposed $(e)_1$ during round α, then

$$S \text{ wins } u \text{ iff } \psi_e(r^*) \leq \psi_e(s^*).$$

If S proposed $(e)_1$ during round α, then

$$S \text{ wins } u \text{ iff } \psi_e(r^*) < \psi_e(s^*).$$

Remarks

(a) We call e the critical number of u, and write $e = \text{crit}(u)$. Notice that $\text{crit}(u)$ exists, and in fact $\text{crit}(u) \leq \langle o, k \rangle$, for any run u of $G_k(p,q)$.

(b) Our convention is that ψ_e takes value ∞ off of A^*, that $\infty \leq \infty$, and that $x < \infty$ iff $x \in OR$. So, for example, if S proposed e_1 then S loses unless $r \in A$. For p and q in $\mathfrak{H}^T A$, let

$$p \leq_k q \text{ iff } S \text{ has a winning strategy (ws) in } G_k(p,q).$$

The next lemma implies that $\leq_k$ is a prewellorder.

Lemma 1. Let $p_0 \in \mathfrak{H}^T A$, and suppose that for all $n \geq 0$, Σ_n is either a ws for F in $G_k(p_n, p_{n+1})$ or a ws for S in $G_k(p_{n+1}, p_n)$.

Then only finitely many Σ_n's are for F.

Proof. Fix a ws τ for I from p_0 in $\mathcal{G}(A,T)$. Assume toward a contradiction that infinitely many Σ_n's are for F. We shall construct runs u_n according to Σ_n s.t. u_n and u_{n+1} agree on a common play $r_{n+1} \subseteq p_{n+1}$ on the p_{n+1} board, and the play $r_0 \subseteq p_0$ on the p_0 board is according to τ. (Notice Σ_n always plays as I on the p_{n+1} board and II on the p_n board.) The definition is by induction on rounds. Suppose we have $u_n \restriction \alpha$ for all n, and consequently $r_n \restriction \omega + \alpha$ for all $n (r_n \restriction \omega = p_n)$. Suppose also that $r_n \restriction \omega + \alpha \in T$ for all n; that is, no board has reached a position in E. We now define round α of the u_n's. Let

$$a_0 = \tau(r_0 \restriction \omega + \alpha)$$

and

$$a_{n+1} = \begin{cases} \Sigma_n(u_n \restriction \alpha^\frown a_n) & \text{if } \Sigma_n \text{ is for } S \\ \Sigma_n(u_n \restriction \alpha) & \text{if } \Sigma_n \text{ is for } F \end{cases}$$

be the αth moves for I given by τ and the Σ_n's. We must now define the proposal phase of $u_n(\alpha)$, for all n. We represent a proposal-acceptance/counterproposal by a pair (b, c) of numbers.

Claim. There are a $t \leq k$, an n_0, and pairs (b_n, c_n) for $n \in \omega$, such that

(i) $u_n \restriction \alpha^\frown\langle a_n, a_{n+1}, b_n, c_n, \rangle$ is according to Σ_n (if Σ_n is for S; otherwise $u_n \restriction \alpha^\frown\langle a_{n+1}, a_n, b_n, c_n\rangle$ is according to Σ_n)

and

(ii) $\langle b_n, c_n\rangle$ settles on freedom for $n < n_0$ and on t for $n \geq n_0$.

and

(iii) for infinitely many n, $\langle b_n, c_n\rangle$ involves Σ_n proposing t.

Proof (Sketch). Let $t_0 \leq k$ be least s.t. infinitely many Σ's which are F-strategies will now propose t_0. Suppose that t_n is given. If all but finitely many Σ's which are S strategies will accept a t_n proposal, stop the induction and set $t = t_n$. Otherwise let $t_{n+1} < t_n$ be such an infinitely many Σ's for S counterpropose t_{n+1} when their opponent proposes t_n.

Since $t_{n+1} < t_n$, we eventually get t. One can check that t works. (If $t_0 = t$, the Σ's verifying (iii) are for F. Otherwise, the Σ's verifying (iii) are for S.)

Now let $t, n_0, \langle(b_n, c_n) | n \in \omega\rangle$ satisfy the claim. We imagine (b_n, c_n) as the proposal phase of $u_n(\alpha)$. We want finally an αth move d_n for II on the p_n board.

Case 1. $t > 0$.

Then let

$$d_n = (t)_0 \qquad \text{for } n \geq n_0$$

and

$$d_n = \begin{cases} \Sigma_n(u_n \restriction \alpha^\frown\langle a_n, a_{n+1}, b_n, c_n, d_{n+1}\rangle) & \text{if } \Sigma_n \text{ is for } S \\ \Sigma_n(u_n \restriction \alpha^\frown\langle a_{n+1}, a_n, b_n, c_n\rangle) & \text{if } \Sigma_n \text{ is for } F \end{cases}$$

for $n < n_0$.

Case 2. $t = 0$.

Let

$$d_n = \begin{cases} \Sigma_n(u_n \restriction \alpha^\frown\langle a_n, a_{n+1}, b_n, c_n, d_{n+1}\rangle) & \text{if } \Sigma_n \text{ is for } S \\ \Sigma_n(u_n \restriction \alpha^\frown\langle a_{n+1}, a_n, b_n, c_n\rangle) & \text{if } \Sigma_n \text{ is for } F. \end{cases}$$

Since infinitely many Σ_n's are for F, this definition makes sense.

Finally, we set

$$u_n(\alpha) = \begin{cases} \langle a_n, a_{n+1}, b_n, c_n, d_{n+1}, d_n\rangle & \text{if } \Sigma_n \text{ is for } S \\ \langle a_{n+1}, a_n, b_n, c_n, d_n, d_{n+1}\rangle & \text{if } \Sigma_n \text{ is for } F. \end{cases}$$

Thus in any case $u_n(\alpha)$ and $u_{n+1}(\alpha)$ agree that $r_{n+1}(\alpha) = \langle a_{n+1}, d_{n+1}\rangle$, while $r_0(\alpha) = \langle a_0, d_0\rangle$ is according to τ.

We can continue to define $u_n(\alpha)$ this way until we reach an α such that $r_i \restriction \omega + \alpha \in E$ for some i. This must happen. Fix the first such α, and fix i. Now since τ is a ws for I and each Σ_n is a ws for F or S in G_k, we see that $r_n \restriction \omega + \alpha \in \Game^{T} A$ for all n. (cf. the payoff for G_k and our convention $x < \infty$ for $x \in OR$.) But then since the first component in any $\psi_e(r^*)$ is $\text{dom}(r)$, $r_n \restriction \omega + \alpha \in E$ for all $n \geq i$.

Now let $e \leq \langle o, k\rangle$ be least such that $e = \text{crit}(u_n)$ for infinitely many $n \geq i$. So $e \leq \text{crit}(u_n)$ for cofinitely many $n \geq i$, so that $\theta^3_{(e)_0}((r_n \restriction \omega + \alpha)^*)$ is eventually constant as $n \to \infty$. Write $r_n = r_n \restriction \omega + \alpha$, and let

$$\beta = \text{eventual value of } \theta^3_{(e)_0}(r_n^*) = |(e)_0|_{F(r_n)}.$$

Thus infinitely many, and hence by construction cofinitely many, u_n's settle on the $(e)_1$ proposal round at β. For infinitely many n, Σ_n is responsible for the proposal. Thus

$$\psi_e(r^*_{n+1}) \leq \psi_e(r^*_n)$$

for cofinitely many n, while

$$\psi_e(r^*_{n+1}) < \psi_e(r^*_n)$$

for infinitely many n. This is a contradiction. ∎

Corollary $\leq_k$ is a prewellorder of $\mathfrak{D}^T A$.

Proof reflexive: if $p \not\leq_k p$, then $p, p, p, \ldots$ violates the lemma
connected: if $p \leq_k q$ and $q \not\leq_k p$, then $pqpqpq \ldots$ violates the lemma
transitive: if $p \leq_k q \leq_k r$, and $p \not\leq_k r$, then $prqprqprq \ldots$ violates the lemma
wellfounded: clear from lemma. ∎

Actually, reflexivity, connectedness, and transitivity could be proved by more direct finite diagrams. Now for $p \in \mathfrak{D}^T A$, let

$$\phi_k(p) = \text{ordinal of } p \text{ in } \leq_k.$$

Lemma 2. $\vec{\phi}$ is a semiscale on $\mathfrak{D}^T A$.

Proof. Let $p_n \to p$ as $n \to \infty, p_n \in \mathfrak{D}^T A$ for all n, and $\forall_i$ ($\phi_i(p_n)$ eventually constant as $n \to \infty$). By thinning the sequence of p_n's we may assume we have a ws Σ_n for S in $G_n(p_{n+1}, p_n)$ for all n. Let τ be a ws for I in $\mathcal{G}(A, T)$ from p.

We need a definition. Suppose $r_n \in T \cup E$ for $n \in \omega$, and for all k

$$|k|_{F(r_n)} \text{ is eventually constant} = \alpha_k,$$

and

$$r_n(\alpha_k) \text{ is eventually constant}$$

as $n \to \infty$. Let $\pi : \beta \to \{\alpha_k | k \in \omega\}$ be the enumeration of $\{\alpha_k | k \in \omega\}$ in increasing order, and define $r : \beta \to \omega$ by

$$r(\gamma) = \text{ eventual value of } r_n(\pi(\gamma)) \text{ as } n \to \infty.$$

We write then

$$r = \lim^*_{n \to \infty} r_n.$$

This notion of "convergence in the codes" is more useful than pointwise convergence in what follows. ($\lim^*_{n \to \infty} r_n = r$ means just that r^*_n converges to r^* in a certain scale.)

We now define by induction on rounds runs u_n of $G_n(p_{n+1}, p_n)$ according to Σ_n. We arrange that u_n and u_{n+1} agree on a common play $r_{n+1} \supseteq p_{n+1}$ for the p_{n+1} board, and that the play $r_0 \supseteq p_0$ on the p_0 board is by τ. So assume that $u_n \upharpoonright \alpha$, hence $r_n \upharpoonright \omega + \alpha$, is given for all n. (Let $r_n \upharpoonright \omega = p_n$.) Let

$$a_0 = \tau(r_0 \upharpoonright \omega + \alpha)$$

and

$$a_{n+1} = \Sigma_n(u_n \restriction \alpha^\frown a_n)$$

be the moves for I on the various boards generated by τ and the Σ_n's. If a_n is eventually constant, say $a_n = a$ eventually, and if $\lim^*_{n\to\infty} r_n \restriction \omega + \alpha = r$ for some r which is a play by σ, let

$$d = \sigma(r^\frown a).$$

Otherwise, let $d = 0$. Now for any n let i_n be the largest i such that $\langle d, i\rangle \leq n$ and $\Sigma_n(u_n \restriction a^\frown a_n{}^\frown\langle d, i\rangle) =$ "accept", if such an i exists. Let $i_n = 0$ otherwise.

Case 1. $i_n \to \infty$ as $n \to \infty$.

Pick n_0 s.t. $i_n > 0$ for $n \geq n_0$. The proposal pair in $u_n(\alpha)$ is (b_n, c_n) where $b_n = \langle d, i_n\rangle$ for $n \geq n_0$ and $b_n = 0 =$ freedom for $n < n_0$, and $c_n =$ accept for all n. Let

$$d_n = \begin{cases} d & \text{if } n \geq n_0 \\ \Sigma_n(u_n \restriction a^\frown\langle a_n, a_{n+1}, b_n, c_n.d_{n+1}\rangle) & \text{if } n < n_0. \end{cases}$$

and set

$$u_n(\alpha) = \langle a_n, a_{n+1}, b_n, c_n, d_{n+1}\rangle$$

Case 2. Otherwise.

Then there is a $\langle d, i\rangle$ such that infinitely many Σ_n's reject $\langle d, i\rangle$. Just as in the claim in Lemma 1, we get a $t < \langle d, i\rangle$, an n_0, and pairs (b_n, c_n) such that

(i) $u_n \restriction \alpha^\frown\langle a_n, a_{n+1}, b_n, c_n\rangle$ is according to Σ_n,

(ii) $\langle b_n, c_n\rangle$ settles on freedom for $n < n_0$, and on t for $n \geq n_0$,

and

(iii) for infinitely many n, (b_n, c_n) involves Σ_n proposing t.

Subcase A. $t > 0$. Then we set

$$d_n = \begin{cases} (t)_0 & \text{if } n \geq n_0 \\ \Sigma_n(u_n \restriction \alpha^\frown\langle a_n, a_{n+1}, b_n, c_n, d_{n+1}\rangle) & \text{if } n < n_0. \end{cases}$$

Subcase B. $t = 0$. Define

$$S = \{\langle d_0 \ldots d_k\rangle | \forall i \leq k(d_i \in \{0,1\}) \text{ and } \forall i < k$$
$$d_i = \Sigma_i(u_i \restriction \alpha^\frown\langle a_i a_{i+1} b_i c_i d_{i+1}\rangle)\}$$

Clearly S is an infinite, finitely branching tree. (This was the reason we restricted II's plays in $\mathcal{G}(A,T)$ to $\{0,1\}$.) Let f be an infinite branch of S, and set $d_i = f(i)$.

Finally, in case 2 we set

$$u_n(\alpha) = \langle a_n, a_{n+1}, b_n, c_n, d_{n+1}, d_n \rangle.$$

We can continue to define $u_n(\alpha)$ this way until we reach, as we must, an α such that $r_i \restriction \omega + \alpha \in E$ for some i. Fix the least such α, and fix i. As before, $r_n \restriction \omega + \alpha \in \quad {}^T A$ for all n, so $r_n \restriction \omega + \alpha \in A$ for all $n \geq i$. Let $u_n = u_n \restriction \alpha$ and $r_n = r_n \restriction \omega + \alpha$.

We claim $\text{crit}(u_n) \to \infty$ as $n \to \infty$. If not, let e be least such that $e = \text{crit}(u_n)$ for infinitely many n. So $e \leq \text{crit}(u_n)$ for cofinitely many n, and $\theta^3_{(e)_0}(r_n^*) = |(e)_0|_{F(r_n)}$ is eventually constant as $n \to \infty$. Let β be this constant value. Then infinitely many u_n's settle on the $(e)_1$ proposal at round β. Thus Case 2 must apply at round β, so by construction cofinitely many u_n's settle on $(e)_1$ at round β, and for infinitely many n, Σ_n is responsible for $(e)_1$. But then

$$\psi_e(r_{n+1})^* \leq \psi_e(r_n^*)$$

for cofinitely many n, while

$$\psi_e(r_{n+1}^*) < \psi_e(r_n^*)$$

for infinitely many n, a contradiction.

Since $\text{crit}(u_n) \to \infty$ as $n \to \infty$, $\psi_e(r_n^*)$ is eventually constant as $n \to \infty$, for all e. Thus r_n^* converges in $\vec{\theta}^3, \vec{\theta}^4$, and $\vec{\theta}^5$, so that

$$\lim^*_{n\to\infty} r_n = r$$

for some $r \supseteq p$ such that $r \in A$. We are done if we show that r is a play according to σ. If not, let β be least so that $r(\beta)$ is not according to σ. Let

$$\beta = |k|_{F(r)}$$

and

$$\gamma = \text{ eventual value of } |k|_{F(r_n)} \text{ as } n \to \infty$$

we claim that

$$\lim^*_{n\to\infty} r_n \restriction \gamma = r \restriction \beta.$$

This follows from the convergence of r_n^* in $\vec{\theta}^0$ and $\vec{\theta}^1$: if $\delta < \gamma$, then the code for δ relative to $F(r_n \restriction \gamma)$ stabilizes as $n \to \infty$ iff the code for δ relative to $F(r_n)$ stabilizes as $n \to \infty$.

Since r^* converges in $\vec{\theta}^2$, we get that $r \restriction \beta \in T$.

Since r_n^* converges in $\vec{\theta}^5$, which is very good, the a_n's defined at round γ are eventually constant $= a$. But then at round γ, $d = \sigma(r \restriction \beta^\frown a)$. Moreover, Case 1 must apply at round γ, since otherwise crit(u_n) has a finite lim inf. Thus $r_n(\gamma) = \langle a, d\rangle$ for all sufficiently large n. Since $\vec{\theta}^5$ is very good, $r(\beta) = \langle a, d, \rangle$. But then $r \restriction \beta + 1$ is according to σ, a constradiction which completes the proof of Lemma 2. ∎

Lemma 2 completes the Proof of Theorem 1, since a well known construction produces as HOD$(\mathbf{R})$ scale on any set which carries a HOD$(\mathbf{R}\ )$ semiscale. ∎

Given a scaled coding F of T, and a pointclass Γ of a set of reals, let

$$\Game^{T,F}\Gamma = \{\Game^T A \mid A^* \in \Gamma\}.$$

The proof of Theorem 1 shows that if T and F are "reasonable" and Γ has the semiscale property, so does $\Game^{T,F}\Gamma$. For example, let $T_n = \{p \in \cup_{\alpha<\omega_1} {}^\alpha\omega \mid \forall\beta \in \text{dom}(p)\ ((L_\beta[p \restriction \beta], \epsilon, p \restriction \beta)$is not Σ_n admissible$\}$, and let F_n be the scaled coding of T_n described at the beginning of this section. Let us write

$$\Game^{\Sigma_n}\Gamma = \Game^{T_n,F_n}\Gamma,$$

and agree that

$$\Gamma - \text{AD}^{\Sigma_n} \text{ iff } \mathcal{G}(A, T_n) \text{ is determined whenever } A^* \in \Gamma.$$

(For "$P(\mathbf{R}) - \text{AD}^{\Sigma_n}$" we write simply "$\text{AD}^{\Sigma_n}$").

Corollary 1. Let $n \geq 1$, and assume $\Gamma - \text{AD}^{\Sigma_n}$, where Γ is closed under $\forall^{\mathbf{R}}$. Suppose all Γ sets admit Γ semiscales. Then all $\Game^{\Sigma_n}\Gamma$ sets admit $\Game^{\Sigma_n}\Gamma$ scales, so that all $\Game^{\Sigma_n}\Gamma$ relations admit $\Game^{\Sigma_n}\Gamma$ uniformizations.

Proof. Suppose $A^* \in \Gamma$. Let us trace through the proof of Theorem 1. We can take $\theta_i^0 \ldots \theta_i^5$ to be Γ norms. We must modify ψ_i slightly to get a Γ norm; let

$$\psi'_i(p^*) = \langle \theta_0^5(p^*), \psi_i(p^*)\rangle.$$

Finally, let $G'_k(p,q)$ be just like $G_k(p,q)$ except that S cannot win $G_k(p,q)$ unless he wins as I on the p board. The proof of Theorem 1 goes through with these modifications (cf. [7], Chapter 6) and yields a $\Game^{\Sigma_n}\Gamma$ semiscale $\vec{\phi}'$ on $\Game^{\Sigma_n}A$. One can easily check that, since $\Game^{\Sigma_n}\Gamma$ is closed under real quantification, the scale of the tree of $\vec{\phi}'$ is in fact a $\Game^{\Sigma_n}\Gamma$ scale. ∎

We would like to point out two curious features of the proof of Theorem 1. First, it handles directly only games where II must play from $\{0,1\}$. Second, the verification that $p_\omega \in \mathfrak{D}^T A$ when $p_n \to p_\omega$ mod $\vec{\phi}$ is indirect: we do not construct a strategy for I in $\mathcal{G}(A,T)$ from p_ω, but instead defeat a strategy for II. Is there a more direct proof avoiding these devices?

§2. Canonical strategies.

We shall construct definable winning strategies for the games of the form $\mathcal{G}(A,T)$, where T admits a scaled coding and A^* a $HOD(\mathbf{R})$ scale. The strategies we define are in some sense "best", as in Moschovakis [6], Chapter 6. However, the games $G_k(p,q)$ are not adequate to evaluate what's best for I; the problem is there is no satisfactory way to decide which k to use in evaluating a given position. So we use the game $G_\omega(p,q)$, which is like $G_k(p,q)$ except that no bound is put on the size of proposals F may make. Now $G_\omega(p,q)$ leads to a probably ill-founded value order on I's possible next moves, but we can avoid that problem by considering directly only games where I must play from $\{0,1\}$. (Curiouser and curiouser!)

Once again, we prove our most general theorem first, then state its sharpened form for games ending at the first Σ_n admissible relative to the play as a corollary.

If $\Sigma_\bullet$ is a strategy for $\mathcal{G}(A,T)$, where T admits a scaled coding, then $\Sigma^* = \{p | p$ is a position according to $\Sigma\}$.

Theorem 2. Assume all clopen games of length ω_1 with HOD ($\mathbf{R}$) payoff are determined. Suppose T admits a scaled coding, and $A \subseteq E$ is such that A^* admits a HOD($\mathbf{R}$) scale. Then if I wins $\mathcal{G}(A,T)$ he wins via a strategy Σ such that Σ^* admits a HOD($\mathbf{R}$) scale.

Proof. We assume without loss of generality that $(p \in T \wedge \delta \in \text{dom}(p)) \Rightarrow (p(\delta)_0 \in \{0,1\} \wedge p(\delta)_1 \in \{0,1\}$; that is, both players must play 0 or 1 in $\mathcal{G}(A,T)$. In order to conform to the notation of §1, we assume we have a fixed $p_0 \in T$ such that $\text{dom}(p_0) = \omega$ and I wins $\mathcal{G}(A,T)$ from p_0 : the canonical Σ we are to construct must win from p_0. This is no loss of generality.

Let $\vec{\rho}, \vec{\theta^0} \dots \vec{\theta^5}$, and $\vec{\psi}$ be the families of norms on A^* defined from a scaled coding of T and a scale on A^* just as in the proof of Theorem 1. For $p, q, \in T$ such that $\text{dom}(p) = \text{dom}(q)$ we define a game $G_\omega(p,q)$. The definition is the same as that of $G_k(p,q)$ for $k \in \omega$, except that: (a) we do not require $\text{dom}(p) = \omega$, or even that p or q are winning positions for I in $\mathcal{G}(A,T)$ (this assumption played no role in the definition of $G_k(p,q)$ anyway), and (b) we allow F to propose any $t \in \omega$, not just $t \leq k$, when it's his turn to propose in $G_\omega(p,q)$. (If S rejects t, he must still counterpropose a $t' < t$.) Now, let for $p \supseteq p_0$,

$$\Sigma(p) = \begin{cases} 0 & \text{if } \forall a \in \{0,1\} \exists b \in \{0,1\} \text{s.t.} S \text{ has a ws in } G_\omega(p^\frown\langle o, a\rangle, p^\frown\langle 1, b\rangle), \\ 1 & \text{otherwise.} \end{cases}$$

The determinacy of G_ω and its symmetry imply that if $\Sigma(p) = 1$, then $\forall a \in \{0,1\}$ $\exists b \in \{0,1\}$ (S has a ws in $G_\omega(p^\frown\langle 1, a\rangle, p^\frown\langle 0, b\rangle)$). Thus

$$\Sigma(p) = i \Rightarrow S \text{ has a ws in } H(p),$$

where $H(p)$ is the game in which F plays a, then S plays b, then F and S play out $G_\omega(p^\frown\langle i, a\rangle, p^\frown\langle 1 - i, b\rangle)$

It is easy to see that Σ^* is of the form $\Game^S B$, for some S admitting a scaled coding and B such that B^* admits a HOD($\mathbf{R}$) scale. By Theorem 1, then, it is enough to show that Σ is a ws for I in $\mathcal{G}(A, T)$ from p_0. So let $q \in E$ be an arbitrary play according to Σ; we want to show $q \in A$.

Fix a ws τ for I in $\mathcal{G}(A, T)$ from p_0. For $\delta \in \text{dom}(q)$, let Σ_δ be a ws for S in $H(q \restriction \delta)$. We respresent a position or completed run of $H(q \restriction \delta)$ by a function $u : \alpha \to \omega$, where $\alpha \geq \delta$ and $u \restriction \delta = q \restriction \delta, u(\delta) = \langle 1 - q(\delta)_0, q(\delta)_0, a, b\rangle$ where a, b are the first two moves of $H(q \restriction \delta)$, and $u(\delta + 1 + \eta)$ codes the (up to) six moves of round η in $G_\omega(q \restriction \delta^\frown\langle q(\delta)_0, a\rangle, q \restriction \delta^\frown\langle 1 - q(\delta)_0, b\rangle)$. If u is a position or run of $H(q \restriction \delta)$, then the lower board of u is r, where $r \restriction \delta = q \restriction \delta, r(\delta) = \langle q(\delta)_0, u(\delta)_2\rangle$, and $r(\delta + 1 + \eta) = \langle u(\delta + 1 + \eta)_1, u(\delta + 1 + \eta)_4\rangle$. Similarly, the upper board of u is s, where $s \restriction \delta = q \restriction \delta, s(\delta) = \langle 1 - q(\delta)_0, u(\delta)_3\rangle$, and $s(\delta + 1 + \eta) = \langle u(\delta + 1 + \eta)_0, u(\delta + 1\eta)_5\rangle$.

Notice that from position u of $H(q \restriction \delta)$ we can recover δ; δ is the least $\alpha \in \text{dom}(u)$ such that $u(\alpha)_0 \neq q(\alpha)_0$. Let us write $\delta = \delta(u)$.

Let us call a sequence $\langle u_\beta | \beta < \gamma\rangle$ a *diagram* if $\gamma \leq \text{dom}(u_0)$ and

(a) for $\beta < \gamma$, u_β is a position in $H(q \restriction \delta)$ according to Σ_δ, where $\delta = \delta(u_\beta)$,

and

(b) letting r_β be the upper board of u_β for $\beta < \gamma$, we have: r_0 according to τ, and $r_{\beta+1}$ is the lower board of u_β for $\beta + 1 < \gamma$.

Our plan is to construct a diagram $\langle u_\beta | \beta < \gamma\rangle$ such that for any limit $\lambda \leq \gamma$, $\text{crit}(u_\beta) \to \omega$ as $\beta \to \lambda$. We also arrange that $r_0 \in E$, that $r_\lambda = \lim^*_{\beta \to \lambda} r_\beta$ if $\lambda < \gamma$ is a limit, and that $q = \lim^*_{\beta \to \gamma} r_\beta$ if γ is a limit, while q is the lower board of $u_{\gamma-1}$ if γ is a successor. The existence of such a diagram easily implies $q \in A$. [$r_0 \in A$ since r_0 is by τ. But then $r_\beta \in A$ for $\beta < \gamma$ by induction: since r_β and $r_{\beta+1}$ are the upper and lower boards of u_β, which is by some $\Sigma_\delta, r_\beta \in A \Rightarrow r_{\beta+1} \in A$. If $\lambda < \gamma$ is a limit, then since $\text{crit}(u_\beta) \to \omega$ as $\beta \to \lambda$, r^*_β converges in $\vec{\theta}^3, \vec{\theta}^4, \vec{\theta}^5$ to some r^* in A^* as $\beta \to \lambda$; moreover $r = r_\lambda$ since $\lim^*_{\beta\to\lambda} r_\beta = r_\lambda$. So $r_\beta \in A$ for $\beta < \gamma$; repeating the argument we get $q \in A$.]

We obtain the desired diagram by means of a sequence $D_\alpha = \langle u^\alpha_\beta | \beta < \gamma_\alpha \rangle$ of diagrams defined by induction on α. We maintain by induction that $\alpha < \sigma \Rightarrow u^\alpha_0 \subseteq u^\sigma_0$. In fact, if $n < \min(\omega, \gamma_\alpha)$ and $\alpha < \sigma$, then $n < \gamma_\sigma$ and $u^\alpha_n \subseteq u^\sigma_n$. (On the other hand, we may well have $\alpha < \sigma$ such that u^α_ω and u^σ_ω are defined and incompatible. We are really building a tree of approximations to the desired u_ω, "continuously" associating to each branch of this tree a tree of approximations to $u_{\omega+\omega}$, etc. This is because we do not know definitely even initial segments of the eventual $r_\omega, r_{\omega+\omega}$, etc., as we build our diagram. For notational simplicity, however, we shall suppress explicit mention of these trees.)

If $\langle u_\beta | \beta < \gamma \rangle$ is any diagram, then $\langle (u_\beta, a_\beta) | \beta < \gamma \rangle$ is an *enlarged diagram* if $a_0 = \tau(r_0)$ and $a_{\beta+1} = \Sigma_\delta(u_\beta{}^\frown\langle a_\beta \rangle)$ for $\beta + 1 < \gamma$ and $\delta = \delta(u_\beta)$. Given such an enlarged diagram, let

$$i_\beta = \begin{cases} \text{the least } t \text{ such that } (t)_0 \in \{0,1\} \text{ and} \\ \Sigma_\delta(u_\beta^\frown\langle a_\beta, a_{\beta+1}, t \rangle) = \text{reject, where } \delta = \delta(u_\beta), \\ \omega, \text{if no such } t \text{ exists.} \end{cases}$$

For $\lambda \leq \gamma$ a limit, we say $\langle (u_\beta, a_\beta) | \beta < \gamma \rangle$ *accepts readily at* λ iff $\lim_{\beta\to\lambda} i_\beta = \omega$.

Given an enlarged diagram $\langle (u_\beta, a_\beta | \beta < \gamma \rangle$ with $\gamma = \beta + 1$, define a_γ by: $a_\gamma = \Sigma_\delta(u_\beta{}^\frown\langle a_\beta \rangle)$, where $\delta = \delta(u_\beta)$.

*Lemma 3.*Suppose $\langle (u_\beta, a_\beta | \beta < \gamma \rangle$ is an enlarged diagram which accepts readily at all limit $\lambda \leq \gamma$. Let $d \in \{0,1\}$. Then there are b_β, c_β, and d_β, for $\beta < \gamma$, such that, setting $d_\gamma = d$,

(a) $(u_\beta{}^\frown\langle a_\beta, a_{\beta+1}, b_\beta, c_\beta, d_{\beta+1}, d_\beta \rangle$ is by $\Sigma_\delta, \delta = \delta(u_\beta)$, for $\beta < \gamma$,

(b) $d_\beta \to d_\lambda$ as $\beta \to \lambda$, for $\lambda \leq \gamma$ a limit,

and

(c) $b_\beta \to \omega$ as $\beta \to \lambda$, for $\lambda \leq \gamma$ a limit
$c_\beta =$ accept, for all but finitely many β.

Proof. By induction on γ. The successor step is easy, so let γ be a limit. Let $\pi : \gamma \xrightarrow{1-1} \omega$ be determined by r_0. Let i_β, for $\beta < \gamma$, be as in the definition of ready acceptance. Let $\eta < \gamma$ be such that $\langle d, o \rangle \leq i_\beta$ for $\eta \leq \beta$ (so $\Sigma_{\delta(u_\beta)}$ accepts $\langle d, o \rangle$ for $\eta \leq \beta$.) Let λ be the largest limit

$\leq \eta$. For $\eta \leq \beta < \gamma$, let

$$b_\beta = \text{the largest } t \leq \max(\pi(\beta), i_\beta) \text{ s.t. } (t)_0 = d$$
$$\text{and } \Sigma_\delta \text{ accepts } t \text{ after } u_\beta{}^\frown\langle a_\beta, a_{\beta+1}, t\rangle,$$
$$\text{where } \delta = \delta(u_\beta),$$

and

$$c_\beta = \text{accept}, \ d_\beta = d.$$

For $\lambda \leq \beta < \eta$, let

$$b_\beta = 0 = \text{freedom}, \qquad c_\beta = \text{accept}$$

and

$$d_\beta = \Sigma\delta(u_\beta{}^\frown\langle a_\beta, a_{\beta+1}, b_\beta, c_\beta, d_{\beta+1}\rangle), \text{ for } \delta = \delta(u_\beta).$$

Finally, the induction hypothesis with $d = d_\lambda$ will give us the desired b_β, c_β and d_β for $\beta < \lambda$. ∎

We shall use tacitly the fact that there is a function which, given a sequence of satisfying the hypothesis of the lemma and a $d \in \{0,1\}$, produces a sequence satisfying its conclusion. (We don't have AC!)

We are ready to define our approximations $D_\alpha = \langle u^\alpha_\beta | \beta < \gamma_\alpha\rangle$ to the desired diagram; the definition is by induction on α.

$\underline{\alpha = 0}$: We may assume q is not by τ; otherwise $q \in A$ and we're done. So let δ be least such that $\tau(q \restriction \delta) \neq q(\delta)_0$. Set

$$\gamma_0 = 1,$$

and

$$u^0_0 = q \restriction \delta^\frown\langle 1 - q(\delta)_0, q(\delta)_0, q(\delta)_1, \Sigma_\delta(q \restriction \delta^\frown\langle 1 - q(\delta)_0, q(\delta)_0, q(\delta)_1\rangle)$$

$\underline{\alpha > 0}$: Set $v_0 = \cup_{\eta<\alpha} u^\eta_0$. We use s_β for the upper board of v_β. If $s_0 \in E$, then our induction on α stops; D_α is undefined. Otherwise, let $a_0 = \tau(s_0)$. We shall define v_β and a_β by induction on β. Suppose we have v_β and a_β for $\beta < \gamma$, and that $\langle(v_\beta, a_\beta)|\beta < \gamma\rangle$ is an enlarged diagram. Suppose also the following four conditions are met.

(1) γ limit $\Rightarrow \langle(v_\beta, a_\beta)|\beta < \gamma\rangle$ accepts readily at γ.

(2) if γ is a limit, then $\lim_{\beta\to\gamma} a_\beta$ and $\lim^*_{\beta\to\gamma} s_\beta$ exists, moreover $p_0 \supseteq \lim^*_{\beta\to\gamma} s_\beta$.

(3) Let $s = \lim^*_{\beta \to \gamma} s_\beta$ if γ is a limit, and s be the lower board of $v_{\gamma-1}$ otherwise. Then $\exists \delta \operatorname{dom}(s) \cap \operatorname{dom}(q) \quad (s(\delta) \neq q(\delta))$; moreover, if δ is the least such ordinal, then $s(\delta)_0 \neq q(\delta)_0$.

(4) For $\theta \in \operatorname{dom}(s_0)$, define t_θ by

$$t_\theta(0) = \theta, \quad t_\theta(\beta+1) = t_\theta(\beta) \text{ for } \beta+1 \leq \gamma,$$

and

$$\begin{aligned} t_\theta(\lambda) = & |e|_{F(r)}, \text{ where } r - \lim^*_{\beta \to \lambda} s_\beta \\ & \text{and } |e|_{F(s_\beta)} = t_\theta(\beta) \\ & \text{for all sufficiently large } \beta < \lambda, \end{aligned}$$

for $\lambda \leq \gamma$ a limit; $t_\theta(\lambda)$ is undefined is no such $e \in \omega$ exists. (One should visualize t_θ as the θth "column" of the diagram $\langle v_\beta | \beta < \gamma \rangle$. Not all columns extend to γ, since not all contribute to $*$-limits all the way down.) Let

$$C_\gamma = \{\theta | t_\theta(\gamma) \text{ is defined}\}.$$

Let s, δ be as in (3). (Our assumptions imply $\operatorname{dom}(s) = \{t_\delta(\gamma) | \theta \in C_\gamma\}$.) Fix θ (unique) such that $\delta = t_\theta(\gamma)$. Then we require that for $\theta \leq \sigma \leq \rho$ and $\sigma, \rho, \in C_\gamma$

$$\gamma < \min(\gamma_\sigma, \gamma_\rho) \text{ (i.e.} u^\sigma_\gamma, u^\rho_\gamma \text{ are defined)},$$

$$u^\sigma_\gamma \subseteq u^\rho_\gamma, \text{ and } s \restriction (t_\sigma(\gamma)+1) \text{ is the upper board of } u^\sigma_\gamma.$$

If conditions (1)-(4) are met, then we set

$$a_\gamma = \lim_{\beta \to \gamma} a_\beta,$$

and

$$v_\gamma = \cup\{u^\sigma_\gamma | \sigma \in C_\gamma \wedge t_\sigma(\gamma) \geq \delta\}.$$

Notice that $\langle (v_\beta, a_\beta) | \beta < \gamma+1 \rangle$ remains an enlarged diagram.

If one of (1)-(4) fails, then v_γ is undefined. This must happen at some $\gamma \leq \sup_{\sigma<\alpha} \gamma_\sigma$ since at least (4) will fail. So suppose γ is at least such that v_γ is undefined. We shall define D_α by taking cases on which of (1)-(4) fail at γ.

Case 1 (1) fails at γ.

So γ is a limit. Set $\gamma_\alpha = \gamma$. Now since (1) fails, we get a $t \in \omega$ and a cofinal $B \subseteq \gamma$ such that for $\beta \in B$, there is a b such that $(b)_0 \in \{0,1\}$ and Σ_δ counterproposes t when F proposes b after $v_\beta{}^\frown\langle a_\beta, a_{\beta+1}\rangle$, where $\delta = \delta(v_\beta)$. Let t be least such that a cofinal $B \supseteq \gamma$ exists, and fix such a B of order type ω.

Subcase A. $t > 0$.

There is an $\eta < \gamma$ such that for $\eta \leq \beta < \gamma$ and $\delta = \delta(v_\beta), \Sigma_\delta$ accepts t after $v_\beta{}^\frown\langle a_\beta, a_{\beta+1}\rangle$; this follows from the minimality of t. Let $\pi : \gamma \xrightarrow{1-1} \omega$, be determined by s_0. For $\eta \leq \beta \leq \gamma$, let

$b_\beta =$ some b witnessing $\beta \in B$, if $\beta \in B$,

$c_\beta =$ (reject, t), if $\beta \in B$,

$b_\beta =$ the largest $i \leq \max(t, \pi(\beta))$ s.t.$(i)_0 = (t)_0$ and Σ_δ accepts i after $v_\beta{}^\frown\langle a_\beta, a_{\beta+1}\rangle$, where $\delta = \delta(v_\beta)$, if $\beta \notin \beta$,

$c_\beta =$ accept, if $\beta \notin B$,

and

$$d_\beta = (t)_0.$$

Let λ be the largest limit ordinal $\leq \eta$. For $\lambda \leq \beta < \eta$, set

$$b_\beta = 0 = \text{freedom}, \ \ c_\beta = \text{accept},$$

and

$$d_\beta = \Sigma_\delta(v_\beta{}^\frown\langle a_\beta, a_{\beta+1}, b_\beta, c_\beta, d_{\beta+1}\rangle), \quad \text{for } \delta = \delta(v_\beta).$$

Finally, since $\langle (v_\beta, a_\beta) | \beta < \gamma\rangle$ accepts readily at all limit $\lambda' \leq \lambda$, we may apply lemma 3 with $d = d_\lambda$ to generate b_β, c_β, and d_β for $\beta < \lambda$. Then let

$$u^\alpha_\beta = v_\beta{}^\frown\langle a_\beta, a_{\beta+1}, b_\beta, c_\beta, d_{\beta+1}, d_\beta\rangle$$

for $\beta < \gamma = \gamma_\alpha$.

Subcase B. $t = 0$.

Let i_β be as in the definition of ready acceptance, and let

$$S = \{\sigma < \gamma | \exists_\beta [\sigma \leq \beta < \sigma + \omega \wedge (\beta \in B \text{ or } i_\beta \leq \max(\langle 0,0\rangle, \langle 1,0\rangle))]\}.$$

Then S has order type ω, as otherwise (1) fails at some $\lambda < \gamma$. Let

$$b_\beta = \begin{cases} \text{some } b \text{ witnessing } \beta \in B, \text{ if } \beta \in B \\ 0, \text{ if } \beta \in S - B \end{cases}$$

and

$$c_\beta = \begin{cases} (\text{reject},0) & \text{if } \beta \in B \\ \text{accept}, & \text{if } \beta \in S - B. \end{cases}$$

Let $\{\sigma_i | i \in \omega\}$ be the increasing enumeration of S, and consider

$$U = \{\langle d_0 \ldots d_k \rangle \in 2^\omega / \forall i < k \text{ (if } \sigma = \sigma_i \text{ and } \delta = \delta(v_\sigma)),$$
$$\text{then } d_i = \Sigma_\delta(v_\sigma{}^\frown\langle a_\sigma, a_{\sigma+1}, b_\sigma, b_\sigma, c_\sigma, d_{\sigma+1}\rangle) \}$$

U is an infinite tree on $\{0,1\}$ so we have an $f \in {}^\omega 2$ such that $f \restriction k \in U$ for all k. Set

$$d_{\sigma_i} = f(i), \quad \text{for } i \in \omega.$$

Finally, suppose $\beta \in \gamma - S$. Let $\sigma \in S$ be least such that $\beta < \sigma$. (σ is a limit.) Set

$$\begin{aligned} b_\beta = &\text{the largest } i \leq \max(i_\beta, \pi(\beta)) \text{s.t.} \\ &(i)_0 = d_\sigma \text{ and } \Sigma_\delta \text{ accepts } i \\ &\text{after } v_\beta{}^\frown\langle a_\beta, a_{\beta+1}\rangle, \text{ where } \delta = \delta(v_\beta), \end{aligned}$$

and

$$c_\beta = \text{accept}, \; d_\beta = d_\sigma.$$

Now, we let, for $\beta < \gamma = \gamma_\alpha$

$$u^\alpha_\beta = v_\beta{}^\frown\langle a_\beta, a_{\beta+1}, b_\beta, c_\beta, d_{\beta+1} d_\beta\rangle.$$

This completes the definition of D_α in Case 1.

Case 2. (1) holds and (2) fails at γ.

Since $\langle (v_\beta), a_\beta | \beta < \gamma\rangle$ accepts readily at all limit $\lambda \leq \gamma$, lemma 3 applied with $d = 0$ yields b_β, c_β, and c_β for $\beta < \gamma$. Set $\gamma_\alpha = \gamma$, and $u^\alpha_\beta = v_\beta{}^\frown\langle a_\beta, a_{\beta+1} b_\beta, c_\beta, d_{\beta+1}, d_\beta\rangle$ for $\beta < \gamma$.

Case 3. (1) and (2) hold but (3) fails at γ.

Let s be as in (3). Let $a = a_\gamma$ if γ is a successor, $a = \lim_{\beta\to\gamma} a_\beta$ otherwise.

Subcase A. $q \subseteq s$, or $\exists\delta \in \mathrm{dom}(s) \cap \mathrm{dom}(q)(s(\delta) \neq q(\delta))$, but $s(\delta)_0 = q(\delta)_0$ for the least such δ.

In this case, proceed exactly as in Case 2.

Subcase B. $s \subsetneq q$ and $q(\mathrm{dom}(s))_0 = a$. In this case apply lemma 3 with $d = q(\mathrm{dom}(s))_1$ to get b_β, c_β and d_β for $\beta < \gamma$. Set $\gamma_\sigma = \gamma$, and $u^\alpha_\beta = v_\beta{}^\frown\langle a_\beta, a_{\beta+1}, b_\beta, c_\beta, d_{\beta+1}, d_\beta\rangle$ for $\beta < \gamma$.

Subcase C. $s \subseteq q$ and $q(\mathrm{dom}(s))_0 = 1 - a$.

This is the only case in which we set up a new board. Let $\gamma_\alpha = \gamma + 1$.

Set

$$u^\alpha_\gamma = s^\frown\langle a, 1-a, q(\delta)_1, \Sigma_\delta(s^\frown\langle a, 1-a, q(\delta)_1\rangle)\rangle,$$

where $\delta = \mathrm{dom}(s)$. Now use the lemma with $d = \Sigma_\delta(s^\frown\langle a, 1-a, q(\delta)_1\rangle)$ to define u^α_β for $\beta < \gamma$.

Case 4. (1)-(3) hold, but (4) fails at γ.

Let $\gamma_\alpha = \gamma$, and apply lemma 3 with $d = 0$ to get u^α_β for $\beta < \gamma$.

This completes the inductive definition of the D_α's. Let $v^\alpha_\beta, s^\alpha_\beta$ be the v_β and s_β occuring in the definitions of $D_\alpha, \alpha > 0$.

Let $M = \langle u_\beta | \beta < \gamma\rangle$ be a diagram with boards $r_\beta, \beta < \gamma$. We call M *good* iff $\lim^*_{\beta\to\lambda} r_\beta$ exists for all limit $\lambda \leq \gamma$, and $\lim^*_{\beta\to\lambda} r_\beta = r_\lambda$ for all limit $\lambda \leq \gamma$. If $M = \langle u_\beta | \beta < \gamma\rangle$ is good, we define t^M_θ for $\theta \in \mathrm{dom}(u_0)$ to be the θth "column" of M, as in condition (4) above: $t^M_\theta(0) = \theta, t^M_\theta(\beta+1) = t^M_\theta(\beta)$ for $\beta + 1 < \gamma$, and $t^M_\theta(\lambda) = |e|_{F(r)}$, where $\lambda \leq \gamma$ is a limit, $r = \lim^*_{\beta\to\lambda} r_\beta$, and $|e|_{F(r)_\beta} = t^M_\theta(\beta)$ for all sufficiently large $\beta < \lambda$. ($\lambda \notin \mathrm{dom}(t^M_\theta$ if no such e exists.) Let also $C^M_\gamma = \{\theta | t^M_\theta(\gamma)$ is defined$\}$.

We now construct the desired diagram. Let $u_0 = \cup\{u^\alpha_0 | u^\alpha_0$ is defined$\}$. Now suppose $M = \langle u_\beta | \beta < \gamma\rangle$ is given, and that

(a) crit $(u_\beta) \to \omega$ as $\beta \to \lambda$, for any limit $\lambda \leq \gamma$,

(b) M is a good diagram,

and

(c) for $\beta < \gamma$, if $\alpha \in C^M_\beta$ and $t^M_\alpha(\beta) \leq \delta(u_\beta)$, then $u^\alpha_\beta = u_\beta \restriction (t^M_\alpha(\beta) + 1)$. (So u^α_β is defined.)

Let $r = \lim^*_{\beta\to\gamma} r_\beta$ if γ is a limit, and let r be the lower board of $u_{\gamma-1}$ otherwise. If $r = q$ then (a) and (b) guarantee that M is the desired diagram. Otherwise, since $r, q \in E$, they are incompatible; let δ be least such that $r(\delta) \neq q(\delta)$. We define u_γ by

$$u_\gamma = \cup\{u_\gamma^\alpha | \alpha \in C_\gamma^M \wedge t_\alpha(\gamma) \geq \delta\};$$

we must see that (a), (b) and (c) remain true. The following claims insure this.

Claim 1. If $\alpha \in C_\gamma^M, \beta < \gamma$, and $t_\alpha^M(\beta) < \delta(u_\beta)$, then $t_\alpha^M(\gamma) < \delta$.

Proof. By induction on $\eta \geq \beta$ we see that $t_\alpha^M(\eta) = t_\alpha^M(\beta)$ and $r_\eta \restriction t_\alpha^M(\eta) = q \restriction t_\alpha^M(\eta)$. If η is a successor, this follows from the definition of "diagram" and "column". If η is a limit, then since $\alpha \in C_\eta^M$, the r_θ-code of $t_\alpha^M(\theta)$ is eventually constant as $\theta \to \eta$. Since r_θ^* converges in $\vec{\theta}^1$ and $\vec{\theta}^2$, this means $r_\eta \restriction t_\alpha^M(\eta) = \lim^*_{\theta\to\eta} r_\theta \restriction t_\alpha^M(\theta) = q \restriction t_\alpha^M(\beta)$.

We can repeat the above argument with γ for η and r for r_η to get $r \restriction t_\alpha^M(\gamma) = q \restriction t_\alpha^M(\gamma)$, so $t_\alpha^M(\gamma) < \delta$.

Claim 2. If $\alpha \in C_\gamma^M$ and $t_\alpha^M(\gamma) = \delta$, then $\gamma_\alpha = \gamma + 1$ and Case 3C held in the definition of D_α. Moreover $r \restriction t_\alpha^M(\gamma) + 1$ is the upper board of u_γ^α.

Proof. By Claim 1, $t_\alpha^M(\beta) \geq \delta(u_\beta)$ for $\beta < \gamma$. So by hypothesis (c) on M,

$$u_\beta^\alpha = u_\beta \restriction t_\alpha^M(\beta) + 1, \text{ for } \beta < \gamma \leq \gamma_\alpha.$$

Also $v_\beta^\alpha = u_\beta \restriction t_\alpha^M(\beta)$, and $s_\beta^\alpha = r_\beta \restriction t_\alpha^M(\beta)$ for $\beta < \gamma$. Now condition (1) holds at γ in the definition of the v_β^α's, as otherwise our construction in Case 1 guarantees that $\text{crit}(u_\beta) \leq \langle e, t\rangle$ for infinitely many $\beta < \gamma$, where e is the eventual r_β code of $t_\alpha^M(\beta)$ and t is as in Case 1. Condition (2) holds: $\lim^*_{\beta\to\lambda} r_\beta \restriction t_\alpha^M(\beta) = r \restriction t_\alpha^M(\gamma)$ exists, so $\lim^*_{\beta\to\gamma} s_\beta^\alpha$ exists. Further, $a_\beta^\alpha = u_\beta^\alpha(t_\alpha^M(\beta))_0 = u_\beta(t_\alpha^M(\beta))_0 = r_\beta(t_\alpha^M(\beta))_0$ for $\beta < \gamma$, and since r_β^* converges in $\vec{\theta}^5$, which is very good, $\lim_{\beta\to\gamma} a_\beta^\alpha = r(t_\alpha^M(\gamma))_0$.

Let s be as in condition (3); then $s = r \restriction (t_\alpha^M(\gamma)) = r \restriction \delta$.

Thus (3) fails at γ in the definition of D_α, and Case 3 holds. Since $s \supseteq r \supseteq q$, 3A cannot hold. Let a be as in Case 3. Then $a = r(\delta)_0$. If $a = q(\text{dom}(s))_0 = q(s)_0$, then the construction in 3B guarantees $r(\delta)_1 = q(\delta)_1$, so $r(\delta) = q(\delta)$, contrary to the definition of δ. Thus 3B cannot hold, and 3C does. The rest of claim 2 is obvious by now.

Claim 3. Let $\alpha \in C_\gamma^M$ and $t_\alpha^M(\gamma) > \delta$; then (1)-(4) hold at γ in the definition of D_α. Moreover, $\text{dom}(u_\gamma)^\alpha = t_\alpha^M(\gamma) + 1$, and $r \restriction t_\alpha^M(\gamma) + 1$ is the upper board of u_γ^α. Finally, if $\eta < \alpha$, $\eta \in C_\gamma^M$, and $t_\eta^M(\gamma) > \delta$, then $u_\gamma^\eta \subseteq u_\gamma^\alpha$.

Proof. This is a tedious induction on α which we leave to the interested reader. The main point is that if $\alpha \in C_\gamma^M$, then letting $N = \langle v_\beta^\alpha | \beta < \gamma \rangle$, $C_\gamma^N = C_\gamma^M \cap \alpha$ and $t_\eta^N = t_\eta^M \restriction \gamma + 1$ for all $\eta < \alpha$.

It is clear from Claim 3 that if we define u_γ as above, then (a), (b), and (c) remain true for $\langle u_\beta | \beta < \gamma + 1 \rangle$.

Lastly, we must see that (a), (b), and (c) are preserved at limit γ. Now (c) is trivial at limits, and for (b) we need only worry that $\lim^*_{\beta \to \gamma} r_\beta$ may not exist; however, this limit must exist granted (a). So it is enough to check (a).

If (a) fails, then we have $\langle e, t \rangle$ such that

$$\langle e, t \rangle = \liminf_{\beta \to \gamma} \mathrm{crit}(u_\beta).$$

Now $r_\beta \in A$ for $\beta < \gamma$ by (a)-(c) below γ, and so $\psi_k(r_\beta)$ converges as $\beta \to \gamma$ for all $k \leq \langle e, t \rangle$. Thus $|e|_{F(r_\beta)}$ is eventually constant, say for $\beta \neq \eta$. Let $t_\alpha^M(\eta) = |e|_{F(r_\eta)}$, so that $t_\alpha^M(\beta) = |e|_{F(r_\beta)}$ for all $\beta \geq \eta$, $\beta < \gamma$. [Though we haven't yet shown $M = \langle u_\beta | \beta < \gamma \rangle$ is good, we can define t_α^M and C_α^M by: $t_\alpha^M = \cup_{\beta < \gamma} t_\alpha^{M \restriction \beta}$ and $C_\alpha^M = \cup_{\beta < \gamma} C_\alpha^{M \restriction \beta}$. Now by (c) below γ we have $u_\beta^\alpha = u_\beta \restriction (t_\alpha^M(\beta) + 1)$ for all $\beta < \gamma$. (In particular v_β^α is defined for $\beta < \gamma$.) If v_γ^α is defined, then $\gamma < \gamma_\alpha$, and then we see from the construction in cases 1-4 that the least proposal in $u_\beta^\alpha(\mathrm{dom}(u_\beta^\alpha) - 1)$ goes to ω as $\beta \to \gamma$. (In this case, our current γ is less than the γ referred to in the case hypothesis.) But this proposal is t cofinally often. So v_γ^α is undefined. If cases 2-4 apply in the definition of D_α, then again the least proposal in $u_\beta^\alpha(\mathrm{dom}(u_\beta^\alpha) - 1)$ goes to ω as $\beta \to \gamma$. So Case 1 applies. But then our construction guarantees that $\Sigma_{\delta(u_\beta)}$ is responsible for the t proposal cofinally often in $u_\beta^\alpha(\mathrm{dom}(u_\beta^\alpha) - 1)$. Setting $i = \langle e, t \rangle$, we get a $\theta < \gamma$ such that

$$\psi_i(r_\eta) \leq \psi_i(r_\beta)$$

for all $\eta \geq \beta \geq \theta$, and

$$\psi_i(r_{\eta+1}) < \psi_i(r_\eta)$$

for infinitely many $\eta \geq \theta$, a contradiction. (The first inequality requires that $\vec{\psi}$ be a scale, not just a semiscale.)

This completes the proof of Theorem 2. ∎

Recall that

$$T_n = \{p \in \cup_{\alpha < \omega_1} {}^\alpha\omega | \forall \beta \in \mathrm{dom}(p)((L_\beta[p \restriction \beta], \epsilon, p \restriction \beta) \text{ is not } \Sigma_n \text{ admissible}\}.$$

Corollary 2. Let $n \geq 1$, and assume $\Gamma - \mathrm{AD}^{\Sigma_n}$, where Γ is closed under $\forall^{\mathbf{R}}$ and has the scale property. Then if I has a ws in $\mathcal{G}(A, T_n)$, where $A^* \in \Gamma$, I has a ws Σ such that Σ^* is $\Game^{\Sigma_n}\Gamma$.

Proof. The proof of this is implicit in the proof of Theorem 2. ∎

§3. An inner model of AD^{Σ_n}

One standard justification for deriving consequences of ZF + AD + DC is that, assuming definable determinacy, this theory has interesting inner models. In particular, if all games of length ω on ω with payoff in $L(\mathbf{R})$ are determined, then $L(\mathbf{R}) = \mathrm{ZF} + AD + DC$. In this section we shall provide an analogous justification for deriving consequeces of ZF + AD^{Σ_n} + DC.

For $\alpha \leq \omega_1$, let $\Gamma - AD^{\alpha}$ be the assertion that all games of length α on ω whose payoff set is in Γ are determined. AD^{α} is the same assertion but with no restriction on payoffs. AD^{ω^2} is the weakest form of full determinacy stronger than that $\mathrm{AD}^{\omega} = \mathrm{AD}$. Blass and Mycielski ([2]) have shown $\mathrm{AD}^{\omega^2} \Leftrightarrow \mathrm{AD}_{\mathbf{R}}$, while Solovay ([7]) has shown that ZF + $\mathrm{AD}_{\mathbf{R}}$ + DC proves Con (ZF + AD + DC). We shall give in detail only the construction of a model of ZF + $AD_{\mathbf{R}}$ + DC; even this seems to involved the machinery of long games and scales developed in §1 and §2. We indicate how to modify the construction in order to get models ZF + AD^{Σ_n} + DC at the end of this section.

We shall rely heavily on work of Woodin ([8]). Woodin constructs "from below" a class M, and shows that if there us any model of ZF + $AD_{\mathbf{R}}$ + DC containing all reals and ordinals, then M is the smallest such model. So our task is to show that some amount of definable determinacy implies $M \models \mathrm{ZF} + \mathrm{AD}_{\mathbf{R}} + \mathrm{DC}$. By Woodin's work on M, it suffices for this to show that every set of reals in M admits a definable scale. We shall show using a "Friedman game" that every $(\underset{\sim}{\Delta}^2_1)^M$ set is $\Game^{\Sigma_2}\underset{\sim}{\Pi}^1_1$. Corollary 2 then implies every set of reals in M is $\Game^{\Sigma_2}\underset{\sim}{\Pi}^1_1$, and Corollary 1 implies that such sets admit definable scales.

In the end, we require $\underset{\sim}{\Delta}^1_2 - \mathrm{AD}^{\Sigma_2}$ to show that $M \models \mathrm{ZF} + \mathrm{AD}_{\mathbf{R}} + \mathrm{DC}$.

In §4 we shall present some evidence that in fact ZFC + HOD($\mathbf{R}$) $- \mathrm{AD}^{\Sigma_1}$ is too weak to construct an inner model of ZF + $\mathrm{AD}_{\mathbf{R}}$ + DC. We believe that $\underset{\sim}{\Delta}^1_2 - \mathrm{AD}^{\Sigma_2}$ is very close to being the weakest definable determinacy hypothesis yielding an inner model of $\mathrm{AD}_{\mathbf{R}}$.

Some terminology: a filter $\mathcal{F}$ on $P_{\omega_1}(X) = \{A \subseteq X | A \text{ is countable}\}$ is *normal* iff $\mathcal{F}$ is closed under diagonal intersection ($A_x \in \mathcal{F}$ for all $x \in X \Rightarrow \{A \in P_{\omega_1}(X) | \forall x \in A(A \in A_x)\} \in \mathcal{F}$) and *fine* iff $\forall x \in X$ $(\{A \in P - \omega_1(X) | x \in A\} \in \mathcal{F})$. A set $C \subseteq P_{\omega_1}(X)$ is a club iff C is closed under countable increasing unions and $\forall A \in P_{\omega_1}(X)\exists B \epsilon C(A \subseteq B)$. The club filter on $P_{\omega_1}(X)$ consists of all $A \subseteq P_{\omega_1}(X)$ such that $C \subseteq A$ for some club C; it is normal and fine. (Normality uses AC.) We say ω_1 *is X-supercompact* iff there is a normal, fine ultrafilter on $P_{\omega_1}(X)$. One of the basic consequences of $AD_{\mathbf{R}}$, due to Solovay, is that the club filter on $P_{\omega_1}(\mathbf{R})$ is a normal ultrafilter, so that ω_1 is $\mathbf{R}$-supercompact [(7)].

We proceed to the main result of this section. It is convenient at this point to add full AC to the metatheory ZF + DC of this paper; this makes possible some simple manipulations of club sets in the proof to follow. (We doubt that AC is actually necessary.)

Theorem 3. (ZFC) Assume $\underset{\sim}{\Delta}^1_2 - AD^{\Sigma_2}$. Then there is an inner model containing all reals and ordinals and satisfying ZF + $AD_{\mathbf{R}}$ + DC+ "ω_1 is $P(\mathbf{R})$-supercompact" + "every set of reals admits a scale."

Proof. We shall define a slight variant of Woodin's model (in order to get $P(\mathbf{R})$-supercompactness.) Let

$$M_0 = V_{\omega+1} = \text{the set of rank} \leq \omega$$

and

$$M_\lambda = \bigcup_{\beta<\lambda} M_\beta, \text{ for } \lambda \text{ a limit.}$$

Now suppose M_α is given. If $M_\alpha \not\models ZF^- +$"$P(P(\mathbf{R}))$ exists", set

$$M_{\alpha+1} = \{a \subseteq M_\alpha | a \text{ is 1st order def. over}(M_\alpha, \epsilon) \text{ from parameters}\}.$$

If $M_\alpha \models ZF^- +$ "$P(P(\mathbf{R}))$ exists" $+ -$ AD, then set

$$M_{\alpha+1} = M_\alpha.$$

(i.e. stop the construction.) Finally, suppose $M_\alpha \models ZF^- +$ "$P(P(\mathbf{R}))$ exists" +AD, and let $\gamma = \theta^{M_\alpha}$ = sup of lengths of prewellorders of $\mathbf{R}$ in M_α.

Case 1. $\text{cof}(\gamma) = \omega$.

Pick any sequence $\langle A_n | n < \omega\rangle$ such that $\forall_n(A_n \in P(\mathbf{R}) \cap M_\alpha)$, but $\langle A_n | n < \omega\rangle \notin M_\alpha$. (By Wadge, this means $\forall B \in (P(\mathbf{R}) \cap M_\alpha)\exists n(B \leq_w A_n)$. Set

$$\begin{aligned} M_{\alpha+1} =&\{a \subseteq M_\alpha | a \text{ is 1st order def.} \\ &\text{from parameters over} \\ &(M_\alpha, \epsilon, \langle A_n | n < \omega\rangle)\}. \end{aligned}$$

(By Wadge, $M_{\alpha+1}$ is independent of the $\langle A_n | n < \omega \rangle$ chosen.)

Case 2. $\mathrm{cof}(\gamma) > \omega$.

Let $X = P(\mathbf{R}) \cap M_\alpha$, and let F_α be the club filter on $P_{\omega_1}(X)$.

Subcase (a). F is not an ultrafilter over $P(P_{\omega+1}(X)) \cap M_\alpha$. (That is, there is an $A \subseteq P_{\omega_1}(X), A \in M_\alpha$, such that neither A nor $P_{\omega_1}(X) - A$ is in F.

Then set

$$M_{\alpha+1} = M_\alpha$$

Subcase (b). Otherwise.

Then set

$$\begin{aligned} M_{\alpha+1} = \{a \subseteq M_\alpha | \, & a \text{ is 1st order def.} \\ & \text{from parameters over} \\ & (M_\alpha, \epsilon, F_\alpha \cap M_\alpha)\}. \end{aligned}$$

This completes the definition of the M_α's. Clearly they constitute an increasing sequence of transitive sets. Set

$$M = \bigcup_{\alpha \in OR} M_\alpha.$$

Let also

$$\mathcal{F}(\alpha, A) \text{ iff Case 2 applied at } \alpha \text{ and} A \in F_\alpha,$$

and $\mathcal{M}_\alpha = (M_\alpha, \epsilon, \mathcal{F} \cap M_\alpha)$.

The point is that $\mathcal{M}_\alpha$ has the information it needs to define $\langle \mathcal{M}_\beta | \beta < \alpha \rangle$. Thus there is a fixed formula defining $\langle \mathcal{M}_\beta | \beta < \alpha \rangle$ over $\mathcal{M}_\alpha$ for all α. There is also a natural sentence expressing "I am an $\mathcal{M}_\alpha$."

Lemma 4. Every $(\underset{\sim}{\Delta}{}^2_1)^M$, set of reals is $\Game^{\Sigma_2} \underset{\sim}{\Pi}{}^1_1$.

Proof. Let S be $(\Delta^2_1)^M$, where we have dropped the real parameter for convenience. Say

$$S(x) \Leftrightarrow M \models \exists A \subseteq \mathbf{R} \; \phi(A, x)$$

and

$$-S(x) \Leftrightarrow M \models \exists A \subseteq \mathbf{R} \; \psi(A, x),$$

where ϕ and ψ have real quantifiers only. We must recursively associate to any $x \in \mathbf{R}$ and AD^{Σ_2} type game G_x such that $S(x)$ iff I wins G_x. For convenience, we shall make the individual moves of G_x reals rather than natural number; for games of the AD^{Σ_2} variety this affects nothing of importance.

Let x be given. We describe the payoff of G_x by specifying the rules of play. The rules governing round α are defined by induction on α.

Round 0. I must play a real coding transitive structure $\mathcal{A}_0$ s.t. $x \in |\mathcal{A}_0|$ and

$$\mathcal{A}_0 \models \text{“I am an } \mathcal{M}_\alpha\text{”} \wedge \exists A \subseteq \mathbf{R}\ \phi(\mathrm{A,x}) \wedge \forall A \subseteq \mathbf{R} - \psi(A,x)$$

II must then play a real coding in a transitive $\mathcal{B}_0$ with $x \in |\mathcal{B}_0|$ and

$$\mathcal{B}_0 \models \text{“I am an } \mathcal{M}_\alpha\text{”} \wedge \exists A \supseteq \mathbf{R}\ \psi(A,x) \wedge \forall A \supseteq \mathbf{R} \rightharpoonup \phi(\mathrm{A,x}).$$

Failure by one of the players is violation of the rules at 0.

Round $\alpha+1$. If neither player has violated the rules at or before α, then we shall have transitive structures $\mathcal{A}_\alpha$ and $\mathcal{B}_\alpha$ at the end of round α. I must now play the code of a transitive $\mathcal{A}_{\alpha+1}$ s.t.

$$\mathbf{R}^{\mathcal{B}_\alpha} \subseteq \mathcal{A}_{\alpha+1}$$

and an elementary

$$i_\alpha : \mathcal{A}_\alpha \to \mathcal{A}_{\alpha+1}$$

with the following properties: Suppose $\mathcal{A}_\alpha \models$ “case occured at stage δ”; that is, $\mathcal{A}_\alpha \models \exists A \mathcal{F}(\delta, A)$. Let

$$C = (P(\mathbf{R}) \cap M_\delta)^{\mathcal{A}_\alpha}.$$

Then for $A \in (P_{\omega_1}(P(\mathbf{R}) \cap M_\delta))^{\mathcal{A}_\alpha}$, I must arrange

$$\mathcal{A}_\alpha \models \mathcal{F}(\delta, A) \text{ iff } i''_\alpha C \in i_\alpha(A).$$

(in particular, $i''_\alpha C \in |\mathcal{A}_{\alpha+1}|$, and is countable in $\mathcal{A}_{\alpha+1}$.

Further, suppose $A_n \in P(\mathbf{R})^{\mathcal{A}_\alpha}$ for all $n < \omega$, and $\langle A_n | n \in \omega\rangle \in |\mathcal{B}_\alpha|$. Then I must arrange

$$\langle i_\alpha(A_n) | n < \omega\rangle \in |\mathcal{A}_{\alpha+1}|.$$

Failure to meet these requirements is a violation of the rules at $\alpha + 1$ by I.

On his $\alpha + 1$st move II must play a $\mathcal{B}_{\alpha+1}$ and a $j_\alpha : \mathcal{B}_\alpha \to \mathcal{B}_{\alpha+1}$ meeting requirements completely symmetric to those on $\mathcal{A}_{\alpha+1}$ and $i_{\alpha+1}$. If he doesn't, II violates the rules at α.

Round λ, λ *limit:* If no one has violated the rules before λ we shall have direct limit systems

$$\langle \mathcal{A}_\alpha, i_{\alpha\beta} | \alpha < \beta < \lambda \rangle$$

and

$$\langle \mathcal{B}_\alpha, j_{\alpha\beta} | \alpha < \beta < \lambda \rangle.$$

Set

$$\mathcal{A}_\lambda = \text{direct limit of} \langle \mathcal{A}_\alpha, i_{\alpha\beta} | \alpha < \beta < \lambda \rangle$$

$$\mathcal{B}_\lambda = \text{direct limit of} \langle \mathcal{B}_\alpha, j_{\alpha\beta} | \alpha < \beta < \lambda \rangle$$

Then I violates the rules at λ if $\mathcal{A}_\lambda$ is illfounded, and II violates the rules at λ if $\mathcal{B}_\lambda$ is illfounded. If there is no violation, we assume $\mathcal{A}_\lambda$ and $\mathcal{B}_\lambda$ are transitive.

The reals played during round λ are meaningless for G_x.

This completes the rules of G_x. The first player to violate these rules loses G_x. In case of a tie, I loses. We shall now show that by the time we reach the first Σ_2 admissible relative to the play, someone must lose.

Claim. Let $p : \lambda \times \omega \to \omega$ be a partial play of G_x, where λ is a limit ordinal. Suppose in p neither player has violated the rules at or before λ. Then $L_\lambda[p]$ is not Σ_2 admissible.

Proof. (Here $\lambda n.p(\alpha, n)$ codes the reals played by I and II at round α in p.) Let p be a counterexample to the claim. Let $\langle \mathcal{A}_\alpha, i_{\alpha\beta} | \alpha < \beta < \lambda \rangle$ and $\langle \mathcal{B}_\alpha, j_{\alpha\beta} | \alpha < \beta < \lambda \rangle$ be the systems produced in p by I and II respectively. Let $\mathcal{A}_\lambda$ and $\mathcal{B}_\lambda$ be their transitive direct limits, and $i_{\alpha\lambda}, j_{\alpha\lambda}$ (for $\alpha < \lambda$) the natural maps.

Now

$$\mathbf{R}^{\mathcal{A}_\lambda} = \bigcup_{\alpha<\lambda} \mathbf{R}^{\mathcal{A}_\alpha} = \bigcup_{\alpha<\lambda} \mathbf{R}^{\mathcal{B}_\alpha} = \mathbf{R}^{\mathcal{B}_\lambda}.$$

On the other hand, $\mathcal{A}_\lambda \models \exists A \subseteq \mathbf{R}\ \phi(A, x) \wedge \forall A \subseteq \mathbf{R} \neg\ \psi(A, x)$ while $\mathcal{B}_\lambda \models \exists A \subseteq \mathbf{R} \psi(A, x) \wedge \forall A \subseteq \mathbf{R} \neg\ \phi(A, x)$, so

$$P(\mathbf{R})^{\mathcal{A}_\lambda} \not\subseteq \mathcal{B}_\lambda$$

and

$$P(\mathbf{R})^{\mathcal{B}_\lambda} \not\subseteq \mathcal{A}_\lambda.$$

Thus we have a $\delta \in \mathrm{OR}^{\mathcal{A}_\lambda} \cap \mathrm{OR}^{\mathcal{B}_\lambda}$ such that

$$M_\delta^{\mathcal{A}_\lambda} = M_\delta^{\mathcal{B}_\lambda} \text{ and } M_{\delta+1}^{\mathcal{A}_\lambda} \neq M_{\delta+1}^{\mathcal{L}_\lambda}.$$

Clearly then, $M_\delta^{\mathcal{A}_\lambda} \models \mathrm{ZF}^- \wedge \mathrm{AD}$. Let

$$\theta = \theta^{(M_\delta^{\mathcal{A}_\lambda})} = \theta^{(M_\delta^{\mathcal{B}_\lambda})}.$$

We claim that both $\mathcal{A}_\lambda$ and $\mathcal{B}_\lambda$ satisfy "cof $(\theta) > \omega$." For suppose e.g. $\mathcal{A}_\lambda$ satisfies " cof$(\theta) = \omega$." Pick $\langle A_n | n < \omega\rangle$ in $|\mathcal{A}|$ such that $\mathcal{A}$ satisfies: $\forall B \in M_\delta \cap P(\mathbf{R}) \exists n (B \leq_{\mathrm{w}} A_n)$. Then for $\alpha < \lambda$ sufficiently large, say $\alpha \geq \alpha_0$, we have a $\langle A_n^\alpha | n < \omega\rangle \in |\mathcal{A}_\alpha|$ s.t.

$$i_{\alpha\lambda}(\langle A_n^\alpha | n < \omega\rangle) = \langle A_n | n < \omega\rangle.$$

Now $A_n \in |\mathcal{B}_\lambda|$ for all n. So for fixed n, we have for all sufficiently large, say $\alpha \geq \alpha_n$, a $B_n^\alpha \in |\mathcal{B}_\alpha|$ s.t.

$$i_{\alpha\lambda}(B_n^\alpha) = A_n.$$

Notice that if α is a limit then $\mathbf{R}^{\mathcal{A}_\alpha} = \mathbf{R}^{\mathcal{B}_\alpha}$ so if α is a limit s.t. both A_n^α and B_n^α are defined, then $A_n^\alpha = B_n^\alpha$. Let

$$S = \{\alpha < \lambda | L_\alpha[p \restriction \alpha] \prec_{\Sigma_1} L_\lambda[p]\}.$$

Then S is club in λ and $L_\lambda[p, S]$ is admissible. By our observations above, for each $n < \omega$ we can find a $\beta_n \in S$ s.t. $\beta_n > \alpha_0$ and $A_n^{\beta_n} \in |\mathcal{B}_{\beta_n}|$. By the stability of β_n, this means $\beta_n \geq \alpha_n$ and $A_n^{\beta_n} = B_n^{\beta_n}$. Since the map $n \longrightarrow \beta_n$ is $\underset{\sim}{\Delta}_1(L_\lambda, [p, S])$, we have a limit γ s.t. $\gamma > \beta_n$ for all n.

Now the requirements on II at round $\gamma + 1$ imply that

$$\langle j_{\gamma,\gamma+1}(B_n^\gamma) | n < \omega\rangle \in |\mathcal{B}_{\gamma+1}|$$

and since $\gamma > \alpha_n$ for all n,

$$j_{\gamma+1,\lambda}(\langle j_{\gamma,\gamma+1}(B_n^\gamma) | n < \omega\rangle) = \langle A_n | n < \omega\rangle.$$

Thus $\langle A_n | n < \omega\rangle \in |\mathcal{B}_\lambda|$, and $\mathcal{B}_\lambda$ satisfies cof $(\theta) = \omega$. Thus both $\mathcal{A}_\lambda$ and $\mathcal{B}_\lambda$ think $M_{\delta+1}$ comes from M_δ via Case 1 using $\langle A_n | n < \omega\rangle$, so $M_{\delta+1}^{\mathcal{A}_\lambda} = M_{\delta+1}^{\mathcal{B}_\lambda}$, a contradiction.

Thus $\mathrm{cof}(\theta) > \omega$ in both $\mathcal{A}_\lambda$ and $\mathcal{B}_\lambda$. Pick $A \in M_\delta^{\mathcal{A}_\lambda} = M_\delta^{\mathcal{B}_\lambda}$ such that

$$\mathcal{A}_\lambda \models \mathcal{F}(\delta, A) \text{ and } \mathcal{B}_\lambda \models \neg\, \mathcal{F}(\delta, A).$$

(Such an A must exist.) Pick α_0 large enough that $i^{-1}_{\alpha_0,\lambda}(A)$, $i^{-1}_{\alpha_0,\lambda}(\delta)$, $j^{-1}_{\alpha_0,\lambda}(A)$, and $j^{-1}_{\alpha_0,\lambda}(\delta)$ all exist. Let

$$\begin{aligned} v_\gamma &= i^{-1}_{\gamma\lambda}(\delta) \\ \mu_\gamma &= j^{-1}_{\gamma\lambda}(\delta) \\ B_\gamma &= i^{-1}_{\gamma\lambda}(A) \\ C_\gamma &= j^{-1}_{\gamma\lambda}(A) \end{aligned}$$

for $\alpha_0 \leq \gamma < \lambda$. Then $\langle (v_\gamma, \mu_\gamma, B_\gamma, C_\gamma) | \gamma < \lambda \rangle$ is Δ_1 over $L_\lambda[p]$.

Now if $X \in P(\mathbf{R}) \cap M_{\nu_\gamma}^{\mathcal{A}_\gamma}$, then there is a $\beta > \gamma$, $\beta \in S$, and a Y in $P(\mathbf{R}) \cap M_{\mu_\beta}^{\mathcal{B}_\beta}$ s.t. $i_{\gamma\lambda}(X) = j_{\beta\lambda}(Y)$. Notice that for $\beta \in S, i_{\gamma\lambda}(X) = j_{\beta\lambda}(Y)$ iff $i_{\gamma\beta}(X) = Y$. By a closure argument then, we can find $\beta \in S, \beta > \alpha_0$, s.t.

$$P(\mathbf{R}) \cap M_{\nu_\beta}^{\mathcal{A}_\beta} = P(\mathbf{R}) \cap M_{\mu_\beta}^{\mathcal{B}_\beta}.$$

Call this set C. Then $i_{\beta,\beta+1}{}''C$ is countable in $\mathcal{A}_{\beta+1}$, and $j_{\beta,\beta+1}{}''C$ is countable in $\mathcal{B}_{\beta+1}$, so

$$i_{\beta+1,\lambda}(i_{\beta,\beta+1}{}''C) = i_{\beta\lambda}{}''C$$

and

$$j_{\beta+1,\lambda}(i_{\beta,\beta+1}{}''C) = j_{\beta\lambda}{}''C.$$

But β is stable, so $j_{\beta\lambda}(X) = i_{\beta\lambda}(X)$ for all $X \in C$, so

$$i_{\beta\lambda}{}''C = j_{\beta\lambda}{}''C.$$

But by our rules

$$\begin{aligned} i_{\beta,\beta+1}{}''C &\in B_{\beta+1} \\ j_{\beta,\beta+1}{}''C &\notin C_{\beta+1} \end{aligned}$$

Hence

$$i_{\beta\lambda}{}''C \in A$$

and

$$j_{\beta\lambda}''C \notin A$$

a contradiction. This proves the claim. ▮

To finish the proof of Lemma 1, we must show that I has a winning strategy in G_x iff $S(x)$. So suppose $S(x)$. Pick an α s.t.

$$M_\alpha \models \mathrm{ZF}^- \wedge \exists A \subseteq \mathbf{R}\ \phi(A,x) \wedge \forall A \subseteq \mathbf{R} \rightharpoonup \phi(A,x)$$

Then we can find a club $\mathcal{C} \subseteq P_{\omega_1}(M_\alpha)$ such that if $P \in \mathcal{C}$, then

$$(P, \epsilon, \mathcal{F} \cap P) \prec \mathcal{M}_\alpha$$

and

$$(A, \delta \in P \wedge \mathcal{F}(\delta, A) \Rightarrow (P(\mathbf{R}) \cap M_\delta \cap P) \in A.$$

[Proof: For each $\delta, A \in M_\alpha$ s.t. $\mathcal{F}(\delta, A)$, pick a club $\mathcal{C}_{\delta,A} \subseteq A$. Define

$$f(\delta, A, Y) = \text{some } P \in \mathcal{C}_{\delta,A} \text{ such that } Y \subseteq P.$$

for $Y \in P_{\omega_1}(P(\mathbf{R}) \cap M_\delta)$. Let $\mathcal{C}$ be the set of all $Q \in P_{\omega_1}(M_\alpha)$ s.t. Q is closed under f and Skolem functions for M_α.]

Now for $P \in \mathcal{C}$, let

$$\pi_p : \mathcal{A}_p \cong (P, \epsilon, \mathcal{F} \cap P)$$

be the inverse of the collapse. Player I should play in G_x so that for all β

$$\mathcal{A}_\beta = \mathcal{A}_{p_\beta}, \text{ for some } P_\beta \in \mathcal{C}$$

and

$$P_\beta \subseteq P_{\beta+1}, P_\lambda = \bigcup_{\beta<\lambda} P_\beta \text{ for } \lambda \text{ limit}$$

and

$$i_{\beta,\beta+1} = \pi^{-1}_{P_{\beta+1}} \circ \pi_{P_\beta}.$$

It is clear that the properties of $\mathcal{C}$ guarantee that he can play this way forever and violate no rules in doing so.

The proof that if $\neg S(x)$ then II has a winning strategy in G_x is entirely symmetric. ∎

By Lemma 4, there must be a game of the form $\mathcal{G}(A, T_2)$ where $A^* \in \underset{\sim}{\Pi}^1_1$, which has no winning strategy in M. (This just means $M \not\models \underset{\sim}{\Pi}^1_1 - \mathrm{AD}^{\Sigma_2}$.) For otherwise the universal $\Game^{\Sigma_2}\Pi^1_1$ set of reals is $(\Delta^2_1)^M$, and hence there is a fixed $(\Delta^2_1)^M$ set universal for $(\underset{\sim}{\Delta}^2_1)^M$ sets, an absurdity. By Theorem 2, then, there is a $\Game^{\Sigma_2}\underset{\sim}{\Delta}^1_2$ set S which is not in M.

Let μ be the club filter on $P_{\omega_1}(\mathbf{R})$. Let Q be a $\Game^{\Sigma_2}\underset{\sim}{\Delta}^1_2$ set which codes a scale on S.

Claim. $L(Q, \mu, R) \models \mathrm{AD}+$ "μ is an ultrafilter."

Proof. Let

$$N_0 = V_{\omega+1}$$

and

$$N_\lambda = \bigcup_{\beta<\lambda} N_\beta, \text{ for } \lambda \text{ limit.}$$

Given N_α such that $N_\alpha \not\models \mathrm{ZF}^-$, let

$$\begin{aligned} N_{\alpha+1} = \{A \subseteq N_\alpha \mid\ & A \text{ is 1st order} \\ & \text{definable over } (N_\alpha, \epsilon, Q) \\ & \text{from parameters.}\} \end{aligned}$$

Given N_α such that $N_\alpha \models \mathrm{ZF}^-$, but $N_\alpha \not\models \mathrm{AD}$ or μ is not an ultrafilter over $P(P_{\omega_1}(\mathbf{R})) \cap N_\alpha$, then set $N_{\alpha+1} = N_\alpha$. Otherwise, let

$$\begin{aligned} N_{\alpha+1} = \{A \supseteq N_\alpha \mid\ & A \text{ is 1st order definable} \\ & \text{over } (N_\alpha, \epsilon, Q, \mu \cap N_\alpha) \\ & \text{from parameters}\}. \end{aligned}$$

Let $N = \cup_{\alpha \in OR} N_\alpha$. An argument similar to the proof of Lemma 4 but much simpler shows that every $(\underset{\sim}{\Delta}^2_1)^N$ set of reals is $\Game^{\omega^2,\mathbf{R}}\underset{\sim}{\Pi}^1_1(Q)$, where $\Game^{\alpha,\mathbf{R}}$ is the game quantifier corresponding to games of length α on $\mathbf{R}$, and $\underset{\sim}{\Pi}^1_1(Q)$ is the least poinclass containing Q and closed under $\forall^{\mathbf{R}}, \cap, \cup$, and continuous substitution. Thus there is a $\underset{\sim}{\Pi}^1_1(Q)$ game of length ω^2 on $\mathbf{R}$ which is not determined in N. Now Martin [4] shows that every such game has a ws which is $\Game^{\omega^2,\mathbf{R}}\underset{\sim}{\Delta}^1_2(P)$, where P is $\Game^{\Sigma_2}\underset{\sim}{\Delta}^1_2$ and codes a scale on Q. (The relevant games of length ω^2 on $\mathbf{R}$ are determined by $\underset{\sim}{\Delta}^1_2 - \mathrm{AD}^{\Sigma_2}$: any game of length $\alpha < \omega_1$ on $\mathbf{R}$ with $\Game^{\Sigma_2}\underset{\sim}{\Delta}^1_2$ payoff is determined.) Thus there is a $\Game^{\omega^2,\mathbf{R}}\underset{\sim}{\Delta}^1_2(P)$, hence a $\Game^{\Sigma_2}\underset{\sim}{\Delta}^1_2$, set not in N. Let B be such a set. Let

$C \subseteq \mathbf{R}, B \leq_w C$; and C codes up ws's for all games of length ω on $\mathbf{R}$ with payoff $\leq_w B$; we can take C to be $\Game^{\Sigma_2}\underset{\sim}{\Delta}^1_2$ by [4].

Subclaim. $L(C, \mathbf{R}) \models$ AD.

Proof. It is enough to see $C^\#$ exists and is $\Game^{\Sigma_2}\underset{\sim}{\Delta}^1_2$. Now every $(\underset{\sim}{\Sigma}^2_1)^{L(C,\mathbf{R})}$ set is $\Game^{\omega,\mathbf{R}}\underset{\sim}{\Pi}^1_1(C)$, so we have a $\Game^{\Sigma_2}\underset{\sim}{\Delta}^1_2$ set D such that $D \notin L(C, \mathbf{R})$. Let α be least such that $L_\alpha(D, \mathbf{R}) \models KP$; then $L_\alpha(D,, \mathbf{R}) \models$ AD since $\Game^{\Sigma_2}\underset{\sim}{\Delta}^1_2$ is closed under "inductive in." Work in $L_\alpha(D, \mathbf{R})$. Then by Wadge, if $X \subseteq \mathbf{R}$ and $X \in L(C, \mathbf{R}) X \leq_w D$. Thus there is a measurable cardinal greater than $\theta^{L(C,\mathbf{R})}$, so $C^\#$ exists. Since $L_\alpha(D, \mathbf{R}) \models C^\#$ exists, $C^\#$ does exist and is in $L_\alpha(D, \mathbf{R})$, hence $\Game^{\Sigma_2}\underset{\sim}{\Delta}^1_2$. ∎

To finish the proof of the claim, let

$$N'_\alpha = N_\alpha^{L(C,\mathbf{R})}.$$

By induction on α we see that $N'_\alpha = N_\alpha, P(\mathbf{R}) \cap N_\alpha \subseteq \{X | X \leq_w B\}$, $N_\alpha \models$ AD, and μ is an ultrafilter over $P(P_{\omega_1}(\mathbf{R})) \cap N_\alpha$. This yields the claim at once. ∎

Now notice that there is a canonical $\pi : \mathbf{R} \xrightarrow{\text{onto}} \{B | B \leq_w Q\}, \pi \in L(Q, \mu, \mathbf{R})$. Suppose now

$$M_\alpha \in L(Q, \mathbf{R}, \mu);$$

then by Wadge $P(\mathbf{R}) \cap M_\alpha \subseteq \{B | B \leq_w Q\}$ so that π induces a canonical

$$\pi_\alpha : \mathbf{R} \xrightarrow{\text{onto}} P(\mathbf{R}) \cap M_\alpha = X_\alpha.$$

Suppose Case 2 occurs in the definition of $M_{\alpha+1}$. For any $\mathcal{A} \subseteq P_{\omega_1}(X_\alpha)$, let

$$\mathcal{A}^* = \{A \in P_{\omega_1}(\mathbf{R}) | \pi''_\alpha A \in \mathcal{A}\};$$

then $\mathcal{A}^*$ is in $L(Q, \mathbf{R}, \mu)$ if $\mathcal{A}$ is. Also, if $A^* \in \mu$, then $\mathcal{A} \in F_\alpha$, while if $P_{\omega_1}(\mathbf{R}) - \mathcal{A}^* \in \mu$, then $P_{\omega_1}(X_\alpha) - \mathcal{A} \in F_\alpha$. [For this reason we seem to need AC in V. Pick $\sigma : X_\alpha \to \mathbf{R}$ such that $\sigma \circ \pi_\alpha = id$. Then if $\mathcal{C} \subseteq P_{\omega_1}(\mathbf{R})$ is club, then $\{A \in P_{\omega_1}(X_\alpha) | \sigma^{-1}(A) \in \mathcal{C}\}$ is club in $P_{\omega_1}(X_\alpha)$; moreover $A \in \mathcal{A}$ iff $\sigma^{-1}(A) \in \mathcal{A}^*$.]

It follows that F_α is an ultrafilter over $P(P_{\omega_1}(X_\alpha)) \cap M_\alpha$; moreover F_α is uniformly-in-M_α definable over $L(Q, \mathbf{R}, \mu)$.

It is now easy to see by induction that $M_\alpha \in L(Q, \mathbf{R}, \mu)$ for all α and that $M_\alpha \neq M_{\alpha+1}$ for all α. In fact, the function $\alpha \longrightarrow M_\alpha$ is definable over $L(Q, \mathbf{R}, \mu)$. Thus M satisfies ZF + AD + "ν is a normal ultrafilter over $P_{\omega_1}(P(\mathbf{R}))$", where ν is the club filter on $P_{\omega_1}(P(\mathbf{R}) \cap M)$ restricted to M. Since $M \models \text{cof}(\theta) > \omega$ by construction, $M \models$ DC (cf. [7]). Finally, every set of reals in M is $\leq_w S$ since $M \supseteq L(Q, \mathbf{R}, \mu)$. Thus each such set has a scale in $L(Q, \mathbf{R}, \mu)$. Woodin [8] shows that this implies each set in M has a scale in M. So $M \models$ "Every set of reals has a scale," and by Martin [5] and Woodin [8], $M \models \forall\alpha < \omega_1 \text{AD}^\alpha$. ∎

Notice that we could have augmented the construction of M by throwing in canonical winning strategies for games of the form $\mathcal{G}(A, T_1), A \in M$. This would give an M^* having the properties of M and satisfying AD^{Σ_1} in addition. [The proof of Lemma 4 must be modified as follows: suppose we have an A in $\mathcal{A}_\alpha \cap \mathcal{B}_\alpha$ such that $\mathcal{A} \models$ I wins $\mathcal{G}(A, T_1)$ and $\mathcal{B}_\alpha \models$ II wins $\mathcal{G}(A, T_1)$. Then the players of G_x take time out to play $\mathcal{G}(A, T_1)$ and produce thereby a $q \in E_{T_1}$. We then require $q \in i_{\alpha,\alpha+1}(A)$ and $q \notin j_{\alpha,\alpha+1}(A)$. With this modification, the proof of Theorem 3 goes through]. A similar argument shows

Theorem 4 .Assume $\underset{\sim}{\Delta}^1_2 - AD^{\Sigma_{n+1}}$, where $n \leq 1$. Then there is an inner model containing all reals and ordinals, and satisfying $ZF + AD^{\Sigma_n} + DC +$ "every set admits a scale" + "ω_1 is $V_{\omega+n+1}$ – supercompact".

In fact, it seems likely that the proof of Theorem 3 will adapt to show: if all clopen games of length ω_1 with HOD($\mathbf{R}$) (or even, "$\underset{\sim}{\Delta}^1_2$ in the codes") payoff are determined, then there is an inner model satisfying ZF + DC + "Every set of reals admits a scale" + "$\mathcal{G}(A, T)$ is determined whenever T admits a scaled coding." If so, then this is nearly as well as one can do in the direction of models of strong forms of full determinacy without modifying the notion of a winning strategy; cf. §5.

§4. A determinacy proof.

The basic problem in this area is to prove that determinacy of the long games we have been considering from ZFC together with some large cardinal hypothesis. One method for doing this involves reducing the long games to games of length ω on higher type objects, as in the Blass-Mycielski proof that $\text{AD}_\mathbf{R}$ implies AD^{ω^2}. In this section we shall present such a reduction for games of the form $\mathcal{G}(A, T_1)$, that is, games ending at the first admissible relative to the play. We shall also prove the determinacy of the length of ω games to which we reduce within the theory ZF + AD + DC + "Every set admits a scale" + "ω_1 is $P(\mathbf{R})$-supercompact". Thus this theory proves AD^{Σ_1}. Now Woodin has recently shown, in ZFC + $\exists\kappa$ (κ is supercompact), that there is an inner model of ZF + AD + DC + "Every set admits a scale" containing all reals and ordinals. It seems likely that his techniques will/do produce a model satisfying "ω_1 is $P(\mathbf{R})$-supercompact" as well. If so, then we have a proof from ZFC + $\exists\kappa$ (κ is supercompact) of $\Gamma - \mathrm{AD}^{\Sigma_1}$, where Γ is the class of sets in Woodin's model (and thus a proof of e.g. $\underset{\sim}{\Pi}^1_1 - \mathrm{AD}^{\Sigma_1}$).

Martin [5] and Woodin [8] have independently shown that ZF + AD + DC + "Every set admits a scale" proves $\forall\alpha < \omega_1(\mathrm{AD}^\alpha)$. Thus $\underset{\sim}{\Pi}^1_1 - \mathrm{AD}^\alpha$, for $\alpha < \omega_1$, follows from the existence of supercompact cardinals. Martin's idea seems likely to yield, with more work, a proof of AD^{Σ_1} from ZF + AD + DC + "Every set admits a scale"; if so, then $\underset{\sim}{\Pi}^1_1 - \mathrm{AD}^{\Sigma_1}$ follows from $ZFC + \exists\kappa$ (κ is supercompact). However, the ideas presented in this section seem more promising than Martin's approach when it comes to proving $\underset{\sim}{\Pi}^1_1$ AD^{Σ_n}, $n \geq 2$, from ZFC + $\exists\kappa$ (κ is a supercompact).

Suppose $T \subseteq \bigcup_{\alpha<\omega_1} {}^\alpha\omega$, and $\pi : T \to \omega$ is such that $\forall p, q \in T$ $(p \subseteq q \Rightarrow \pi(p) \neq \pi(q))$; then we shall call π a *continuous coding of* T. Our determinacy proof actually applies to game of the form $\mathcal{G}(A, T)$, where T admits a continuous coding. It is an easy exercise to show that T_1 admits a continuous coding, as do the trees for the games ending at the first recursively inaccessible, recursively Mahlo , . . . , relative to the play. It is also easy to see that T_2, the tree for games ending at the first Σ_2 admissible relative to the play, does not admit a continuous coding.

We begin with the higher type games to which we shall reduce our long games. Let X be any set, and $A \subseteq {}^\omega X$. By $G_{\mathcal{A}}$ we mean the game length ω on X with payoff set $\mathcal{A}$. We call $G_{\mathcal{A}}$ determined if one of the players has a winning quasi-strategy, or wqs. We call $\mathcal{A}$ *Suslin* if for some ordinal κ and tree T on $X \times \kappa$, $\mathcal{A} = p[T] = \{f \in {}^\omega X | \exists g \in {}^\omega\kappa \forall n (f \restriction n, g \restriction n) \in T\}$. We call $\mathcal{A}$ co-Suslin if ${}^\omega X - \mathcal{A}$ is Suslin. The author learned the proof of the following lemma from A.S. Kechris; it is due to him. By OD(X) we mean the class of set ordinal definable from finitely many elements of X.

Lemma 5. Assume ZF + AD + DC. Suppose ω_1 is X-supercompact, as witnessed by the ultrafilter μ. Suppose that $\mathcal{A} \subseteq {}^\omega X$ is Suslin and co-Suslin, as witnessed by the trees T and U. Then there is a OD($\{\mu, T, U\}$) wqs for $G_{\mathcal{A}}$.

Proof. For $C \in P_{\omega_1}(X)$, let $G^c_{\mathcal{A}}$ be just like $G_{\mathcal{A}}$, except that the player's moves are restricted to lie in C. Then $G^c_{\mathcal{A}}$ is determined by AD. For any C such that I wins $G^c_{\mathcal{A}}$, we can define a canonical "best" wqs Σ_c for I in $G^c_{\mathcal{A}}$, as in Moschovakis [6], Chapter 6. (In order to make the strategy canonical, we must make it "quasi.") Σ_c is OD($\{T, C\}$) uniformly in C. Similarly, if II wins $G^c_{\mathcal{A}}$ we get a canonical wqs τ_c for II which is OD($\{U, C\}$), uniformly in C.

Now suppose I wins G^c_a for $\mu - a.e.C$. Define Σ for I in G_a by:

$$p \text{ is according to } \Sigma \text{ iff for } \mu - a.e.C,\ p \text{ is according to } \Sigma_c.$$

Σ is OD($\{T, \mu\}$), and it is easy to check that Σ is a wqs for I in $G_{\mathcal{A}}$.

Similarly, if II wins $G^c_{\mathcal{A}}$ for $\mu - a.e.C$, we get an OD($\{U, \mu\}$) wqs τ for II in $G_{\mathcal{A}}$. ∎

The most natural reduction of continuously coded long games to games of length ω takes us to games on $\mathbf{R} \times P(\mathbf{R})$. Unfortunately there is a very simple $\mathcal{A} \subseteq {}^{\omega}(\mathbf{R} \times P(\mathbf{R}))$ which is not Suslin in models of AD, namely $\mathcal{A} = \{(\vec{x}, \vec{A}) | \lambda_n \cdot x_n(0) \in A_0\}$. Thus we must take care to make Lemma 5 applicable. We do this by reducing to a game on $\mathbf{R} \times \mathcal{S}$, where

$$\mathcal{S} = \{(\bar{\phi}, \bar{\psi}) | \bar{\phi} \text{ and } \bar{\psi} \text{ are scales, and } \text{proj}(\bar{\phi}) = \mathbf{R} - \text{proj}(\bar{\phi})\}.$$

Here proj($\bar{\phi}$) is the projection of $\bar{\phi}$, that is, the common domain of the norms $\phi_i, i \in \omega$. For $g \in {}^{\omega}\mathcal{S}$, let proj($g$) $= h$, where $h(i) = \text{proj}(g(i)_0)$ for all $i \in \omega$. For $\mathcal{A} \subseteq {}^{\omega}(\mathbf{R} \times \mathcal{S})$, let

$$\text{Proj}(\mathcal{A}) = \{(f, h) | \exists g((f, g) \in \mathcal{A} \wedge h = \text{proj}(g))\};$$

thus proj($\mathcal{A}$) $\subseteq {}^{\omega}(\mathbf{R} \times P(\mathbf{R}))$. We say that $\mathcal{A} \subseteq {}^{\omega}(\mathbf{R} \times \mathcal{A})$ is *projection-invariant* iff whenever proj($\mathcal{B}$) = proj($\mathcal{A}$), then $\mathcal{B} \subseteq \mathcal{A}$. Now one can easily modify the example of the last paragraph to obtain an $\mathcal{A} \subseteq {}^{\omega}(\mathbf{R} \times \mathcal{S})$ which is not Suslin in models of AD; however there us no simple projection-invariant $\mathcal{A}$ with this property.

Conjecture. There is an inner model of ZF + AD + DC + "Every projection-invariant $\mathcal{A} \subseteq {}^{\omega}(\mathbf{R} \times \mathcal{S})$ is Suslin".

Such a model must of course satisfy "Every set of reals admits a scale." The model M of §3 may verify the conjecture; the author is not sufficiently in command of Woodin's [8] to decide.

Fortunately, we shall need the conjecture only for reasonably simple $\mathcal{A}$. Let $B \subseteq \mathbf{R}$ and $\mathcal{C} \subseteq {}^{\omega}(\mathbf{R} \times P(\mathbf{R}))$; we call $\mathcal{C}$ $B-$*projective* iff $\mathcal{C}$ is the smallest class of relations containing the basic relations

$$R(\vec{x}^1 \ldots \vec{x}^n, \vec{A}) \text{ iff } x^m_i \in A_j,$$

$$S(\vec{x}^1 \ldots \vec{x}^n, \vec{A}) \text{ iff } x^m_i \in B,$$

and

$$Q(\vec{x}^1 \dots \vec{x}^n, \vec{A}) \text{ iff } x_i^m(j) = l,$$

for all i, j, l, m, in ω, and closed under countable intersection, countable union, complement, and quantification over ${}^\omega\mathbf{R}$. The proof of the following lemma is implicit in that of Moschovakis' 2nd periodicity theorem, so we omit it.

Lemma 6. Assume ZF + AD + DC. Suppose $B \subseteq \mathbf{R}$ is Suslin and co-Suslin via the trees T and U. Let $\mathcal{A} \subseteq {}^\omega(\mathbf{R} \times \mathcal{S})$ be projection-invariant, and suppose proj$(\mathcal{A})$ is B-projective. Then $\mathcal{A}$ is Suslin via a tree in OD$(\{T, U\})$.

Theorem 5. Assume ZF + AD + DC + "Every set of reals admits a scale" + ω_1 is $P(\mathbf{R})$-supercompact". Then $\mathcal{G}(A, T)$ is determined whenever T admits a continuous coding. In particular, AD^{Σ_1} holds.

Proof. Let π be a continuous coding of T. There is a scaled coding F of T induced by π. Let $\hat{\pi} = \{(p^*, n) | \pi(p) = n\}$. Let $B \subseteq \mathbf{R}$ be such that $\hat{\pi}, T^*, E_T^*, A^*, C$, and their complements all have scales Wadge reducible to B. (Here $C = C(i, j, k, y)$ is as the definition of "scaled coding.") Let μ be a normal, fine ultrafilter on $P_{\omega_1}(\mathbf{R} \times \mathcal{S})$; μ exists since we can map $\mathbf{R} \times \mathcal{S}$ into $P(\mathbf{R})$ in a 1-1 way.

Since we have AD + "Every set has a scale", $P(\mathbf{R}) \not\subseteq \mathrm{OD}(\{S\} \cup \mathbf{R})$ for any set S. Thus for any set S there is an $A \subseteq \mathbf{R}$ such that any set of reals in OD$(\{S\} \cup \mathbf{R})$ has a scale Wadge reducible to A. Let

$$H_0 = \mathrm{OD}(\{B\} \cup \{\mu\} \cup \mathbf{R})$$

and

$$H_{n+1} = \mathrm{OD}(\{\mu\} \cup \{A\} \cup \mathbf{R}), \text{ where } A \text{ Wadge-minimal such that every set of reals in } H_n \text{ has a scale } \leq_w A.$$

We now describe a game G^* on $\mathbf{R} \times \mathcal{S}$ auxilliary to $\mathcal{G}(A, T)$. The game G^* is quite similar to the auxilliary game in Martin's proof of Borel determinacy in [3]. A set $\Sigma \subseteq \cup_{\alpha < \omega_1} {}^\alpha\omega$ is a tree if $\forall p \in \Sigma \forall \beta \in \mathrm{dom}(p)(p \restriction \beta \in \Sigma)$. Play in G^* is divided into rounds; before beginning a round we have a G^*-position and an associated $\mathcal{G}$ position. For convenience we assume $p \in E \Rightarrow \mathrm{dom}(p)$ is a limit.

Round 0. We begin with G^* and $\mathcal{G}$-positions $\emptyset$.

(a) I plays $a \in \omega$, then II plays $b \in \omega$. Let e be least in $\omega - \{\pi(\langle a, b\rangle)\}$; we call e critical for round 0.

(b) I now plays a $(\bar{\phi}, \bar{\psi})$ in $S \cap H_1$ such that $\mathrm{proj}(\bar{\phi}) = \Sigma$ is a subtree of $T(\Sigma = \emptyset$ is o.k.). II may either accept or reject Σ.

II accepts: I is now obliged to keep the $\mathcal{G}$-position in Σ henceforth. II is obliged to insure that $e \neq \pi(p \upharpoonright \beta)$ for any $\beta \leq \mathrm{dom}(p)$ and any $\mathcal{G}$-position p reached henceforth.

II rejects: II must plays a $q \in \Sigma$ extending $\langle a, b\rangle$ and such that $e = \pi(q)$. No one incurs any obligations.

The new $\mathcal{G}$ position is $\langle a, b\rangle$ if II accepts and q if II rejects.

Round n, $n > 0$: We have a G^* position r with associated $\mathcal{G}$ position p.

(a) I plays $a \in \omega$, II plays $b \in \omega$. Let e be least in $\omega - \{\pi((p^\frown\langle a, b\rangle) \upharpoonright \beta) \,|\, \beta \leq \mathrm{dom}(p) + 1\}$; e is critical for round n.

(b) I now plays a $(\bar{\phi}, \bar{\psi})$ in $S \cap H_{(n+1)}$ such that $\mathrm{proj}(\bar{\phi}) = \Sigma$ is a (set of reals coding a) subtree of T. II may either accept or reject Σ.

II accepts: I is now obliged to keep the $\mathcal{G}$ position in Σ henceforth. II is obliged to insure that if q is a $\mathcal{G}$ position later reached, then $e \neq \pi(q \upharpoonright \beta)$ for all $\beta \leq \mathrm{dom}(q)$.

II rejects: II must play a $q \in \Sigma$ extending $p^\frown\langle a, b\rangle$ and such that $e = \pi(q)$. No obligations are incurred.

The new $\mathcal{G}$ position is $p^\frown\langle a, b\rangle$ is II accepts, and q if II rejects.

The first player to fail to meet any one if his obligations loses G^*. If both players fail first at the same round, then the one violating the obligation incurred earliest loses, with I's obligation incurred at round n coming after those incurred by II at rounds $m < n$ and before those incurred by II at rounds $m \geq n$.

Suppose both players meet all obligations incurred in G^*. After ω moves they have produced a $\mathcal{G}$ position p. If $p \in E$, then I wins G^* iff $p \in A$. Otherwise, notice that $\pi(p)$ was critical at some round, and II accepted at that round, so that p violates II's obligation incurred at that round (and no earlier II obligations). So if $p \notin E$, and $\pi(p)$ was critical at round n, then I wins G^* iff p violates no I obligation incurred at some round $m \leq n$.

Clearly we may regard G^* as a game on $\mathbf{R} \times S$. Since the relation $y \in H_n$ is $\mathrm{OD}(\{B, \mu\})$, there is a tree U_1 on $\mathbf{R} \times \mathcal{S}$ such that $U_1 \in \mathrm{OD}(\{B, \mu\})$ and

$$r \in [U_1] \text{ iff } r \text{ is a play of } G^* \text{ in which I loses at no finite stage.}$$

Similarly, there is an $\mathrm{OD}(\{B, \mu\})$ tree U_2 on $\mathbf{R} \times \mathcal{S} \times \omega$ such that

$$r \in p[U_2] \text{ iff } r \text{ is a play of } G^* \text{ in which}$$
$$\text{II loses at some finite stage.}$$

Finally, there is a projection invariant $\mathcal{B} \subseteq {}^{\omega}(\mathbf{R} \times \mathcal{S})$ such that $\mathrm{proj}(\mathcal{B})$ is $\langle \hat{\pi}, T^*, E^*, A^*C \rangle-$ projective and whenever r is a play of G^* where no one loses at a finite stage,

$$r \in \mathcal{B} \text{ iff } r \text{ is a win for I at stage } \omega.$$

It follows from Lemma 6 that $\mathcal{B}$ is Suslin via some $\mathrm{OD}(\{B, \mu\})$ tree U_3. But then

$$r \text{ is a win for I in} G^* \text{iff} (r \in [U_1] \wedge$$
$$(r \in p[U_2] \text{ or } r \in p[U_3]))$$

so that $G^* = G_{\mathcal{A}}$, where $\mathcal{A}$ is Suslin via an $\mathrm{OD}(\{B, \mu\})$ tree. Similarly, $\mathcal{A}$ is co-Suslin via an $\mathrm{OD}(\{B, \mu\})$ tree, so Lemma 5 gives a wqs for G^* which is $\mathrm{OD}(\{B, \mu\})$.

Case 1. I had a wqs in G^*.

We shall construct a ws Γ for I in $\mathcal{G}(A, T)$. Notice that there is a set of reals not in $\cup_n H_n$ by DC, and thus there us a map from $\mathbf{R}$ onto $\cup_n H_n \cap \mathcal{S}$. By uniformization we can thus convert I's wqs in G^* into a full ws; call it $\hat{\Gamma}$.

We define Γ from $\hat{\Gamma}$ by associating inductively to each p we reach following Γ a G^* position $\hat{p}$ which is according to $\hat{\Gamma}$ and whose associated $\mathcal{G}$ position is p. We arrange that in $\hat{p}$ II has met all his obligations.

Suppose we have reached $p \in T$ by following Γ, and we have the associated $\hat{p}$ according to $\hat{\Gamma}$. Let $\hat{\Gamma}(\hat{p}) = a$; then set $\Gamma(p) = a$. Now suppose II responds with b in $\mathcal{G}(A, T)$.

Case A. $p^\frown\langle a, b \rangle$ meets all II obligations incurred in $\hat{p}$.

In this case, let $(\bar{\phi}, \bar{\psi}) = \hat{\Gamma}(\hat{p}^\frown\langle a, b \rangle)$, and set

$$\widehat{(p^\frown\langle a, b \rangle)} = \hat{p}^\frown\langle a, b, (\bar{\phi}, \bar{\psi}), \text{ accept} \rangle.$$

Case B. Otherwise

Let the first II obligation violated by $p^\frown\langle a,b\rangle$ be incurred at round n (i.e., in $\hat{p}(n)$). Let $\hat{p}(n) = \langle c,d,(\bar{\phi},\bar{\psi})\text{ accept}\rangle$. Set

$$\widehat{(p^\frown\langle a,b,\rangle)} = \hat{p}\restriction n^\frown\langle c,d,(\bar{\phi},\bar{\psi}),\text{reject},p^\frown\langle a,b\rangle\rangle.$$

(Notice $p^\frown\langle a,b\rangle \in \text{proj}(\bar{\phi})$, as otherwise $\hat{p}^\frown\langle a,b\rangle$ is a loss for $\hat{\Gamma}$.

In either case, the relation between p and $\hat{p}$ still holds between $p^\frown\langle a,b\rangle$ and $\widehat{(p^\frown\langle a,b\rangle)}$.

Finally, we must define $\hat{p}$ for p a position of limit length reached by following Γ. Let $\lambda = \text{dom}(p)$. Notice that if n is least such that $\widehat{(p\restriction\beta)}(n) \neq \widehat{(p\restriction\beta+1)}(n)$, then $\widehat{(p\restriction\beta)}(n)_3 = \text{accept}$ and $\widehat{(p\restriction\beta+1)}(n)_3 = \text{reject}$. So $q(n) = \lim_{\beta\to\lambda}\widehat{(p\restriction\beta)}(n)$ exists for all n. If $p \in E$, set $\hat{p} = q$. In this case $\hat{p}$ is by $\hat{\Gamma}$, so $p \in A$, and we have verified that Γ wins. If $p \notin E$, let $e = \pi(p)$. Then e was critical at some round, say round n, of q. Let $q(n) = \langle c,d,(\bar{\phi},\bar{\psi}),\text{accept}\rangle$. Set

$$\hat{p} = q\restriction n^\frown\langle c,d,(\bar{\phi},\bar{\psi}),\text{reject},p\rangle.$$

Since q was by $\hat{\Gamma}$, which wins for I in G^*, $p \in \text{proj}(\bar{\phi})$. Thus $\hat{p}$ is a position according to $\hat{\Gamma}$ in which II meets all his obligations, with associated position p, as desired.

This defines Γ. We have also verified Γ wins $\mathcal{G}(A,T)$ for I.

Case 2. II has a wqs in G^*.

We shall construct a ws Γ for II in $\mathcal{G}(A,T)$. Let $\hat{\Gamma}$ be a wqs for II in G^* which is $\text{OD}(\{B,\mu\})$.

Let $\sigma : \mathbf{R} \xrightarrow{\text{onto}} \{q|q \text{ is a } G^* \text{ position}\}$. If $\sigma(x) = q$ we call x an index of q. We shall define Γ by associating inductively to each $\mathcal{G}$ position p we reach the following Γ in G^* position $\hat{p}$ according to $\hat{\Gamma}$, together with a real indexing $\hat{p}$. We arrange that in $\hat{p}$ I has met all his obligations and that p is the $\mathcal{G}$-position associated to $\hat{p}$.

Suppose we have reached p by following Γ, and have $\hat{p}$ and x such that $\sigma(x) = \hat{p}$. Suppose I now plays $a \in \omega$ in $\mathcal{G}(A,T)$. Let $b \in \omega$ be least such that $\hat{p}^\frown\langle a,b\rangle$ is by $\hat{\Gamma}$; set $\Gamma(p^\frown\langle a\rangle) = b$. Now let $q = p^\frown\langle a,b\rangle$; we want to define $\hat{q}$ and an index for $\hat{q}$.

Case A. q violates no I-obligation in $\hat{p}$.

In this case, let

$$\begin{aligned}\Sigma = \{r \in \cup_{\alpha<\omega_1} {}^{\alpha}\omega \mid &\text{if } \beta \leq \text{dom(r), then no position of the form}\\ &\hat{p}^\frown\langle a, b, (\bar{\phi}, \bar{\psi}), \text{reject}, r \restriction \beta\rangle,\\ &\text{with}(\bar{\phi}, \bar{\psi}) \in \mathcal{S} \cap H_{n+1}, n = \text{dom}(\hat{p}), \text{is in accord with } \hat{\Gamma}\}.\end{aligned}$$

Then $\Sigma \in H_n$, where $n = \text{dom}(\hat{p})$, since $\hat{p} \in H_n$ and $\hat{\Gamma} \in H_n$. So there is a $(\bar{\phi}, \bar{\psi}) \in \mathcal{S} \cap H_{n+1}$ such that $\Sigma = \text{proj}(\bar{\phi})$. Now $\hat{\Gamma}$ must accept Σ after $\hat{p}^\frown\langle a, b\rangle$ by the definition of Σ, so we set

$$\hat{q} = \hat{p}^\frown\langle a, b, (\bar{\phi}, \bar{\psi}), \text{accept}\rangle,$$

where $(\bar{\phi}, \bar{\psi}) \in \mathcal{S} \cap H_{n+1}$, $\Sigma = \text{proj}(\bar{\phi})$, and the index of $\hat{q}$ is obtained from x by the appropriate uniformizing function. (We ought to have fixed at the outset a function $h : \mathbf{R} \to \mathbf{R}$ such that if $\sigma(x) = \hat{p}$ has the properties above, and Σ is defined as above, then $\sigma(h(x, a)) = \hat{q}$ is related to Σ and $\hat{p}$ as above.)

Case B. Otherwise

Let n be least such that the I-obligation incurred in $\hat{p}(n)$ is violated by q. Then $\hat{p}(n) = \langle c, d, (\bar{\phi}, \bar{\psi}), \text{accept}\rangle$, where $\text{proj}(\bar{\phi}) = \Sigma$ is as in Case A, and $q \notin \Sigma$ while $q \restriction \beta \in \Sigma$ for $\beta < \text{dom}(q)$. By definition of Σ, there is a $(\bar{\theta}, \bar{\rho}) \in \mathcal{S} \cap H_{n+1}$, such that $\hat{p} \restriction n^\frown\langle c, d, (\bar{\theta}, \bar{\rho}), \text{reject}, q\rangle$is in accord with $\hat{\Gamma}$. Using uniformization, we can pick such a $(\bar{\theta}, \bar{\rho})$ together with an index of

$$\hat{q} = \hat{p} \restriction n^\frown\langle c, d, (\bar{\theta}, \bar{\rho}), \text{reject}, q\rangle.$$

Finally, suppose we reach a position p of limit length following Γ; we must define $\hat{p}$ and an index of same. As in Case 1, $q(n) = \lim_{\beta\to\text{dom}(p)} \widehat{(p \restriction \beta)}(n)$ exists for all n, so that q is a completed run of G^* by $\hat{\Gamma}$. If $p \in E$, then since p is the run of $\mathcal{G}(A, T)$ associated to $q, p \notin A$, so that Γ has won. If $p \notin E$, then let $\pi(p) = e$, and let n be such that e is critical at round n in q. Since $\hat{\Gamma}$ wins for II, q violates some I obligation incurred in $q(m)$ for some $m \leq n$. Then let

$$\hat{p} = q \restriction m^\frown\langle c, d, (\bar{\theta}, \bar{\rho}), \text{reject}, q\rangle,$$

together with an index of $\hat{p}$, be obtained as in Case B above.

This completes the definition of Γ and the verification that it wins for II. ∎

There is a natural strategy for iterating the reduction achieved in Theorem 5, and thereby proving $\underset{\sim}{\Pi}^1_1 - \text{AD}^{\Sigma_n}$ (assuming, ultimately, ZFC + $\exists\kappa$ (κ is supercompact).) It seems one must strengthen the hypothesis "Every set has a scale" of Theorem 5 (as well as requiring ω_1 be P_{ω_1} ($V_{\omega+\omega}$-supercompact) in order to do this. The first step of the strengthening is given by the conjecture mentioned in this section.

The proof of the Kechris-Woodin theorem of §3 adapts at once to show

Corollary 3. ZF + AD + DC "Every set has a scale" + "ω_1 is $P(\mathbf{R})$-supercompact" provides the consistency of $\mathrm{ZFC} + \mathrm{HOD}(\mathbf{R}) - \mathrm{AD}^{\Sigma_1}$.

Thus $\mathrm{HOD}(\mathbf{R}) - \mathrm{AD}^{\Sigma_1}$ would not have sufficed to produce the model M of §3.

§5. Questions.

The reader who has arrived at this section by slogging through the intermediate ones will probably have his own list of questions by now. We list here only a few salient ones.

Ad §1, the natural question is whether definable scales extend still further. For $p \in \cup_{\alpha<\omega_1} {}^{\alpha}\omega$, let

$$N_p = \{f \in {}^{\omega_1}\omega | p \subseteq f\},$$

and call a set $A \subseteq {}^{\omega_1}\omega$ open-$\underset{\sim}{\Pi}^1_1$ if there is a $\underset{\sim}{\Pi}^1_1$ set A^* of codes for elements of $\cup_{\alpha<\omega_1} {}^{\alpha}\omega$ such that

$$A = \cup\{N_p | \exists x \in A^* (x \text{ codes } p)\}.$$

If CH holds, then $\Game^{\omega_1}$ (closed-$\underset{\sim}{\Pi}^1_1$) is just $\underset{\sim}{\Sigma}^2_1$, and by Moschovakis' argument, $\underset{\sim}{\Pi}^2_1$ has the prewell-ordering property. Does some form of definable determinacy, e.g. $\mathrm{HOD}(\mathbf{R}) - \mathrm{AD}^{\omega_1}$, imply that all $\Game^{\omega_1}$ (open-$\underset{\sim}{\Pi}^1_1$) sets of reals admit definable scales?

Ad §2, does every length ω_1 game with open-$\underset{\sim}{\Pi}^1_1$payoff won by I have a definable winning strategy (assuming e.g. $\mathrm{HOD}(\mathbf{R}) - \mathrm{AD}^{\omega_1}$?). There is a simple such game won by II for which II has no definable winning strategy. It is part of the folklore: I and II take a nap until, at some move α, I awakens with a start and asks II to produce $x \in \mathrm{WO}$ such that $|x| = \alpha$. If I never awakens he loses, otherwise II must comply or lose. For all we know, however, every length ω_1 game with $\mathrm{HOD}(\mathbf{R})$ payoff has a definable *pseudo-strategy,* where S is a pseudo-strategy winning G_A for (say) I if

$$S : \bigcup_{\alpha<\omega_1} ({}^{\alpha}\omega \times \mathrm{WO}_{\alpha}) \to \omega,$$

where $\mathrm{WO}_{\alpha} = \{x \in \mathrm{WO} || x| = \alpha\}$, and whenever $g(\alpha) \in \mathrm{WO}_{\alpha}$ for all $\alpha < \omega_1$,

$$\forall_{\alpha}(f(\alpha)_0 = S(f \restriction \alpha, g(\alpha))) \Rightarrow f \in A,$$

for all $f \in {}^{\omega_1}\omega$. Does HOD(**R**) $-$ AD$^{\omega_1}$ guarantee HOD(**R**) pseudo-strategies for length ω_1 with HOD(**R**) payoff? (The notion of a pseudo-strategy is due independently to Woodin.)

Ad §3, notice that the folklore example above shows ZF + AD + (open-$\boldsymbol{\Pi}^1_1$) - AD$_{\omega_1}$ is inconsistent. A slight modification shows that ZF + AD implies the existence of a non-determined "clopen-$\boldsymbol{\Pi}^1_1$" game of length ω_1. [At any move β before he awakes, I must pay for continuing to nap by playing an $x \in WO_\beta$. Notice that, in contrast to the games admitting scaled codings, the game is not clopen in V, only in models of AD.] Thus §3 goes nearly as far as possible in producing inner models with full winning strategies for games not limited by definability of the payoff. Can we produce stronger inner models using pseudo-strategies?

Another question suggested by §3: what is the consistency strength of AD$_{\mathbf{R}}$+ "θ is regular", or even AD$_{\mathbf{R}}$ + cof$(\theta) > \omega_1$, vis-a-vis the long game hierarchy? Does $\boldsymbol{\Delta}^1_2$ AD$^{\Sigma_2}$ give us a model of AD$_{\mathbf{R}}$ + cof$(\theta) > \omega_1$?

Finally, §4 suggests the obvious question: does ZFC + $\exists\kappa$ (κ is supercompact) prove $\boldsymbol{\Pi}^1_1$ $-$ AD$^{\Sigma_n}$ for all $n < \omega$? Woodin has shown in ZFC + $\exists\kappa$ (κ is *supercompact*) that all $⅁^{\omega_1}$ (closed-$\boldsymbol{\Pi}^1_1$) sets are Lebesgue measurable. This suggests that in fact (open-$\boldsymbol{\Pi}^1_1$) $-$AD$^{\omega_1}$ might follow from ZFC + $\exists\kappa$ (κ is supercompact).

We conclude by mentioning the strongest form of determinacy we know (other than the inconsistent forms). There are three parameters entering into the description of a class of games: complexity of payoff, complexity of individual moves, and length. The known limitations in each direction are:

(a) *payoff:* there is a non-determined game on $\{0,1\}$ of length ω. (Gale-Stewart)

The natural response is to limit the payoffs considered to OD(**R**).

(b) *moves:* there us a non-determined game on $P(\mathbf{R})$ of length ω whose payoff is OD. [I plays a non-determined $A \subseteq {}^{\omega}\omega$, then I and II play out G_A.] (folklore)

The natural response is to limit the moves to be OD(**R**).

(c) *length:* there is a non-determined game on $\{0,1\}$ with OD payoff of length $\omega_1 + \omega$. [In his first ω_1 moves I must describe an uncountable $A \subseteq {}^{\omega}\omega$ with no perfect subset. In the next ω moves, I and II play the perfect set game for A.] (Galvin, Laver)

The natural (most generous) response is to limit the intermediate positions to be OD(**R**). (This subsumes our response to (b)). This leads to a "maximum determinacy" principle, or MD.

MD. Let G be a game on OD($\mathbf{R}$) which ends as soon as the players reach a position which is not OD($\mathbf{R}$), with the winner declared according to some $OD(\mathbf{R})$ payoff condition. Then G is determined.

Superficially, anyway, MD allows games of any ordinal length. Is MD consistent? Is it good for anything?

REFERENCES

[1] H. Becker, *A property equivalent to the existence of scales,* Transactions of the AMS, vol. 287, 1985, pp. 591-612.

[2] Andreas Blass, *The equivalence of two atrong forms of determincay,* Proc. of AMS, vol. 52, 1975, pp. 373-376.

[3] D.A. Martin, *Borel determinacy,* Ann. of Math. (2) vol. 102, 1975, pp. 363-371.

[4] D.A. Martin, *The real game quantifier propagates scales,* Cabal Seminar 79-82, Springer Lecture Notes in Math.,

[5] D.A. Martin, unpublished.

[6] Y.N. Moschovakis, *Descriptive set theory,* North Holland, Amsterdam, 1980.

[7] R.M. Solovay, *The independence of DC from AD,* Cabal Seminar 76-77, Springer Lecture Notes in Math., vol. 689, pp. 171-189.

[8] W.H. Woodin, unpublished.

"AD + UNIFORMIZATION" IS EQUIVALENT TO "HALF $AD_{\mathbf{R}}$"

Alexander S. Kechris*
Department of Mathematics
California Institute of Technology
Pasadena, CA 91125

Let ω be the set of natural numbers and $\mathbf{R} = 2^\omega$ the set of "reals". As usual AD is the assertion that every game on ω is determined and $AD_{\mathbf{R}}$ is the assertion that every game on $\mathbf{R}$ is determined. Let us also consider an intermediate principle $AD_{\mathbf{R}}^{1/2}$, which is the assertion that every game in which one of the players plays in ω and the other in $\mathbf{R}$ is determined. Such games appear often in applications instead of full games on $\mathbf{R}$, see for instance [K].

Clearly (in ZF) $AD_{\mathbf{R}} \Rightarrow AD_{\mathbf{R}}^{1/2} \Rightarrow AD$. One other immediate consequence of $AD_{\mathbf{R}}^{1/2}$ is UNIFORMIZATION, which is the assertion: If $S \subseteq \mathbf{R} \times \mathbf{R}$, there is $F : \mathbf{R} \to \mathbf{R}$ with $\exists y\, S(x,y) \Rightarrow S(x, F(x))$, i.e. every relation on the reals can be uniformized. We prove here that conversely AD + UNIFORMIZATION imply $AD_{\mathbf{R}}^{1/2}$, i.e. we have the following

Theorem. *Assume* ZF + DC. *Then*

$$AD + \text{UNIFORMIZATION} \iff AD_{\mathbf{R}}^{1/2}.$$

The relationship between $AD_{\mathbf{R}}$ and $AD_{\mathbf{R}}^{1/2}$ has not been fully understood. Woodin (unpublished) has shown (using also results of Becker [B]) that (over ZF + DC) $AD_{\mathbf{R}}$ and $AD_{\mathbf{R}}^{1/2}$ are equiconsistent, but it is not known if $AD_{\mathbf{R}}^{1/2}$ implies $AD_{\mathbf{R}}$ (over ZF + DC again).

The proof of the above theorem is "local", so it yields results about various classes of games. For any pointclass Γ let $\Gamma - AD$ and $\Gamma - AD_{\mathbf{R}}^{1/2}$ denote the determinacy of the corresponding games in Γ and let Γ-UNIFORMIZATION be the assertion that every relation in Γ can be uniformized by a function (whose graph is) in Γ. Call a pointclass Γ *nice* if Γ contains all the Borel sets and is closed under Borel substitutions, conjunctions and disjunctions, existential quantification over $\mathbf{R}$ and negation. Then we have

Corollary (of the proof). *Assume* ZF + DC, *and let* Γ *be a nice pointclass. Then*

$$\Gamma - AD + \Gamma - \text{UNIFORMIZATION} \Rightarrow \Gamma - AD_{\mathbf{R}}^{1/2}.$$

(Of course we cannot invert the arrow here unless we assume that games $A \subseteq \mathbf{R} \times \mathbf{R}^\omega$ in Γ have also strategies in Γ).

*Research partially supported by NSF Grant DMS-8416349.

For example for Γ = Projective since by Moschovakis' Theorem (see [M]) $\Gamma-$ AD $\Rightarrow$ Γ-UNIFORMIZATION, using the usual notation PD and $\mathrm{PD}_{\mathbf{R}}^{1/2}$ in this case, we have:

$$\mathrm{ZF} + \mathrm{DC} \vdash \mathrm{PD} \iff \mathrm{PD}_{\mathbf{R}}^{1/2}.$$

Similarly for $\Gamma = (\boldsymbol{\Delta}_1^2)^{L(\mathbf{R})}$ (in which case again $\Gamma-$ AD $\Rightarrow$ Γ- UNIFORMIZATION, by a theorem of Martin-Steel, (see [MS])) we have in ZF + DG

$$(\boldsymbol{\Delta}_1^2)^{L(\mathbf{R})} - \mathrm{AD} \iff (\boldsymbol{\Delta}_1^2)^{L(\mathbf{R})} - \mathrm{AD}_{\mathbf{R}}^{1/2}.$$

Recall that $(\boldsymbol{\Delta}_1^2)^{L(\mathbf{R})}-$AD $\iff$ $L(\mathbf{R})-$ AD (see [KS]), but of course $L(\mathbf{R})$ -UNIFORMIZATION fails.

We give now the proof of the theorem. Assume AD + UNIFORMIZATION. Let $A \subseteq \mathbf{R} \times \mathbf{R}^\omega$ be given and consider the corresponding game G_A:

I	$\alpha(0)$		$\alpha(1)$		$\cdots$	α
II		x_0		x_1		$\vec{x}$

$$\alpha(i) \in \omega,\ x_i \in \mathbf{R};\ \text{I wins iff } (\alpha, \vec{x}) \in A.$$

We have to show that this game is determined.

The proof is based on a method of "countable approximations" of games with (some) moves in $\mathbf{R}$, which in different variations (depending on the choice of "approximation") is quite useful in several applications.

For each real y consider the following "countable approximation" $G_A^{(y)}$ of G_A:

I	$\alpha(0)$		$\alpha(1)\cdots$		α
II		x_0		$x_1\cdots$	$\bar{x}$

$\alpha(i) \in \omega$; $x_i \in \mathbf{R}$, $x_i \leq_T y$ (i.e. x_i are recursive in y); I wins iff $(\alpha, \bar{x}) \in A$.

This is basically a game on ω, so by AD it is determined for each $y \in \mathbf{R}$. If II has a winning strategy in $G_A^{(y)}$ for some y, clearly II has a winning strategy in G_A (and conversely). So assume I has a winning strategy in $G_A^{(y)}$, for all $y \in \mathbf{R}$. We will prove that I has a winning strategy in G_A.

By UNIFORMIZATION let F be a function that assigns to each real y a winning strategy $\tau_y \equiv F(y)$ for I in $G_A^{(y)}$. (Thus $\tau_y : \{x : x \leq_T y\}^{<\omega} \to \omega$.)

Let us recall here the following result of Martin, (a strengthening of) which will be crucial in our argument.

Theorem (Martin). *Assume ZF + DC + AD. Let* $A \subseteq \mathbf{R}$. *If A is cofinal in the Turing degrees, i.e.* $\forall x \in \mathbf{R}\ \exists y \in A(x \leq_T y)$, *then A contains a pointed perfect set (i.e. there is a perfect binary tree T such that* $x \in [T] \Rightarrow T \leq_T x$ *and* $[T] \subseteq A$).

Proof. Consider the game

I	$x(0)$		$x(1)$		...
II		$y(0)$		$y(1)$	

$$x(i), y(i) \in \{0,1\}; \text{ II wins iff } x \leq_T y \ \&\ y \in A.$$

If I had a winning strategy, II could play any $y \in A$ with $\tau \leq_T y$ to defeat this strategy. So II must have a winning strategy σ. For each sequence of moves $u(0) \cdots u(n)$ of I let $\sigma * u(0) \cdots \sigma * u(n)$ be the corresponding sequence of moves for II following σ. Fix also a real $\hat{\sigma} \in 2^\omega$ with $\sigma \equiv_T \hat{\sigma}$.

Define now the following perfect binary tree T_1:

Let $u_\emptyset = \emptyset$. Then let $u_{(0)}$, $u_{(1)}$ be the first two binary sequences whose even part agrees with $\hat{\sigma}$ (when $u \in 2^{<\omega} \cup 2^\omega$ we say that the even part of u agrees with $x \in 2^\omega$ is $u(2i) = x(i)$ for $2i <$ length (u)), and are such that $\sigma * u_{(0)}$, $\sigma * u_{(1)}$ are incompatible. Such $u_{(0)}$, $u_{(1)}$ evidently exist since for $x \in 2^\omega$ $x \leq_T \sigma * x$ thus $x \mapsto \sigma * x$ is not constant on any subset of 2^ω containing reals of arbitrary high Turing degree. Let then $u_{(00)}$, $u_{(01)}$ be the first two binary sequences extending $u_{(0)}$ whose even part agrees with $\hat{\sigma}$ and are such that $\sigma * u_{(00)}$, $\sigma * u_{(11)}$ are incompatible, and similarly define $u_{(10)}$, $u_{(11)}$, etc. Let T_1 consist of all the subsequences of some u_t, $t \in 2^{<\omega}$.

Finally let

$$T = \{\sigma * u : u \in T_1\}.$$

Clearly T is a perfect binary tree, and $[T] \subseteq A$. Also $T \leq_T \sigma$. Let now $y \in [T]$. Then $y = \sigma * x$, for some $x \in [T_1]$. So the even part of x agrees with $\hat{\sigma}$, thus $\sigma \leq_T x$. But also (by the rules of the game) $x \leq_T y$ $\therefore$ $\sigma \leq_T y$ and thus $T \leq_T y$, i.e. T is pointed. ⊣

From Martin's Theorem we can obtain easily the following

Lemma. *Assume ZF + DC + AD. Let T be a pointed perfect tree and let* $F : [T] \to \omega$. *Then for each real x there is pointed perfect tree* T' *with* $T' \subseteq T$, $x \leq_T T'$ *and* $F|[T']$ *constant.*

Proof. First note that for any $G : 2^\omega \to \omega$ there is a pointed perfect S with $G|[S]$ constant. This follows from the fact that by Turing Determinacy some $G^{-1}[\{n\}]$ must be unbounded and the preceding theorem.

Let now $h : 2^\omega \to [T]$ be the canonical homeomorphism and let $G = F \circ h : 2^\omega \to \omega$. The let T_1 be pointed perfect tree with $G|[T_1]$ constant. By an easy standard fact find $T_1' \subseteq T_1$ pointed perfect tree with $x \leq_T T_1'$. Let now T' be the image of T_1' under h. If $z \in [T']$ then $h^{-1}(z) \in [T_1'] \therefore T_1' \leq_T h^{-1}(z) \therefore T_1' \leq_T z \oplus T$ (note that $h \leq_T T$) and so $T' \leq_T z \oplus T$. But $z \in [T]$ so $T \leq_T z$ thus $T' \leq_T z$, i.e. T' is pointed. Moreover letting $z =$ leftmost branch of T_1' we see that $x \leq_T T_1' \leq_T z \oplus T \leq_T z \leq_T T'$. Finally F is clearly constant on $[T']$ and we are done. $\dashv$

We are finally ready to define a strategy for I in G_A:

Let $F_0 : \mathbf{R} \to \omega$ be given by $F_0(y) = \tau_y(\emptyset)$. Choose a pointed perfect tree T_0 such that $F_0|[T_0]$ is fixed, say with value k_0. I plays k_0 as his first move in G_A. Assume now II plays x_0. Find $T_0^{x_0} \subset T_0$ perfect pointed such that $x_0 \leq_T T_0^{x_0}$. Then consider the function $F_1^{x_0} : [T_0^{x_0}] \to \omega$ given by $F_1^{x_0}(y) = \tau_y((x_0))$. (Note that for $y \in [T_0^{x_0}]$ we have $x_0 \leq_T y$, thus $\tau_y((x_0))$ is defined.) By the preceding lemma find $T_1^{x_0} \subseteq T_0^{x_0}$ pointed perfect tree with $x_0 \leq_T T_1^{x_0}$ and $F_1|[T_1^{x_0}]$ constant, say with value k_1. Then II plays next in G_A this k_1. (Note that by UNIFORMIZATION we can choose $T_1^{x_0}$ for each x_0 as asserted.) Let next II play x_1 in G_A. Find $T_1^{x_0,x_1} \subseteq T_1^{x_0}$ pointed perfect tree with $x_0 \oplus x_1 \leq_T T_1^{x_0,x_1}$. Then consider the function $F_2^{x_0,x_1} : [T_1^{x_0,x_1}[\to \omega$ given by $F_2^{x_0,x_2}(y) = \tau_y((x_0,x_1))$. Again this is well defined since for $y \in [T_1^{x_0,x_1}]$, $x_0, x_1 \leq_T y$. Then find $T_2^{x_0,x_1}$ pointed perfect tree with $x_0 \oplus x_1 \leq_T T_2^{x_0,x_1}$ and $F_2|[T_2^{x_0,x_1}]$ constant, say with value k_2. I plays next this k_2, etc.

Thus if II plays successively $x_0, x_1, \ldots$, we have found pointed perfect sets

$$T_0 \supseteq T_1^{x_0} \supseteq T_2^{x_0,x_1} \supseteq T_3^{x_0,x_1,x_2} \supseteq \cdots$$

with $x_0 \oplus x_1 \oplus \cdots \oplus x_{n-1} \leq_T T_n^{x_0,x_1,\ldots,x_{n-1}}$, and such that $\tau_y((x_0,\ldots,x_{n-1}))$ is fixed on $T_n^{x_0,x_1,\ldots,x_{n-1}}$ with value say k_n. I plays then $k_0, k_1, \ldots$. We claim that this is a winning strategy for I in G_A, i.e. letting $\vec{k} = (k_0, k_1, \ldots)$, $\vec{x} = (x_0, x_1, \ldots)$ we have $(\vec{k},\ \vec{x}) \in A$. To see this note that $\cap_n [T_n^{x_0,x_1,\ldots,x_{n-1}}] \neq \emptyset$ and pick $y \in \cap_n [T_n^{x_0,x_1,\ldots,x_{n-1}}]$. Then for each n, $x_0 \oplus \cdots \oplus x_{n-1} \leq_T y$ i.e. $x_n \leq_T y$ for each n. Moreover $k_n = \tau_y((x_0 \cdots x_{n-1}))$, i.e. this is a run of the the game $G_A^{(y)}$ in which I followed τ_y, thus $(\vec{k},\ \vec{x}) \in A$ and we are done.

One final remark: One can easily obtain from the preceding proof definability estimates for $\mathrm{AD}_{\mathbf{R}}^{1/2}$ games in terms of definability estimates for uniformization. For example assuming PD and letting $A \subseteq \mathbf{R} \times \mathbf{R}^\omega$ be a projective game in which I has a winning strategy, then I has a projective winning strategy. (A strategy for I in a map $F : \mathbf{R}^{<\omega} \to \omega$, i.e. essentially a set of reals.) Of course finer level-by-level estimates and lightface versions can be also easily extracted but we will not spell them out here.

References

[B] H. Becker, *A property equivalent to the existence of scales,* Trans. Amer. Math. Soc., 287 (1985), 591-612.

[K] A. S. Kechris, *A coding theorem for measures,* this volume.

[KS] A. S. Kechris and R. M. Solovay, *On the consistency strength of determinacy hypotheses,* Trans. Amer. Math. Soc., 290 (1985), 179-211.

[MS] D. A. Martin and J. R. Steel, *The extend of scales in $L(\mathbf{R})$,* Cabal Seminar 79-81, Lecture Notes in Mathematics, Vol. 1019, Springer-Verlag, 86-96.

[M] Y. N. Moschovakis, *Descriptive set theory,* North Holland, 1980.

A CODING THEOREM FOR MEASURES

Alexander S. Kechris[1]
Department of Mathematics
California Institute of Technology
Pasadena, CA 91125

1. Introduction. Assuming ZF + DC + AD Moschovakis (see [M1]) has shown that if there is a surjection $\pi : \mathbf{R} \longrightarrow \lambda$ from the reals ($\mathbf{R} = \omega^\omega$ in this paper) onto an ordinal λ, then there is a surjection $\pi^* : \mathbf{R} \longrightarrow p(\lambda)$ from the reals onto the power set of λ. Let us denote by $\beta(\lambda)$ the set of ultrafilters on λ. The question was raised whether there is an analog of Moschovakis' Theorem for $\beta(\lambda)$, i.e. if there is a surjection from $\mathbf{R}$ onto λ, is there one from $\mathbf{R}$ onto $\beta(\lambda)$? Martin showed that this cannot be proved in ZF + DC + AD alone because if $V = L(\mathbf{R})$ and $\lambda = \delta^2_1$, there is no surjection of $\mathbf{R}$ onto $\beta(\lambda)$. We prove in this paper that if one strengthens the determinacy hypothesis from AD to $\text{AD}^{1/2}_{\mathbf{R}}$ (see [K1]), then this question has a positive answer. Our main theorem is then

Theorem. *Assume* ZF + DC + $\text{AD}^{1/2}_{\mathbf{R}}$. *If there is a surjection from* $\mathbf{R}$ *onto an ordinal* λ, *then there is a surjection from* $\mathbf{R}$ *onto* $\beta(\lambda)$.

This result can be rephrased as follows. By a result of Kunen (for a proof see [K2]) $\beta(\lambda)$ is wellorderable, so it has a definite cardinality, which we denote also by $\beta(\lambda)$. Since the sup of the ordinals onto which we can map the continuum is denoted by Θ, the preceding theorem says:

Assuming ZF + DC + $\text{AD}^{1/2}_{\mathbf{R}}$, if $\lambda < \Theta$ then $\beta(\lambda) < \Theta$.

Combining the above result with Moschovakis' Theorem one can actually obtain a stronger statement which implies both. Recall first the standard fact that in ZF + DC + AD every ultrafilter is countably complete. Denote then for each orginal λ by $q(\lambda)$ the set of countably complete filters on λ. This contains $\beta(\lambda)$ but also contains $p(\lambda)$ by the natural identification of $A \subseteq \lambda$ with the filter $\hat{A} = \{X \subseteq \lambda : A \subseteq X\}$. We have now the following

Corollary. *Assume* ZF + DC + $\text{AD}^{1/2}_{\mathbf{R}}$. *Then if* $\lambda < \Theta$ *there is a surjection of* $\mathbf{R}$ *onto* $q(\lambda)$.

Actually in Moschovakis' Theorem one obtains, for each $\pi : \mathbf{R} \longrightarrow \lambda$, estimates of the complexity of $A^\pi = \{x \in \mathbf{R} : \pi(x) \epsilon A\}$, for A $\subseteq \lambda$, in terms of $\leq_\pi$ where $x \leq_\pi y \Leftrightarrow \varphi(x) \leq \varphi(y)$, which are very useful in various applications. Similarly we obtain definability estimates for the

[1]Research partially supported by NSF Grants MCS-8117804 and DMS-8416349

complexity of ultrafilters. As an immediate consequence we have various local versions of the main theorem of which we mention as an example the following:

Corollary. *Assume* ZF + DC + AD *(this is all we need here). Then*

(i) *If* $\lambda < \delta^1_\omega = \sup\{\delta^1_n : n < \omega\}$, *then* $\beta(\lambda) < \delta^1_\omega$.

(ii) *If* $\lambda < \kappa^{KL}$ *(= the Kleene ordinal of the continuum), then* $\beta(\lambda) < \kappa^{KL}$. *Similarly for* $\kappa^{\mathbf{R}}$ *(for the definition of* κ^{KL}, $\kappa^{\mathbf{R}}$ *see for instance* [K2]).

(iii) *If* $\lambda < (\delta^2_1)^{L(\mathbf{R})}, \beta(\lambda) < (\delta^2_1)^{L(\mathbf{R})}$.

Finally let us mention another corollary concerning the absoluteness of ultrafilters for certain inner models of AD .

Corollary. *Assume* ZF + DC + AD *and let* M *be an inner model of* ZF + DC + $\mathrm{AD}^{1/2}_{\mathbf{R}}$ *containing* $\mathbf{R}$. *Let* $\lambda < \Theta^M$. *Then every ultrafilter on* λ *is in* M. *In particular, if* λ *is measurable, then* $M \models$ *"λ is measurable".*

Actually a finer version of this result is possible which implies the following (note that $L(\mathbf{R}) \models \neg\, \mathrm{AD}^{1/2}_{\mathbf{R}}$).

Corollary. *Assume* ZF + DC + AD. *If* $\lambda < \Theta^{L(\mathbf{R})}$, *then every ultrafilter on* λ *belongs to* $L(\mathbf{R})$, *thus if* λ *is measurable,* $L(\mathbf{R}) \models$ *"λ is measurable".*

2. A game for coding ultrafilters. Let $C \subseteq \mathbf{R}, \pi : C \longrightarrow p(\lambda)$, λ an ordinal and write for simplicity $X_\alpha \equiv \pi(\alpha)$. Consider then the following game $G \equiv G_\pi$, which is a "coded" version of a "cut-and-choose" game on λ:

I	α_0		α_1	$\alpha_j \in \mathbf{R}$
				$\cdots$
II		i_0	i_1	$i_j \in \{0,1\}$

I wins iff : (i) $\forall n(\alpha_n \in C)$ and (ii)

$$\cap\{X_{\alpha_n} : i_n = 0\} \cap \cap\{\sim X_{\alpha_n} : i_n = 1\} \neq \emptyset\,.$$

We claim first that, assuming ZF + DC + AD, I has no winning strategy in G. Indeed if τ was such a strategy, towards a contradiction, let for every $x \in 2^\omega$ $\alpha^x_0, \alpha^x_1, \ldots$ be I's moves following τ when II plays $x(0), x(1), \ldots$ .Put

$$f(x) = \text{least } \xi[\xi \in \cap \{X_{\alpha_n^x} : x(n) = 0\} \cap$$
$$\cap \{\sim X_{\alpha_n^x} : x(n) = 1\}].$$

Then $f : 2^\omega \to \lambda$ is an injection, i.e. there is a wellordering of 2^ω, which violates AD.

A strategy F for player II in this game is a map $F : \mathbf{R}^{<\omega} \to \{0,1\}$. We now have the key.

Lemma. *If F is a winning strategy for player II in G, and $\mathcal{U} \subseteq p(\lambda)$ is an ultrafilter, then there are $\alpha_0, \alpha_1, \ldots, \alpha_{n-1} \in C$ such that for all $\alpha \in C$,*

$$X_\alpha \in \mathcal{U} \Leftrightarrow F(\alpha_0 \ldots \alpha_{n-1}, \alpha) = 1 .$$

In particular each such $\mathcal{U}$ is completely determined by $F \mid \mathbf{R}^{n+1}$.

Proof. Call a sequence $(\alpha_0, i_0, \alpha_1, i_1, \ldots, \alpha_{m-1}, i_{m-1})$ *$\mathcal{U}$-good* if it is a finite run of the game G in which II follows $F, \alpha_0 \ldots \alpha_{m-1} \in C$ and $\forall j \leq m-1(X_{\alpha_j} \in \mathcal{U} \Leftrightarrow i_j = 0)$. The empty sequence is $\mathcal{U}$-good by convention. If every $\mathcal{U}$-good sequence has a $\mathcal{U}$-good proper extension we can obtain an infinite run $(\alpha_0, i_0, \alpha_1, \ldots)$ of G in which II followed F, $\alpha_0, \alpha_1, \ldots \in C$ and $X_{\alpha_j} \in \mathcal{U} \Leftrightarrow i_j = 0$, for all j. Then clearly $\cap\{X_{\alpha_n} : i_n = 0\} \cap \cap\{\sim X_{\alpha_n} : i_n = 1\} \in \mathcal{U}$ (recall that $\mathcal{U}$ is countably complete), so this intersection is non-$\emptyset$, and I won, a contradiction.

So let $(\alpha_0, i_0, \ldots, \alpha_{n-1}, i_{n-1})$ be a maximal $\mathcal{U}$-good sequence. Let $\alpha \in C$. Then $(\alpha_0, i_0, \ldots, \alpha_{n-1}, \alpha, F(\alpha_0, \ldots, \alpha_{n-1}, \alpha))$ is not $\mathcal{U}$-good, i.e.

$$X_\alpha \in \mathcal{U} \Leftrightarrow F(\alpha_0 \ldots \alpha_{n-1}, \alpha) = 1$$

and we are done. ⊣

3. On real-integer games. Note that the preceding game G_π is a game in which one player plays reals and the other integers, so it is not necessarily determined using AD alone. We denote the determinacy of such games by $\mathrm{AD}_{\mathbf{R}}^{1/2}$. (In [K1] it is shown that in ZF + DC, $\mathrm{AD}_{\mathbf{R}}^{1/2}$ is equivalent to UNIFORMIZATION.) So we have immediately

Theorem 3.1. *Assume* ZF + DC + $\mathrm{AD}_{\mathbf{R}}^{1/2}$. *If $\lambda < \Theta$, then $\beta(\lambda) < \Theta$.*

Proof. Using the notation of §2, with $C = \mathbf{R}, \pi : \mathbf{R} \longrightarrow \lambda$ map each $n+1-$ tuple $\vec{\alpha} = (\alpha_0, \ldots \alpha_{n-1}, \alpha_n)$ in $\mathbf{R}^{n+1}$ to $V_{\vec{\alpha}} = \{X_{\alpha_n} : F(\vec{\alpha}) = 1\}$. Then

$$\beta(\lambda) \subseteq \{V_{\vec{\alpha}} : \vec{\alpha} \in C^{n+1}, \quad n = 0, 1, 2, \ldots\} .$$

So there is a subjection of $\mathbf{R}$ onto $\beta(\lambda)$. ⊣

We also have easily the result about countably complete filters.

Corollary 3.2. *Assume* ZF + DC + $AD_{\mathbf{R}}^{1/2}$. If $\lambda < \Theta$, *then there is a surjection from* $\mathbf{R}$ *onto* $q(\lambda)$, *the set of countably complete filters on* λ.

Proof. By a result of Kunen, if I is a proper countably complete filter on λ then there is an ultrafilter $\mathcal{U}$ on λ extending I. (For a proof see [S], p.148). Thus if I is a proper countably complete filter then $I = \cap\{\mathcal{U} : \mathcal{U}$ in an ultrafilter on λ containing I $\}$. Thus $p(\beta(\lambda))$ can be mapped onto $q(\lambda)$ and since $\beta(\lambda) < \Theta$, $\mathbf{R}$ can be mapped onto $q(\lambda)$. $\dashv$

4. The complexity of ultrafilters. In the notation of §2 again let for each ultrafilter $\mathcal{U}$ on λ.

$$\mathcal{U}^* \equiv \mathcal{U}^*_\pi = \{\alpha \in C : X_\alpha \in \mathcal{U}\} .$$

It follows that for some $\alpha_0 \ldots \alpha_{n-1} \in C$, and all $\alpha \in C$,

$$\alpha \in \mathcal{U}^* \Leftrightarrow F(\alpha_0 \ldots \alpha_{n-1}, \alpha) = 1,$$

for any winning strategy F for II in G_π. Thus the complexity of $\mathcal{U}^*$ depends (beyond that of C, π) on the complexity of $F \mid \mathbf{R}^m, m = 1, 2, \ldots$ (each of which is essentially a set of reals) for any winning strategy F for II.

There are certain real-integer games which can be proved determined in AD only. In those instances we get reasonably good estimates for the winning strategies which give us corresponding estimates for ultrafilters. Here is a relevant result.

Theorem 4.1. *(Woodin, unpublished.) Assume* ZF + DC + AD. *If* $A \subseteq \mathbf{R}^\omega \times \omega^\omega$ *is co-Souslin, then the (real-integer) game corresponding to* A *is determined.*

Here $A \subseteq \mathbf{R}^\omega \times \omega^\omega$ is co-Souslin iff $A' \subseteq \omega^\omega \times \omega^\omega$ given by $A'(\alpha, \beta) \Leftrightarrow A(\{(\alpha)_n\}, \beta)$ is co-Souslin, where $\alpha \mapsto \{(\alpha)_n\}$ is a recursive 1 - 1 correspondence between ω^ω and $\mathbf{R}^\omega$.

We give below an alternative version and proof (motivated by the ideas of [K1]) of that result, which also gives the definability estimates we want.

Theorem 4.2. *Assume* ZF +DC + AD. *Let* $A \subseteq \mathbf{R}^\omega \times \omega^\omega$ *and suppose* $\sim A$ *(viewed as a subset of* $\omega^\omega \times \omega^\omega$ *as above) carries a scale* $\{\varphi_n\}$ *such that each corresponding relation* $\leq^*_{\varphi_n}, <^*_{\varphi_n}$ *belongs to a pointclass* Γ_n, *where* $\Gamma_0 \subseteq \Gamma_1 \subseteq \cdots \subseteq \Gamma_n \subseteq \cdots$ *and* Γ_n *is closed under* $\wedge, \vee, \exists m \leq n, \forall m \leq n$ *and recursive substitutions. Consider the game* G

I	α_0		α_1		$\cdots$	$\alpha_j \in \mathbf{R}$
II		i_0		i_1	$\cdots$	$i_j \in \mathbf{R}$

II wins iff $(\vec{a}, \vec{i}) \notin A$.

If I *has no winning strategy in* G, *then* II *has a winning strategy* $F : \mathbf{R}^{<\omega} \to \omega$ *such that for each* $n, F \mid \mathbf{R}^n \notin \Game^2\Gamma_n$. *(Here* $\Game$ *is the game quantifier, see [M1], and* $\Game^2 = \Game\Game$*).*

Proof. Fix a Turingg degree d. Then clearly I does not have a winning strategy in the game where he plays reals $\alpha_i \leq_T d$. So by the Third Periodicity Theorem of Moschovakis [M1], II has a winning strategy F_d in this restricted game, such that for each n the relation

$$\alpha_i \leq_T d \ \wedge \ F_d(\alpha_0 \ldots \alpha_{n-1}) = i$$

is in $\Game\Gamma_n$ uniformly on d_1 i.e. the relation

$$R_n(x, \alpha_0 \ldots \alpha_{n-1}, i) \Leftrightarrow \alpha_i \leq_T x \wedge F_{[x]_T}(\alpha_0 \ldots \alpha_{n-1}) = i$$

is in $\Game\Gamma_n$. Define now $F(\alpha_0 \ldots \alpha_{n-1}) = i \Leftrightarrow$ for a cone of d's, $F_d(\alpha_0 \ldots \alpha_{n-1}) = i$. Clearly $F \mid \mathbf{R}^n \in \Game\,(\Game\Gamma_n) = \Game^2\Gamma_n$ and F is a winning strategy for II in the game G. ⊣

Let us mention now some specific applications. We assume ZF + DC + AD below.

First let $\lambda < \delta^1_\omega$ be a projective ordinal, let $\varphi : \mathbf{R} \longrightarrow \lambda$ be a projective norm and using the Moschovakis' Coding Lemma (see [M1]) let in the notation of §2 $C \subseteq \mathbf{R}$, $\pi : C \longrightarrow p(\lambda)$ be also projective. (For π this means that the relation "$\varphi(x) \in \pi(y)$" is projective.) Then for any ultrafilter $\mathcal{U}$ on λ, $\mathcal{U}^*$ is projective (actually in some fixed level of the projective hierarchy) and $\beta(\lambda) < \delta^1_\omega$.

One can of course state finer level-by-level versions of this result. However the recent work of S. Jackson (see for example [J]) provides a fairly complete analysis of ultrafilters on projective ordinals which provides much more accurate estimates for the definability of ultrafilters.

Similarly if $\lambda < K^{KL}$ every ultrafilter on λ is Kleene recursive in ${}^3E$ and a real (in the codes), and $\beta(\lambda) < \kappa^{KL}$. If $\lambda < \kappa^{\mathbf{R}}$ then every ultrafilter on λ is hyperprojective in the codes and

$\beta(\lambda) < \kappa^{\mathbf{R}}$. For $\lambda = \kappa^{\mathbf{R}}$ itself we have that each ultrafilter on $\kappa^{\mathbf{R}}$ is $\sum_m^*$ for some m (depending on $\mathcal{U}$). For the definition of the classes $\sum_m^*$ see [M2]. So $\beta(\kappa^{\mathbf{R}}) < \delta_\omega^*$. Martin has showed that also $\beta(\kappa^{\mathbf{R}}) \geq \delta_\omega^*$ so in fact $\beta(\kappa^{\mathbf{R}}) = \delta_\omega^*$.

Finally if $\lambda < (\delta_1^2)^{L(\mathbf{R})}$ then every ultrafilter on λ is $(\Delta_1^2)^{L(\mathbf{R})}$ in the codes and $\beta(\lambda) < (\delta_1^2)^{L(\mathbf{R})}$.

5. Absoluteness of ultrafilters. The following fact is immediate from the analysis in §2.

Corollary 5.1. *Assume* ZF + DC + AD. *Let* M *be an inner model of* ZF + DC + AD *containing* $\mathbf{R}$, *let* $\lambda < \Theta^M$ *and let* C, π *(as in §2) be in* M. *If* F *is a winning strategy for* II *in* G_π *such that* $F \mid R^n \in M$ *for all* $n \in \omega$, *then every ultrafilter on* λ *is in* M.

In particular we have

Corollary 5.2. *Assume* ZF + DC + AD. *Let* M *be an inner model of* ZF + DC + AD *containing* $\mathbf{R}$ *and let* $\lambda < \Theta^M$. *Then every ultrafilter on* λ *is in* M *and thus, if* λ *is measurable,* $M \models$ "λ *is measurable*".

Finally we can obtain this result for $L(\mathbf{R})$ itself.

Corollary 5.3. *Assume* ZF + DC + AD. *Let* $\lambda < \Theta^{L(\mathbf{R})}$. *Then every ultrafilter on* λ *is in* $L(\mathbf{R})$, *and if* λ *is measurable,* $L(\mathbf{R}) \models$ "λ *is measurable*".

Proof. If $V = L(\mathbf{R})$ there is nothing to prove. If $V \neq L(\mathbf{R})$ then by [SV], $\mathbf{R}^\#$ exists. Then by a result of Solovay (unpublished - see however [MS], p. 93) every $A \in L(\mathbf{R})$ admits a scale $\{\varphi_n\}$ with $\leq_{\varphi_n}^*$ $<_{\varphi_n}^*$, in $L(\mathbf{R})$, thus by 4.2 the hypothesis of 5.1 is satisfied and we are done. ⊣

References

[J] S. Jackson, *AD and the projective ordinals, this volume.*

[K1] A.S. Kechris, "AD + *UNIFORMIZATION*" *is equivalent to "HALF* $\mathrm{AD}_{\mathbf{R}}$," this volume.

[K2] ________, *Determinacy and the structure of* $L(\mathbf{R})$, Proc. Symp. in Pure Math., Amer. Math. Soc., 42 (1985), 271-283.

[MS] D.A. Martin and J.R. Steel, *The extend of scales in* $L(\mathbf{R})$, Cabal Seminar 79-81, Lecture Notes in Math., Vol. 1019, Springer-Verlag, 1983,86-96.

[M1] Y.N. Moschovakis, *Descriptive Set Theory,*North Holland, 1980.

[M2] Y.N. Moschovakis, *Scales on conductive sets,* Cabal Seminar 79-81, ibid, 77-85.

[S] R.M. Solovay, *A* Δ_3^1 *coding of the subsets of* ω_ω, Cabal Seminar 76-77, Lecture notes in Math., Vol. 689, Springer-Verlag, 1978, 133-150.

[SV] J.R. Steel and R. Van Wesep, *Two consequences of determinacy consistent with choice,* Trans. Amer. Math. Soc., 272 (1982), 67-85.

SUBSETS OF $\aleph_1$ CONSTRUCTIBLE FROM A REAL

Alexander S. Kechris[1]
Department of Mathematics
California Institute of Technology
Pasadena, CA 91125

The purpose of this paper is to give a necessary and sufficient condition for a subset of $\aleph_1$ to be constructible from a real in terms of structural properties of the code set of A, valid under the condition that an appropriate measurable cardinal exists. This can be combined with recent results of Woodin to provide upper bounds for the consistency strength, of theories of the form ZFC + $\forall x \in \omega^\omega (x^\# \text{ exists})+$ "every subset of $\aleph_1$ with code set in Γ is constructible from a real," for various pointclasses Γ.

For each $A \subseteq \aleph_1$ let

$$A^* = \{w \in WO : |w| \in A\},$$

where WO = the set of reals coding wellorderings of ω and for $w \in WO$ we put $|w|$ = the ordinal of the wellordering coded by w. The *code set* of A is then

$$\langle A \rangle = \{0\hat{\ }w : w \in A^*\} \cup \{1\hat{\ }w : w \in (\aleph_1 - A)^*\}$$

i.e. the disjoint union of A^* and $(\aleph_1 - A)^*$.

A set of reals $P \subseteq \omega^\omega$ is called *Souslin* if there is a tree T on $\omega \times \lambda$, λ an ordinal, with $P = p[T] = \{x \in \omega^\omega : \exists f \in \lambda^\omega (x, f) \in [T]\}$. If the tree T can be taken to be *homogeneous* we call P *homogeneously Souslin.* For the definition of homogeneous trees see [K]. If finally $P = p[R] = \{x : \exists y (x, y) \in R\}$, where $R \subseteq \omega^\omega \times \omega^\omega$ and R is homogenously Souslin we call P *weakly homogeneously Souslin.*

We have now the following characterization.

Theorem. (ZF + DC).

i) *If $\aleph_1$ is measurable, then $A \subseteq \aleph_1$ is constructible from a real iff $\langle A \rangle$ is Souslin.*

---

[1]Research partially supported by NSF Grants MCS-8117804 and DMS-8416349

ii) *If there exists a measurable cardinal, then $A \subseteq \aleph_1$ is constructible from a real iff $\langle A \rangle$ is weakly homogeneously Souslin.*

Let us mention now some consequences. Woodin (unpublished) has shown that

Con(ZFC + $\exists$ infinitely many strong cardinals)

$\Rightarrow$ Con(ZFC + Every projective set is Souslin),

so that we have

Con(ZFC + $\exists$ infinitely many strong cardinals)

$\Rightarrow$ Con(ZFC + $\forall x(x^{\#}$ exists) + Every projective subset of $\aleph_1$ is constructible for a real).

(Recall that κ is *strong* if $\forall\lambda\exists j : V \to M$ $(crit(j) = \kappa$ & $V_\lambda \subseteq M))$. On the other hand it is known that "ZFC + $\forall x(x^{\#}$ exists) + Every projective subset of $\aleph_1$ is constructible from a real" is (consistency-wise) quite strong, stronger at least than the large cardinals for which core models have been constructed (which includes at least κ with $o(\kappa) = \kappa^+$).

Another consequence is that from the consistency of "ZF + DC + Every set of reals is Souslin + $\aleph_1$ is measurable" one obtains the consistency of "ZFC +$*$" – see [W] for this principle and its consequences.

We proceed now to prove the theorem. For technical convenience and without loss of generality we will assume that $A \subseteq \aleph_1 - \omega$.

First we need some preliminaries. Let $\tau_0, \tau_1, \tau_2, \ldots$ be a 1 - 1 enumeration of $\omega^{<\omega}$, with $\tau_0 = \emptyset$, $lh(\tau_i) \leq i$, $\tau_i \supsetneqq \tau_j \Rightarrow i > j$. For $\sigma \in \omega^{<\omega}$, $lh(\sigma) = n$ let

$$T_\sigma = \{\tau_i \in \omega^{<\omega} : i < n \wedge \forall j(\emptyset \neq \tau_j \subseteq \tau_i \Rightarrow \sigma(j) = 0\} \cup \{\emptyset\}.$$

Then T_σ is a finite tree on ω and $\sigma \subseteq \sigma' \Rightarrow T_\sigma \subseteq T_{\sigma'}$. For $\sigma \neq \emptyset$, define the following ordering $<_\sigma$ on n :

$$\begin{aligned} i <_\sigma j \Leftrightarrow\ & (i, j < n \wedge \tau_i, \tau_j \notin T_\sigma \wedge i < j) \vee \\ & (\tau_i \notin \tau_\sigma \wedge \tau_j \in \tau_\sigma) \vee (\tau_i, \tau_j \in \tau_\sigma \wedge \tau_i <_{KB} \tau_j), \end{aligned}$$

where $<_{KB}$ is the Kleene-Brouwer ordering on $\omega^{<\omega}$. Clearly 0 is the top element of $<_\sigma$ and $\sigma \subseteq \sigma' \Rightarrow <_\sigma \subseteq <_{\sigma'}$ (since for $i < n = lh(\sigma)$, $\tau_i \in T_\sigma \Leftrightarrow \tau_i \in T_{\sigma'}$ in view of $lh(\tau_i) \leq \tau_i$). For $\alpha \in \omega^\omega$ let $<_\alpha = \bigcup_n <_{\alpha \restriction n}$. Then $<_\alpha$ is the Kleene-Brouwer ordering of $T_\alpha = \{\tau_i \in \omega^{<\omega} : \forall j(\emptyset \neq \tau_j \subseteq \tau_i \Rightarrow \alpha(j) = 0)\} \cup \{\emptyset\}$, transferred to ω via the coding, with the rest of ω thrown in at the bottom with its usual ordering. Put

$$WO = \{\alpha : T_\alpha \text{ well founded}\},$$
$$|\alpha| = \text{ rank of 0 in } <_\alpha .$$

Thus $\{|\alpha| : \alpha \in WO\} = \omega_1 - \omega$. Let

$$SH = \{(\sigma, u) \in \omega^{<n} \times \text{ORD}^{<n} \mid u : n \to \text{ORD} \wedge u \text{ is order preserving for } <_\sigma\}.$$

Then clearly

i) $WO = p[SH] = p[SH \restriction \kappa]$, for any $\kappa \geq \aleph_1$.

ii) $(\sigma, u) \in SH \wedge lh(\sigma) = n \Rightarrow u(0) > u(1) \wedge u(0) > u(2) \wedge \cdots \wedge u(0) > u(n-1)$.

iii) $(w, f) \in [SH] \Rightarrow |w| \leq f(0)$.

iv) If $\pi_\sigma : n \to n$ is the permutation defined by

$$i <_\sigma j \Leftrightarrow \pi_\sigma(i) < \pi_\sigma(j),$$

then

$$(\sigma, u) \in SH \Leftrightarrow \exists v \in [\text{ORD}]^n \quad (u = v \circ \pi_\sigma).$$

So if κ is a measurable cardinal and μ_n is the n-fold cartesian product of a normal measure μ on κ, then $SH \restriction \kappa$ is homogeneous with homogeneity measures $\mu_\sigma = (\pi_\sigma)_*(\mu_n)$.

Finally if $P \subseteq \omega^\omega$ is any Π^1_1 set and $F : \omega^\omega \to \omega^\omega$ is a Lipschitz function such that $F^{-1}[WO] = P$, consider the tree induced by F i.e.

$$(\sigma, u) \in SH_F \Leftrightarrow (F(\sigma), u) \in SH.$$

Then $P = p[SH_F]$ and SH_F inherits all the above properties of SH.

Given $\emptyset \neq A \subseteq \aleph_1 - \omega$ consider its Solovay game S_A :

I	II	II wins iff
x	y, α	

$$x \in WO \Rightarrow y \in WO \wedge |x| \leq |y| \wedge \forall n[(\alpha)_n \in WO] \wedge$$
$$A \cap (|y|+1) \subseteq \{|(\alpha)_n| : n \in \omega\} \subseteq A.$$

By the usual boundedness I cannot have a winning strategy. Also if II has a winning strategy, then A is constructible from a real.

Proof of i). Assume $\aleph_1$ is measurable. If $A \subseteq \aleph_1 - \omega$ is constructible from a real, then using $\forall x \in \omega^\omega(x^\# \text{ exists})$ it is easy to check that $\langle A \rangle$ is Σ^1_2, so it is Souslin.

Conversely assume that $\langle A \rangle$ is Souslin. Let

$$Q(v,\alpha) \Leftrightarrow \forall n[(\alpha)_n \in WO] \wedge v \in WO \wedge$$
$$\forall z \leq_T v[z \in A^* \wedge |z| \leq |v| \Rightarrow \exists k(|z| = |(\alpha_k)|] \wedge \forall n[(\alpha)_n \in A^*],$$

thus

$$Q(v,\alpha) \Leftrightarrow \forall n[(\alpha)_n \in WO] \wedge v \in WO \wedge A \cap (|v|+1) \subseteq \{|(\alpha)_n| : n \in \omega\} \subseteq A.$$

Moreover since $\langle A \rangle$ is Souslin, so is Q, say $Q = p[T]$, T a tree on $\omega \times \lambda$.

Let also

$$R(w,v) \Leftrightarrow w, v \in WO \wedge |w| \leq |v|.$$

Then R is Π^1_1 so let F be Lipschitz with $F^{-1}[WO] = R$ and let $S' = SH_F \restriction \aleph_1$. Then $R = p[S']$ and S' is homogeneous with homogeneity measures $\mu'_{\sigma,\tau}$; $\sigma, \tau \in \cup_n(\omega^n \times \omega^n)$. Let also $S = SH \restriction \aleph_1$, so that $p[S] = WO$ and S has homogeneity measures μ_σ, $\sigma \in \omega^{<\omega}$.

Now consider the following closed game, motivated by a game of Martin (see [K],§7):

I	II	$w, v, \alpha \in \omega^\omega; f, g \in \aleph_1^\omega, h \in \lambda^\omega;$
w, f	v, g, α, h	

II wins iff

$$\forall n \geq 1[(w \restriction n, f \restriction n) \in S \Rightarrow$$
$$(w \restriction n, v \restriction n, g \restriction n) \in S' \wedge (v \restriction n, \alpha \restriction n, h \restriction n) \in T].$$

This game is of course determined. We want to show that I cannot have a winning strategy.

Assume he did, towards a contradiction, and call it G.

His first move by G is a_0, ξ_0 with $(a_0, \xi_0) \in S$. Choose then $v_0 \in WO$ with $v_0 \in WO$ and $|v_0| \geq \xi_0$ and then choose α_0 with $Q(v_0, \alpha_0)$ and h_0 with $(v_0, \alpha_0, h_0) \in [T]$. We can then use the homogeneity of S' (and the fact that after his first move I following G has to play ordinals $< \xi_0$, as long as II plays in the appropriate trees, thus the countable additivity of the homogeneity measures on S' sufficies) to find a g_0 such that v_0, g_0, α_0, h_0 defeats G. The details of this kind of argument are spelled out in a similar situation in [K], p. 68 (see proof of direction ($\Rightarrow$)).

So II has a winning strategy. Then consider the game

I	II	$w, v, \alpha \in \omega^\omega; f \in \aleph_1^\omega;$
w, f	v, α	

II wins iff

$(w, f) \in [S] \Rightarrow v \in WO \wedge |w| \leq |v| \wedge Q(v, \alpha)$.

Clearly II has a winning strategy in that game too, call if F. (F is obtained by a winning strategy of II in the preceding closed game by forgetting about the g, h, which are witnesses that $|w| \leq |v|$ and $Q(v, \alpha)$.

Using F define finally the following strategy Σ for II in the Solovay game S_A :

$$\Sigma(w \restriction n+1) = (v \restriction n+1, \alpha \restriction n+1),$$

where on some set $X(w \restriction n+1)$ of $\mu_{w \restriction n+1}$-measure 1 and for all $f \restriction n+1 \in X(w \restriction n+1)$, $F(w \restriction n+1, f \restriction n+1) = (v \restriction n+1, \alpha \restriction n+1)$. In other words Σ is obtained by integrating F over the homogeneity measures for S.

We claim that this is winning for II thereby completing the proof. Indeed assume w, v, α have been played with I playing w and v, α determined by Σ. If $w \notin WO$ there is nothing to prove. Else $w \in WO$, so by homogeneity find f with $f \restriction n+1 \in X(w \restriction n+1)$, for all n. Then clearly w, f, v, α is a run of the preceding game in which II followed F, thus since $(w, f) \in [S]$ we have $v \in WO \wedge |w| \leq |v| \wedge Q(v, \alpha)$, i.e. II won this run of the Solovay game and we are done.

Proof of ii). Let κ be the least measurable cardinal. Assume now $\langle A \rangle$ is weakly homogeneously Souslin. Since weakly homogeneously Souslin sets are closed under conjunctions, disjunctions and countable intersections it follows that the relation Q we defined before is weakly homogeneously Souslin, say

$$(v, \alpha) \in Q \Leftrightarrow \exists \beta \exists h (v, \alpha, \beta, h) \in [T],$$

where T is a tree on $\omega \times \lambda$, T homogeneous with homogeneity measures $\mu''_{\sigma,\tau,\rho}, \sigma, \tau, \rho \in \cup_n(\omega^n \times \omega^n \times \omega^n)$. Let now $S = SH \restriction k$, $S' = SH_F \restriction k$ with homogeneity measures $\mu_\sigma, \mu'_{\sigma,\tau}$.

Consider again the game

I	II	II wins iff
w, f	v, g, α, β, h	$\forall n \geq 1[(w \restriction n, f \restriction n) \in S \Rightarrow$
		$(w \restriction n, v \restriction n, g \restriction n) \in S' \wedge (v \restriction n, \alpha \restriction n, \beta \restriction n, h \restriction n) \in T]$.

It will be of course enough to show I has no winning strategy. Say he had one and call it G, towards a contradiction.

Define now a strategy

$$\sigma(v \restriction n, \alpha \restriction n, \beta \restriction n) = (w \restriction n+1, f \restriction n+1)$$

so that for each n, $(w \restriction n+1, f \restriction n+1) \in S$ as follows:

$$\sigma(\emptyset, \emptyset, \emptyset) = (w(0), f(0)) \stackrel{\text{def}}{=} G(\emptyset, \emptyset, \emptyset, \emptyset, \emptyset) \in S'.$$

Given $(v \restriction 1, \alpha \restriction 1, \beta \restriction 1)$ consider $S'(w \restriction 1, f \restriction 1)$ and $T(v \restriction 1, \alpha \restriction 1, \beta \restriction 1)$. Then for each $h \restriction 1 \in T(v \restriction 1, \alpha \restriction 1, \beta \restriction 1)$ and $g \restriction 1 \in S'(w \restriction 1, v \restriction 1)$ consider $G(v \restriction 1, g \restriction 1, \alpha \restriction 1, \beta \restriction 1, h \restriction 1)$. We claim that on a set of measure 1 of $(h \restriction 1, g \restriction 1)$ in the product measure $\mu''_{v \restriction 1, \alpha \restriction 1, \beta \restriction 1} \times \mu'_{w \restriction 1, v \restriction 1}$ we have that $G(v \restriction 1, g \restriction 1, \alpha \restriction 1, \beta \restriction 1, h \restriction 1)$ is fixed say with value $(w \restriction 2, f \restriction 2) \in S$, which we define to be our $\sigma(v \restriction 1, \alpha \restriction 1, \beta \restriction 1)$. This is because for $(w \restriction 2, f \restriction 2) \in S$, $f(1) < f(0) < \kappa$ and the product measure is countably complete thus κ-complete since κ is the least measurable.

Proceeding similary we define $\sigma(v \restriction n, \alpha \restriction n, \beta \restriction n)$. Put now

$$\begin{aligned}\sigma'(v \restriction n, \alpha \restriction n, \beta \restriction n) &= w \restriction n+1 \\ &\stackrel{\text{def}}{=} \text{the first coordinate of } \sigma(v \restriction n, \alpha \restriction n, \beta \restriction n),\end{aligned}$$

and let $\sigma'(v, \alpha, \beta) = \cup_n \sigma'(v \restriction n, \alpha \restriction n, \beta \restriction n)$. Then clearly $\sigma'(v, \alpha, \beta) \in WO$ for all v, α, β, thus by boundedness find $\xi_0 < \aleph_1$ with

$$\sup\{|\sigma'(v, \alpha, \beta)| : v, \alpha, \beta \in \omega^\omega\} < \xi_0.$$

Pick now $v_0 \in WO$ with $\xi_0 < |v_0|$ and α_0 with $Q(v_0, \alpha_0)$. Then pick β_0 with $(v_0, \alpha_0, \beta_0) \in$

$p[T]$. Let X_{n+1} be sets of $\mu''_{v_0\restriction n+1,\alpha_0\restriction n+1,\beta_0\restriction n+1}$-measure 1 so that if $h\restriction n+1\in X_{n+1}$ then for a set of $\mu'_{w_0|n+1,v_0|n+1}$-measure 1 worth of $g\restriction n+1$ we have

$$\begin{aligned}&G(v_0\restriction n+1,g\restriction n+1,\alpha_0\restriction n+1,\beta_0\restriction n+1,h\restriction n+1)\\&\qquad=\sigma(v_0\restriction n+1,\alpha_0\restriction n+1,\beta_0\restriction n+1)=(w_0\restriction n+2,f_0\restriction n+2).\end{aligned}$$

By the homogeneity of T find h_0 with $h_0\uparrow n+1\in X_{n+1}$, all n. Then

$$\begin{aligned}&G(v_0\restriction n+1,g\restriction n+1,\alpha_0\restriction n+1,\beta_0\restriction n+1,h_0\restriction n+1)\\&\qquad=\sigma(v_0\restriction n+1,\alpha_0\restriction n+1,\beta_0\restriction n+1)\end{aligned}$$

for $g\restriction n+1\in Y_{n+1}$, where Y_{n+1} has measure 1 in $\mu'_{w_0\restriction n+1,v_0\restriction n+1}$. Since $w_0=\sigma'(v_0,\alpha_0,\beta_0)$, we have $|w_0|\le\xi\le|v_0|$, and using the homogeneity of S' we can find g_0 with $g_0\restriction n+1\in Y_{n+1}$ for all n, i.e.

$$\begin{aligned}&G(v_0\restriction n+1,g_0\restriction n+1,\alpha_0\restriction n+1,\beta_0\restriction n+1,h_0\restriction n+1)\\&\qquad=\sigma(v_0\restriction n+1,\alpha_0\restriction n+1,\beta_0\restriction n+1)=(w_0\restriction n+2,f_0\restriction n+2)\end{aligned}$$

for all n. Then if II plays v_0, g_0, α_0, β_0, h_0 he defeats G and we are done.

References

[K] A.S. Kechris, *Homogeneous trees and projective scales,* Cabal Seminar 77-79, Lecture Notes in Mathematics, Vol. 839, Springer-Verlag, 1981, 33-74.

[W] W.H. Woodin, *Some consistency results in* ZFC *using* AD, Cabal Seminar 79-81, Lecture Notes in Mathematics, Vol. 839, Springer-Verlag, 1981, 33-74.

AD AND THE PROJECTIVE ORDINALS

Steve Jackson
Department of Mathematics
California Institute of Technology
Pasadena, CA 91125

Acknowledgements

The author wishes to thank D. A. Martin for many helpful conversations during the research for this manuscript. This work is an outgrowth of the calculation of $\boldsymbol{\delta}^1_5$ (the case $n = 0$ of this paper) which the author completed while working with Martin at U.C.L.A.. Aside from independently discovering many of the basic techniques and methods of use there, it was a few basic discoveries of Martin (such as [Ma]) which provided the impetus for this research.

I. Introduction. The purpose of this paper is to calculate an upper-bound for $\boldsymbol{\delta}^1_{2n+5}$, $n \geq 0$, assumingcertainbasic inductive assumptions concerning the lower projective ordinals. In a later paper, we will verify the lower bound for $\boldsymbol{\delta}^1_{2n+5}$ and establish the inductive assumptions at the next level.

The case $n = 0$ appears in [J], and may be obtained as a special case of the results in this paper.

We assume the reader is familiar with the results in [Ke] and [Ma], although the reader may take as "axioms" the results needed. Indeed, the following paper is essentially self-contained, except for a knowledge of the homogeneous tree construction, which is only used indirectly.

For background on the projective ordinals as well as their significance in descriptive set theory, we refer the reader to [Mo].

We work in AD + DC throughout, although the inductive hypotheses at the lower levels suffice.

II. Definitions and preliminary results. In this section we define two families of measures; one being essentially the general measures allowed in the homogeneous tree construction, and the other a family of canonical measures. This is the only point in this paper where we use the homogeneous tree construction; the reader not familiar with it may take on faith the fact that our family captures the most general such measure.

We first introduce our main inductive hypotheses:

I_{2n+1} : $\boldsymbol{\delta}^1_{2n+1}$ has the strong partition relation, $\boldsymbol{\delta}^1_{2n+3}$ has the weak partition relation, and $\boldsymbol{\delta}^1_{2n+3} = \aleph_{\tau(2n+1)}$, where $\tau(0) = 1$ and $\tau(k+1) = \omega^{\tau(k)}$.

K_{2n+3}: a) for any measure μ on Ξ_{2n+3}, = the predecessor of $\boldsymbol{\delta}^1_{2n+3}$, and any $\alpha < \boldsymbol{\delta}^1_{2n+3}$, $J_\mu(\alpha) < \boldsymbol{\delta}^1_{2n+3}$ where J_μ denotes the embedding from the ultrapower by the measure μ.

b) For any $g : \boldsymbol{\delta}^1_{2n+3} \to \boldsymbol{\delta}^1_{2n+3}$, and any normal measure V on $\boldsymbol{\delta}^1_{2n+3}$, there is a tree T on $\boldsymbol{\delta}^1_{2n+3}$ such that for some measure one set A with respect to V, and all $\alpha \in A$, $g(\alpha) \leq |T|(\sup J(\alpha))|$, where the sup ranges over embeddings corresponding to measures on Ξ_{2n+3} which occur in the homogeneous tree construction on a $\boldsymbol{\Pi}^1_{2n+2}$ complete set.

We remark that for $n = 0$, I_{2n+1} and $K_{2n+3}(a)$ are well-known theorems of determinancy, and $K_{2n+3}(b)$ is a theorem of Martin's – see [Ma].

We assume I_{2n+1} and K_{2n+3} for the remainder of this paper, and establish the upper bound for $\boldsymbol{\delta}^1_{2n+5}$, along with some additional auxiliary results.

We introduce the family of canonical measures:

We let W_1^n = the n-fold product of the ω-cofinal normal measure on ω_1. We identify the domain of W_1^n with an ordinal by ordering the n-tuples $(\alpha_1, ..., \alpha_n)$ by α_n first, then α_{n-1}, etc.

We define $S_1^{1,n}$ from the strong partition relation on ω_1 as follows: We let $<^n$ denote the ordering on n-tuples of ordinals $(\alpha_1, ..., \alpha_n)$, where $\alpha_1 < \cdots < \alpha_n$ defined by $(\alpha_1, ..., \alpha_n) <^n (\beta_1, ..., \beta_n)$ if $(\alpha_n, \alpha_1, ..., \alpha_{n-1}) <^L (\beta_n, \beta_1, ..., \beta_{n-1})$, where $<^L$ denotes lexicographic ordering. A set A has measure one w.r.t. $S_1^{1,n}$ if there is a c.u.b. subset of ω_1, C, such that for all $F :<^n \to C$, order-preserving, non-normal, and of uniform cofinality ω, the ordinal represented by F w.r.t. W_1^n is in A. By uniform cofinality ω we mean that there is a function F_2 from the tuples $(\alpha_1, ..., \alpha_{n-1}, m, \alpha_n)$, where $m \in \omega$, order-preserving into ω_1 with the order given by lexicographic ordering on $(\alpha_n, \alpha_1, ..., \alpha_{n-1}, m)$, and induces F in the sense that $F(\alpha_1, ..., \alpha_n) = \sup_{m \in \omega} F_2(\alpha_1, ..., \alpha_{n-1}, m, \alpha_n)$ for all $\alpha_1, ..., \alpha_n$.

In general, for a given order-type, we say that functions from that order-type into the ordinals which are order-preserving, non-normal, and of uniform cofinality ω (with the obvious meaning) are of the correct type.

We assume $W^m_{2n'-1}$, and $S^{\ell,m}_{2n'-1}$ have been defined for $n' \leq n$, $1 \leq \ell \leq 2^{n'} - 1$, and all m, and $W^m_{2n'-1}$ is a measure on $\boldsymbol{\delta}^1_{2n'-1}$, $S^{\ell,m}_{2n'-1}$ is a measure on $\Xi_{2n'+1}$. We then let W^m_{2n+1}, be the measure induced by the weak partition relation on $\boldsymbol{\delta}^1_{2n+1}$, functions $f : K_m \to \boldsymbol{\delta}^1_{2n+1}$

of the correct type, where K_m = the ordinal on which $S^{2^n-1,m}_{2n-1}$ is a measure, and the measure $S^{2^n-1,m}_{2n-1}$. That is, A has measure one w.r.t. W^m_{2n+1} if there is a c.u.b. subset of $\boldsymbol{\delta}^1_{2n+1}$, C, such that for all $f : K_m \to C$ of the correct type, $[f]_{S^{2^n-1,m}_{2n-1}} \in A$.

We let $S^{\ell,m}_{2n+1}$ for $2 \leq \ell \leq 2^{n+1} - 1$ be the measure induced by the strong partition relation on $\boldsymbol{\delta}^1_{2n+1}$, functions $F : \boldsymbol{\delta}^1_{2n+1} \to \boldsymbol{\delta}^1_{2n+1}$ of the correct type, and the measure $\mu^{\theta^{\ell,m}}_{2n+1}$ = the measure induced by the weak partition relation on $\boldsymbol{\delta}^1_{2n+1}$, functions $f : \theta^{\ell,m} \to \boldsymbol{\delta}^1_{2n+1}$ of the correct type, and the measure $R^{\ell,m}$ on $\theta^{\ell,m}$, where $R^{\ell,m}$ enumerates (w.r.t.ℓ) the measures W^m_1, $S^{1,m}_1, ..., W^m_{2n-1}$, $S^{1,m}_{2n-1}, ..., S^{2^n-1,m}_{2n-1}$. That is, A has measure one w.r.t. $S^{\ell,m}_{2n+1}$ ($\ell > 1$) if there is a c.u.b. subset of $\boldsymbol{\delta}^1_{2n+1}$, C such that for any $F : \boldsymbol{\delta}^1_{2n+1} \to C$ of the correct type $[F]_{\mu^{\theta^{\ell,m}}_{2n+1}} \in A$.

We frequently use the abbreviated versions of the above definitions.

For $\ell = 1$, we let $<^m$ denote the ordering on m-tuples of ordinals $< \boldsymbol{\delta}^1_{2n+1}$, $(\alpha_1, ..., \alpha_m)$, defined by $(\alpha_1, ..., \alpha_m) <^m (\beta_1, ..., \beta_m)$ if $(\alpha_m, \alpha_1, ..., \alpha_{m-1}) <^L (\beta_m, \beta_1, ..., \beta_{m-1})$, where $<^L$ again denotes lexicographic ordering. We let $S^{1,m}_{2n+1}$ be the measure induced by the strong partition relation on $\boldsymbol{\delta}^1_{2n+1}$, functions $F : <^m \to \boldsymbol{\delta}^1_{2n+1}$ of the correct type, and the m-fold product of the ω-cofinal normal measure on $\boldsymbol{\delta}^1_{2n+1}$.

From I_{2n+1}, it follows that $S^{\ell,m}_{2n+1}$, W^m_{2n+1} are defined.

If v is a measure we let $\theta_v = \theta(v)$ be the ordinal on which v is a measure (and bounded subsets have measure zero). When there is no danger of confusion, we use v, θ_v interchangeably, and speak of $\alpha \in v$.

We let $R_{2n+1} = \cup_m W^m_{2n+1} \cup \cup_{\ell,m} S^{\ell,m}_{2n+1}$, and let $R^{\ell,m}_{2n+1}$ denote $S^{\ell,m}_{2n+1}$ if $\ell > 1$ and either $S^{\ell,m}_{2n+1}$ or W^m_{2n+1} if $\ell = 1$.

This completes the definition of the canonical family of measure $R^{\ell,m}_{2n+1}$.

We now define a more general collection of measures $\mathcal{R}_{2n+1} = \mathcal{W}_{2n+1} \cup \mathcal{S}_{2n+1}$, where the measures in $\mathcal{W}_{2n+1}$ are measures on tuples of ordinals $< \boldsymbol{\delta}^1_{2n+1}$, and the measures in $\mathcal{S}_{2n+1}$ are measures on tuples of ordinals $< \Xi_{2n+3}$. For $v \in \mathcal{R}_{2n+1}$, we let θ_v denote the tuple of ordinals on which v is a measure. By coding tuples, we may think of v as a measure on an ordinal, which will also denote by θ_v, which should cause no confusion. We proceed by induction on n, and we simultaneously define embeddings $\Pi^{v^j}_{v^i} : \theta_{v^j} \to \theta_{v^i}$, for certain $v^j, v^i \in \mathcal{R}$.

<u>$n = 0$</u>. $\mathcal{W}_1$ consists of the measures $\mathcal{W}^m_1$ on ω^m_1, where a permutation π^m of $\{1, ..., m\}$ is associated with each such measure. We identify a measure in $\mathcal{W}_1$ with a measure on an ordinal

by identifying $(\alpha_1, ..., \alpha_m)$ with its order type in the ordering on these tuples where we order first by α_{π_1} then α_{π_2} etc., where $\pi^m = (\pi_1, \pi_2, ..., \pi_m)$.

If $\mathcal{V}_1 \in W_1$ with permutation π^m, and $\mathcal{V}_2 \in \mathcal{W}_1$ with permutation π^{m+1} and the first m elements of π^{m+1} are ordered (as integers) as the elements of π^m, then we define $\pi_{v^1}^{v^2} = \pi_1^2$ by $\pi_1^2(\beta_1, ..., \beta_{m+1}) = (\beta_1, ..., \hat{\beta}_k, ..., \beta_{m+1})$ where k is the last element in the permutation π^{m+1}.

S_1 consists of products of basic measures in S_1, where we proceed to define the basic measures in S_1. For fixed n, we consider tuples of the form $S = \langle \alpha_1, i_1, \alpha_2, i_2, ..., \alpha_m, i_m \rangle$, where $m \le n$, $\alpha_1, ..., \alpha_m < \omega_1$, $1 \le i_1 \le n_1$ for some integer n_1, etc. We also assume that for each $(i_1, ..., i_k)$, where $k < n$, we have a permutation $\pi_{(i_1,...,i_k)}$ of a size $k+1$ subset of $\{1, ..., n\}$, beginning with n, such that if $(\vec{i})$ extends $(\vec{j})$, then $\pi_{(\vec{i})}$ extends $\pi_{(\vec{j})}$. We then require that for $k < n$, and fixed $(i_1, ..., i_k)$ that $(\alpha_1, ..., \alpha_{k+1})$ are ordered as the integers in $\pi_{(j_1,...,j_k)}$. We order the tuples s by the Kleene-Browner ordering (i.e. $S_1 <^s S_2$ if S_1 is less at the least point of disagreement, or if S_1 extends S_2). The measure in S_1 then, is the measure on tuples $(\cdots, \beta^{(i_1,...,i_m)}, \cdots)$ indexed by indices $(i_1, ..., i_m)$, where $m \le n$ as above, defined using the strong partition relation on ω_1, functions $F :<^s \to \omega_1$ of the correct type, where for fixed F, $(i_1, ..., i_m)$, $\beta^{(i_1,...,i_m)}$ is represented with respect to the m-fold product of the ω-cofinal normal measure on ω_1 by the function $k(\gamma_1, ..., \gamma_m) = F(\langle \gamma_{j_1}, i_1, \gamma_{j_2}, i_2, ..., \gamma_{j_m}, i_m \rangle)$, where $(\gamma_{j_1}, ..., \gamma_{j_m})$ is the permutation of $(\gamma_1, ..., \gamma_m)$ ordered as $\pi_{(i_1,...,i_{m-1})}$.

We define Π on basic measures in S_1 as follows: We let $\mathcal{V}_1, \mathcal{V}_2$ be basic measures in S_1 as above, with $\mathcal{V}_1$ corresponding to n_1, $\mathcal{V}_2$ corresponding to n_2, and $n_1 < n_2$. We require that if $(i_1, ..., i_k)$ is an allowed index as in the definition of $\mathcal{V}_1$, then it is allowed in $\mathcal{V}_2$, and for $1 \le k \le n_1 - 1$ the integers in $\pi^{\mathcal{V}^1}_{(i_1 \cdots i_k)}$ and $\pi^{\mathcal{V}^1}_{(i_1 \cdots i_k)}$ are ordered similarly. In this case, the tuples S_1 as in the definition of $\mathcal{V}_1$ are tuples S_2 allowed in $\mathcal{V}_2$. Hence, an $F_2 :<^{s_2} \to \omega_1$ of the correct type induces an $F_1 :<^{s_1} \to \omega_1$ of the correct type. This, in turn, induces the map $\Pi_1^2 = \Pi_{\mathcal{V}_1}^{\mathcal{V}_2}$ from $\theta_{\mathcal{V}^2}$ into $\theta_{\mathcal{V}^1}$. It follows readily that if $A \subseteq \theta_{\mathcal{V}^1}$ has measure one w.r.t. $\mathcal{V}^1$ then for almost all $\theta < \theta_{\mathcal{V}^2}$ w.r.t. $\mathcal{V}^2$, $\Pi_1^2(\theta) \in A$.

We extend Π to products of basic measures in S_1 componentwise.

It also follows readily that Π extended to products has the same property above.

$\underline{n > 1}$. We assume $\mathcal{W}_{2m+1}$, S_{2m+1} have been defined for $m < n$, and define $\mathcal{W}_{2n+1}$, S_{2n+1}.

For $\mathcal{V}_1$, $\mathcal{V}_2$, measures in $\cup_{m<n} \mathcal{W}_{2m+1} \cup \cup_{m<n} S_{2m+1}$, we define $\mathcal{V}_2 < \mathcal{V}_1$ if $\Pi_1^2 : \theta_{v_2} \to \theta_{v_1}$ is defined.

Definition of $\mathcal{W}_{2n+1}$: We define an $n - *$ tuple $T = T^n$ to be a function with domain certain tuples of integers $(i_1, ..., i_m)$, $m \leq n$ and satisfying the following properties:

1) $i_1 \in$ dom T for $1 \leq i_1 \leq n_1$, for some integer n_1.

2) If $(i_1, ..., i_\ell) \in$ dom T, then $(i_1, ..., i_k) \in$ dom T for $k \leq \ell$.

3) T associates measures to certain of the tuples $(i_1, ..., i_k)$, where $k < n$, in its domain, which we denote by $v^{(i_1,...,i_k)}$, where $v^{(i_1,...,i_k)} \in \cup_{m<n} \mathcal{R}_{2m+1}$. We require that if $v^{(i_1,...,i_{k+1})}$ is defined, then so is $v^{(i_1,...,i_k)}$, and $v^{(i_k,...,i_{k+1})} < v^{(i_1,...,i_k)}$.

4) If $v^{(i_1,...,i_k)}$ is defined, then $(i_1, ..., i_{i+1}) \in$ dom T for $1 \leq i_{k+1} \leq n_{(i_1,...,i_k)}$ for some integer $n_{(i_1,...,i_k)}$, and if $(i_1, ..., i_{k+1}) \in$ dom T, then $v^{(i_1,...,i_k)}$ is defined.

For each $n - *$ tuple T, we define a corresponding ordering $<^T$ as follows: We let $\Pi^{(i_1,...,i_{k+1})}_{(i_1,...,i_k)}$ denote the embedding for $v^{(i_1,...,i_{k+1})}$, $v^{(i_1,...,i_k)}$, where defined. The domain of $<^T$ consists of ordinals $\alpha^{(i_1,...,i_k)}$ indexed by indices $(i_1, ..., i_k) \in$ dom T, where $\alpha \in \theta^{(i_1,...,i_{k-1})}$ $(= \theta_{v^{(i_1,...,i_{k-1})}})$, if $k > 1$. For $k = 1$, we take $\alpha = 1$, and allow $1^{(i_1)} \in$ dom $<^T$ for some (but not necessarily all – we think of this as being coded in T) $i_1 \in$ dom T. We set $\alpha^{(i_1,...,i_k)} <^T \beta^{(j_1,...,j_\ell)}$ provided

$$\begin{aligned}&\langle i_1, \Pi^{(i_1,...,i_{k-1})}_{(i_1)}(\alpha),\ i_2, \Pi^{(i_1,...,i_{k-1})}_{(i_1,i_2)}(\alpha), ..., \alpha, i_k\rangle \\ &<^{KB} \langle j_1, \Pi^{(j_1,...,j_{\ell-1})}_{(j_1)}(\beta), j_2,\ \Pi^{(j_1,...,j_{\ell-q})}_{(j_1,j_2)}, ..., \beta,\ j_\ell\rangle,\end{aligned}$$

where $<^{KB}$ denotes the Kleene-Brouwer ordering of these sequences. We use here the notation (defined inductively by)

$$\begin{aligned}\Pi^{(i_1,...,i_{k-1})}_{(i_1,...,i_m)} &= \Pi^{i_1,...,i_{m+1})}_{(i_1,...,i_m)} \circ \Pi^{(i_1,...,i_{k_1})}_{(i_1,...,i_{m+1})}, \text{ where} \\ \Pi^{(i_1,...,i_{k-1})}_{(i_1,...,i_{k-1})} &= \text{ identity.}\end{aligned}$$

We let $\mu^{<^T}$ be the measure on tuples $(\cdots, \eta^{(i_1,...,i_k)}, \cdots)$ of ordinals $< \boldsymbol{\delta}^1_{2n+1}$ induced by the weak partition relation on $\boldsymbol{\delta}^1_{2n+1}$, functions $f\ :<^T \rightarrow\ \boldsymbol{\delta}^1_{2n+1}$ of the correct type, and the measures $v^{(i_1,...,i_k)}$. That is, A has measure one if there is a c.u.b. subset of $\boldsymbol{\delta}^1_{2n+1}$, C, such that for any $f\ :<^T \rightarrow\ C$ of the correct type, $(\cdots, \eta^{(i_1,...,i_k)}, \cdots) \in A$, where $\eta^{(i_1,...,i_k)}$ is represented w.r.t. $v^{(i_1,...,i_{k-1})}$ by the function $f^{(i_1,...,i_k)}\ :\ \theta^{(i_1,...,i_{k-1})} \rightarrow \boldsymbol{\delta}^1_{2n+1}$ defined by $f^{(i_1,...,i_k)}(\beta) = f(\beta^{(i_1,...,i_k)})$, for $k > 1$, and if $k = 1$ and $i_1 \in$ dom $<^T$, then $\eta^{(i_1)} = f(1^{i_1})$.

We define W_{2n+1} to consist of the measures of the form $\mu^{<^T}$ corresponding to all $n - *$ tuples T.

Def. of Π on $\mathcal{W}_{2n+1}$. If T_1 is an $n - *$ tuple, and T_2 an $(n + 1) - *$ tuple, we say that T_2 extends T_1 provided:

1) $(i_1,...,i_k) \in$ dom T_1, iff $(i_1,...,i_k) \in$ dom T_2, for $k \le n$ and $v_{T_2}^{(i_1,...,i_{k-1})} = v_{T_1}^{(i_1,...,i_{k-1})}$.

2) If $(i_1,...,i_{n+1}) \in$ dom T_2, then $v_{T_2}^{(i_1,...,i_n)} < v_{T_1}^{(i_1,...,i_{n-1})}$.

3) If $1^{i_1} \in$ dom $<^{T_1}$ then $1^{i_1} \in$ dom $<^{T_2}$, and conversely.

We write $T_2 < T_1$ for T_2 extending T_1.

For $T_2 < T_1$, and the corresponding measures $\mu^{<^{T_2}}$, $\mu^{<^{T_1}}$ on θ_2, $\theta_1 (= (\delta^1_{2n+1})^{c_1}$, $(\delta^1_{2n+1})^{c_2}$ for come $c_1, c_2)$, we define $\Pi^2_1 : \theta_2 \to \theta_1$ as follows: If $(\cdots, \eta^{(i_1,...,i_k)}, \cdots)$ is a tuple in θ_2, $(i_1,...,i_k) \in$ dom T_2, we set $\Pi^2_1(\cdots, \eta^{(i_1,...,i_k)}, \cdots) = (\cdots, \eta^{(i_1,...,i_k)}, \cdots)$ where $(i_1,...,i_k)$ now ranges over the indices in dom T_1. We note that if $(\cdots, \eta^{(i_1,...,i_k)}, \cdots)$ is represented by $f :<^{T_2} \to \delta^1_{2n+1}$ (which holds almost everywhere w.r.t. $\mu^{<^{T_2}}$), then f induces a function $\bar{f} :<^{T_1} \to \delta^1_{2n+1}$, since $<^{T_1}$ is a subordering of $<^{T_2}$, and $\bar{f}$ then represents $\Pi^2_1(\cdots, \eta^{(i_1,...,i_k)}, \cdots)$.

We also note that if $A \subseteq \theta_1$ has measure one w.r.t. $\mu^{<^{T_1}}$, then for almost all $\vec{\eta}$ w.r.t. $\mu^{<^{T_2}}$, $\Pi^2_1(\vec{\eta}) \in A$.

Def. of S_{2n+1}. We fix an $n-*$ tuple T, and associated order in $<^T$. We define, relative to T, a collection C to be a set $C = \cup_{1 \le m \le n} C_n$ where:

1) C_1 consists of a_0 sequences for some integer a_0, of the form $C_1 = (i_1^{(1)},...,i_{k(1)}^{(1)}),...,C_{a_0} = (i_1^{(a_0)},...,i_{k(a_0)}^{(a_0)})$, of integers between 1 and n_1 (= the number of $(i_1) \in$ dom T).

2) C_ℓ consists of tuples of the form $C_{k_1,...,k_\ell}$, where $C_{k_1,...,k_{\ell-1}} \in C_{\ell-1}$. Each $C_{k_1,...,k_\ell}$ is a sequence of tuples of the form $C_{k_1,...,k_\ell} = (\cdots, (i_1,...,i_\ell), \cdots)$, where each tuple $(i_1,...,i_\ell) \in C_{k_1,...,k_\ell}$, extends $(i_1,...,i_{\ell-1}) \in C_{k_1,...,i_{\ell-1}}$ and $(i_1,...,i_\ell) \in$ dom T.

Corresponding to T, C, we consider sequences S of ordinals $< \delta^1_{2n+1}$ and integers of the form $S = \langle \alpha_0, I_1, \alpha_1,...,\alpha_{j_1}, I_2,...,\alpha_{j_1,j_2},..., I_n,...,\alpha_{j_1,...,j_n},...\rangle$ or an initial segment of such a tuple, where:

1) $\alpha_0 < \delta^1_{2n+1}$

2) $1 \le I_1 \le a_0 =$ the number of elements in C_1

3) $j_1 =$ the number of elements in C_{I_1}.

4) $\alpha_1,...,\alpha_{j_1} < \delta^1_{2n+1}$

5) In general, $(I_1,...,I_k)$ for $k \le n$ is an index corresponding to an element of C.

6) The ordinals $\alpha_{j_1,...,j_k}$ following I_k are indexed by indices $(j_1,...,j_k)$ corresponding to

elements of $C_{I_1,\dots,I_k}$.

7) $\alpha_{j_1,\dots,j_k} < j_{v^{(j_1,\dots,j_k)}}(\alpha_0)$ = the utrapower of α_0 by the measure $v^{(j_1,\dots,j_k)}$. (If $n_1 \in$ dom $<^T$, it would be enough to require $\alpha_{j_1},\dots,\alpha_{j_k} < \alpha_0$ here.)

We let $<^T_C$ denote the Kleene-Brouwer ordering on these tuples. In the definition of S, we assume the ordering on indices $(j_1,\dots,j_k)$ is the same as their occurrence in $C_{(I_1,\dots,I_k)}$.

We then let $\mathcal{S}_{2n+1}{}^T_C$ be the measure induced by the strong partition relation that is on $\boldsymbol{\delta}^1_{2n+1}$, functions $F :<^T_C\to \boldsymbol{\delta}^1_{2n+1}$ of the correct type, and the measure $\mu^{<^T}$, that is, $\mathcal{S}_{2n+1}{}^T_C$ is a measure on tuples of ordinals $(\cdots,\alpha_{(I_1,\dots,I_k)},\cdots)$, indexed by indices $(I_1,\dots,I_k)$ corresponding to indices in $\mathcal{C}$, where A has measure one w.r.t. $\mathcal{S}_{2n+1}{}^T_C$ if there is a c.u.b. $C \subseteq \boldsymbol{\delta}^1_{2n+1}$ such that for all $F :<^T_C\to C$ of the correct type, $(\cdots,\alpha^F_{(I_1,\dots,I_k)},\dots,) \in A$, where $\alpha^F_{(I_1,\dots,I_k)}$ is represented w.r.t. $\mu^{<^T}$ by a function $g(\cdots,\xi_{(i_1,\dots,i_k)},\cdots)$ defined as follows: if $h :<^T\to \boldsymbol{\delta}^1_{2n+1}$ represents the tuple $(\cdots,\xi_{(i_1,\dots,i_k)},\cdots)$ w.r.t. the measures $v^{(i_1,\dots,i_k)}$, then we set $g(\cdots,\xi_{(i_1,\dots,i_k)},\cdots)$ $= F(S) = F(S^h_{(I_1,\dots,I_k)})$, where $S = \langle\alpha_0, I_1, \alpha_1,\dots,\alpha_{j_1}, I_2,\dots,\alpha_{j_1,j_2},\dots,I_k,\dots,\alpha_k,\dots,\alpha_{j_1,\dots,j_k},\cdots\rangle$ where

1) $\alpha_0 = \underset{\text{a.e.}}{\sup}(\text{range } h)$, where $\underset{\text{a.e.}}{\sup}$ denotes the almost everywhere sup w.r.t. the measures $v^{(i_1,\dots,i_k)}$. If h is of the correct type everywhere, this is the same as sup(range h).

2) $\alpha_{j_1,\dots,j_\ell}$ is represented w.r.t. $(v^{j_1,\dots,j_{\ell-1}})$ by the function $h^{(j_1,\dots,j_\ell)} : \theta_{v^{(j_1,\dots,j_{\ell-1})}} \to \boldsymbol{\delta}^1_{2n+1}$ induced by h; that is $h^{(j_1,\dots,j_\ell)}(\gamma) = h(\gamma^{(j_1,\dots,j_\ell)})$, for $\gamma \in \theta_{V^{(j_1,\dots,j_{\ell-1})}}$.

This is well defined. We let $\mathcal{S}_{2n+1}$ consist of the r-fold products of measures of the form $\mathcal{S}_{2n+1}{}^T_C$. We refer to the $\mathcal{S}_{2n+1}{}^T_C$ as the basic measures in $\mathcal{S}_{2n+1}$.

Definition of Π on $\mathcal{S}_{2n+1}$. We again define Π component-wise, so it is enough to define it on basic measures of the form $\mathcal{S}_{2n+1}{}^T_C$.

We consider measures of the form $\mathcal{S}_{2n+1}{}^{T_1}_{C_1}$, $\mathcal{S}_{2n+1}{}^{T_2}_{C_2}$ which satisfy the following:

1) T_2 is an $(n+1)-*$ tuple which extends T_1.

2) $\mathcal{C}_1$, $\mathcal{C}_2$ are collections defined relative to T_1, T_2 respectively.

3) $\mathcal{C}_2$ extends $\mathcal{C}_1$ in the sense that if $C_{(I_1,\dots,I_k)} \in \mathcal{C}_1$, then $C_{(I_1,\dots,I_k)} \in \mathcal{C}_2$.

We define the embedding $\Pi^{(T_2,C_2)}_{(T_1,C_1)}$ corresponding to $\mathcal{S}_{2n+1}{}^{T_2}_{C_1}$, $\mathcal{S}_{2n+1}{}^{T_2}_{C_1}$ in this case. If a tuple $(\dots,\alpha_{(I_1,\dots,I_k)},\dots)$, $k \leq n+1$ is given in the measure space $\mathcal{S}_{2n+1}{}^{T_2}_{C_2}$, and is represented by $F :<^{T_2}_{C_2}\to \boldsymbol{\delta}^1_{2n+1}$, then F induces an $\bar{F} :<^{T_1}_{C_1}\to \boldsymbol{\delta}^1_{2n+1}$, by restriction, representing a tuple $(\cdots,\beta_{(I_1,\dots,I_k)},\cdots)$, $k \leq n$, with respect to $\mu^{<^{T_1}}$. We set $\Pi^{(T_2,C_2)}_{(T_1,C_1)}(\cdots,\alpha_{(I_1,\dots,I_k)},\cdots) =$

$(\cdots,\beta_{(I_1,\ldots,I_k)},\cdots)$.

It follows readily that is well-defined on a measure one set w.r.t. $\mathcal{S}_{2n+1}{}^{T_2}_{C_2}$. We take Π to be zero off this measure one set.

This completes the definitions of the measures R_{2n+1}, $\mathcal{R}_{2n+1}$, and the embeddings Π.

We define an equivalence relation $\sim$ on the measures in R_{2n+1} by induction as follows: We set $\mathcal{V}_1 \sim \mathcal{V}_2$ provided one of the following holds:

1) $\mathcal{V}_1 = S^{1,m_1}_{2n+1}$ and $\mathcal{V}_2 = S^{1,m_2}_{2n+1}$.

2) $\mathcal{V}_1 = W^{m_1}_{2n+1}$, $\mathcal{V}_2 = W^{m_2}_{2n+1}$

3) $\mathcal{V}_1 = S^{\ell_1,m_1}_{2n+1}$, $\mathcal{V}_2 = S^{\ell_2,m_2}_{2n+1}$, where $\ell_1,\ell_2 > 1$, in which case $\ell_1 = \ell_2$.

A useful notation is to let, for $\mathcal{V} = S^{\ell,m}_{2n+1}$, $\ell > 1$, $v(S^{\ell,m}_{2n+1})$ denote the measure v such that $S^{\ell,m}_{2n+1}$ is defined from the strong partition relation on $\boldsymbol{\delta}^1_{2n+1}$, functions $F : \boldsymbol{\delta}^1_{2n+1} \to \boldsymbol{\delta}^1_{2n+1}$ of the correct type, and the measure μ^v = the measure induced by the weak partition on $\boldsymbol{\delta}^1_{2n+1}$, fucntions $f : \theta_v \to \boldsymbol{\delta}^1_{2n+1}$ of the correct type and the measure v.

We note that (by induction) 3 above is equivalent to $v(S^{\ell_1,m_1}_{2n+1}) \sim v(S^{\ell_2,m_2}_{2n+1})$.

We prove some basic lemmas which will be used frequently. We recall we are assuming I_{2n+1} throughout.

Lemma 1. *If $\mathcal{V}_1, \mathcal{V}_2 \in R_{2n+1}$ and $\mathcal{V}_1 \not\sim \mathcal{V}_2$, then* $\mathrm{cof}(\theta_{\mathcal{V}_1}) \neq \mathrm{cof}(\theta_{\mathcal{V}_2})$.

Proof. If not, we fix $\mathcal{V}_1 \not\sim \mathcal{V}_2$, such that $\mathrm{cof}(\theta_{\mathcal{V}_1}) = \mathrm{cof}(\theta_{\mathcal{V}_2})$.

We consider first the case where $\mathcal{V}_1 = S^{\ell_1,m_1}_{2n+1}$, $\mathcal{V}_2 = S^{\ell_2,m_2}_{2n+1}$, where $\ell_1,\ell_2 > 1$ and $\ell_1 \neq \ell_2$. We note that $\theta_{\mathcal{V}_1}, \theta_{\mathcal{V}_2}$ are always limit ordinals. We fix $H : \theta_{\mathcal{V}_1} \to \theta_{\mathcal{V}_2}$ monotonicially increasing and cofinal. We consider the following partition.

$\mathcal{P}$: We partition pairs of functions $F_1, F_2 : \boldsymbol{\delta}^1_{2n+1} \to \boldsymbol{\delta}^1_{2n+1}$ of the correct type, where we require that if $f_1 : v(\mathcal{V}_1) \to \boldsymbol{\delta}^1_{2n+1}$, $f_2 : v(\mathcal{V}_2) \to \boldsymbol{\delta}^1_{2n+1}$, then if $\sup_{\text{a.e.}} f_1 < \sup_{\text{a.e.}} f_2$, $F_1([f_1]_{v(\mathcal{V}_1)}) < F_2([f_2[_{v(\mathcal{V}_2)})$ and similarly for $\sup_{\text{a.e.}} f_2 < \sup_{\text{a.e.}} f_1$. If $\sup_{\text{a.e.}} f_1 = \sup_{\text{a.e.}} f_2$ (hence one of f_1, f_2 is not of the correct type a.e., since $\mathrm{cof}\theta_{v(\mathcal{V}_1)} \neq \mathrm{cof}\theta_{v(\mathcal{V}_2)}$ by induction), we require $F_1([f_1]) < F_2([f_2])$. In using the strong partition relation on $\boldsymbol{\delta}^1_{2n+1}$, we think of the pair F_1, F_2 as being coded by a single function of the correct type. We then partition such F_1, F_2 (or more precisely the single function coding them) according to whether or not $H([F_1]_{\mu^{v_1}}) < [F_2]_{\mu^{v_2}}$, where $v_1 = v(\mathcal{V}_1)$, $v_2 = v(\mathcal{V}_2)$.

We then claim that on the homogeneous side of the partition, the property stated in partition $\mathcal{P}$ holds. We suppose not, and fix a c.u.b. $C \subseteq \boldsymbol{\delta}^1_{2n+1}$ homogeneous for the contrary side. We assume that C is closed under j_{v_1}, j_{v_2}, the ultrapowers from the measures v_1, v_2 (we use $K_{2n+1}(a)$ here), and C consists of limit ordinals. We fix $F_1 : \boldsymbol{\delta}^1_{2n+1} \to C$ of the correct type with $F_1(\alpha) > \alpha$ for all α, and fix $F_2 : \boldsymbol{\delta}^1_{2n+1} \to C$ of the correct type with $[F_2] > H([F_1])$ and $F_2(\alpha) > \alpha$ for all α.

We let C' be a c.u.b. subset of the closure points of $C(= \{\beta \in C :$ for all $\alpha < \beta$, β is the βth element of C after $\alpha\})$ which is closed under F_1, F_2 in the sense that for all $\beta \in C'$, if $\alpha < \beta$ and $f_1 : \theta_{v_1} \to \alpha$ then $F_1([f_1]_{v_1}) < \beta$, and similarly for F_2.

We then claim that there are functions F_1', F_2' satisfying:

1) $F_1' = F_1$, $F_2' = F_2$, a.e. w.r.t. μ^{v_1} and μ^{v_2} respectively.

2) F_1', F_2' have range a subset of C and are of the correct type.

3) F_1', F_2' are ordered as in the statement of $\mathcal{P}$, everywhere.

We define F_1', F_2' as follows: If $\theta \in C'$ and cof $\theta =$ cof θ_{v_1}, then we set $F_1'([f]) = F_1([f])$ for any $f : \theta_{v_1} \to \boldsymbol{\delta}^1_{2n+1}$ s.t. $\sup_{\text{a.e.}} f = \theta$. For such θ we set $F_2'(\theta) = N_C(\sup_\beta F_1(\beta))$, where the sup ranges over $\beta < \boldsymbol{\delta}^1_{2n+1}$ represented w.r.t. v_1 by $f : \theta_{v_1} \to \boldsymbol{\delta}^1_{2n+1}$ s.t. $\sup_{\text{a.e.}} f = \theta$, and $N_c(\gamma) =$ the ωth element of C after γ. If $\theta \in C'$ and cof $\theta =$ cof θ_{v_2}, then we set $F_2'([f]) = F_2([f])$ for $f : \theta_{v_2} \to \boldsymbol{\delta}^1_{2n+1}$ with $\sup_{\text{a.e.}} f = \theta$, and we set $F_1'(\theta) = N_c(\sup_\beta F_2(\beta))$ similarly.

For other θ, we proceed inductively: We fix $f_1 : \theta_{v_1} \to \boldsymbol{\delta}^1_{2n+1}$, $f_2 : \theta_{v_2} \to \boldsymbol{\delta}^1_{2n+1}$ representing θ w.r.t. v_1, v_2, and we let $\theta_1 = \sup_{\text{a.e.}} f_1$, $\theta_2 = \sup_{\text{a.e.}} f_2$. We then set $F_1'(\theta) = N_c(\max(\sup_\alpha F_1'(\alpha), \sup_\beta F_2'(\beta)))$, where α ranges over ordinals represented by $\bar{f}_1 : \theta_{v_1} \to \boldsymbol{\delta}^1_{2n+1}$ with $[\bar{f}_1]_{v_1} < [f_1]_{v_1}$, and β ranges over ordinals represented by $\bar{f}_2 : \theta_{v_2} \to \boldsymbol{\delta}^1_{2n+1}$ with $\sup_{\text{a.e.}} \bar{f}_2 < \theta_1$. We similarly define $F_2'(\theta) = N_c(\max(\sup_\alpha F_1'(\alpha), \sup_\beta F_2'(\beta))$, where now α ranges over ordinals represented by $\bar{f}_1 : \theta_{v_1} \to \boldsymbol{\delta}^1_{2n+1}$ with $\sup_{\text{a.e.}} \bar{f}_1 \leq \theta_2$, and β is represented by $\bar{f}_2 : \theta_{v_2} \to \boldsymbol{\delta}^1_{2n+1}$ with $[\bar{f}_2]_{v_2} < [f_2]_{v_2}$. This is well-defined.

We claim that F_1', F_2' satisfy the above properties. Property 1 follows since $F_1' = F_1$ for all $f : \theta_{v_1} \to C'$ of the correct type, and similarly for F_2'. Property 2 is immediate from the definition of F_1', F_2'. Property 3 follows from the definition of F_1', F_2', the fact that $F_1'(\alpha) > \alpha$, $F_2'(\alpha) > \alpha$, and the definition of C_1' which implies that for $f_1 : \theta_{v_1} \to C'$ of the correct type, $F_1'([f_1]) = F_1([f_1]) \geq \sup_{\text{a.e.}} f_1 = \beta \in C'$, and hence $> F_2'([f_2])$ for all $f_2 : \theta_{v_2} \to \boldsymbol{\delta}^1_{2n+1}$ with $\sup_{\text{a.e.}} f_2 < \beta$, and property 3 follows readily. From property 1, we have that $[F_2'] > H([F_1'])$, which contradicts properties 2,3, and the definition of C. Hence, on the homogeneous side of the partition, the property stated in partition $\mathcal{P}$ holds.

We fix a c.u.b. $C \subseteq \delta^1_{2n+1}$ homogeneous for $\mathcal{P}$. We define F_2 as follows: for $\alpha < \delta^1_{2n+1}$ represented by $f_2 : \theta_{v_2} \to \delta^1_{2n+1}$ where $\sup_{\text{a.e.}} f_2 = \theta$, we set $F_2(\alpha) = N_c(\sup_{\alpha'} F_2(\alpha'))$, where α' ranges over ordinals represented by $\bar{f}_2$, where $\sup_{\text{a.e.}} \bar{f}_2 = \theta$, and $[\bar{f}_2]_{v_2} < [f_2]'_{v_2}$ provided this sup is non-empty, and otherwise we set $F_2(\alpha) = N^2_c(\theta)$.

We then claim for almost all $[F_1]_{\mu^{v_1}}$, $H([F_1]) < [F_2]$, contradicting the fact that H is cofinal, and establishing Lemma 1.

We fix $F_1 : \delta^1_{2n+1} \to C$ of the correct type. To prove the claim, it is enough to find F'_1, F'_2 satisfying:

1) $F'_1 = F_1$, $F'_2 = F_2$ a.e. w.r.t. μ^{v_1}, μ^{v_2} respectively.

2) F'_1, F'_2 have range a subset of C, and are of the correct type.

3) F'_1, F'_2 are ordered as in $\mathcal{P}$.

The construction of F'_1, F'_2 proceeds similarly to the above, starting with a c.u.b. $C' \subseteq \delta^1_{2n+1}$ contained in the closure points of C, and such that $\alpha \in C'$, cof α = cof θ_{v_2}, we have that if $f : \theta_{v_1} \to \alpha$ is of the correct type, then $F_1([f]_{v_1}) < \alpha$. It is here we use the fact that cof $\theta_{v_1} \neq$ cof θ_{v_2}. We also use N^2_c in the above definition, since for $\alpha \in C'$ (in which case $F'_2([f_2]) = F_2([f_2])$ for $\sup_{\text{a.e.}} f_2 = \alpha$), we take $F'_1([f_1]) = N_c(\alpha)$, for f_1 the constant function α. For $\alpha \notin C'$, the definition proceeds as above. This completes the proof of Lemma 1 in the case considered.

If $\ell_1 = 1$, $\ell_2 > 1$ (or vice-versa), the argument is similar.

If $\mathcal{V}_1 = W^m_{2n+1}$, $\mathcal{V}_2 = S^{\ell_2,m_2}_{2n+1}$ (or vice-versa), the argument is again similar: we fix an $H : \delta^1_{2n+1} \to \theta_{\mathcal{V}_2}$, cofinal, monotonically increasing, and consider the partition $\mathcal{P}$: we partition pairs of functions f_1, F_2, $f_1 : \theta_{v_1} \to \delta^1_{2n+1}$ of the correct type, where $W^n_{2n+1} = \mu^{v_1}_{2n+1}$, and $F_2 : \delta^1_{2n+1} \to \delta^1_{2n+1}$ of the correct type, where sup $f_1 <$ inf F_2, according to whether or not $H([f_1]_{v_1}) < [F_2]_{\mu^{v_2}}$. It follows by a similar argument that on the homogeneous side of the partition, the above property holds. By another similar argument, we contradict the cofinality of H.

If $\mathcal{V}_1 \in \cup_{m<n} R_{2m+1}$, and $\mathcal{V}_2 = W^m_{2n+1}$, Lemma 1 follows from the regularity of δ^1_{2n+1}, and if $\mathcal{V}_2 = S^{\ell_2,m_2}_{2n+1}$, we contradict the existence of such an H by the δ^1_{2n+1}-additivity of the measure S^{ℓ_2,m_2}_{2n+1}, $\ell_2 \geq 1$.

If $\mathcal{V}_1, \mathcal{V}_2 \in \cup_m R_{2n+1}$, the lemma follows by induction.

This completes the proof of Lemma 1.

We refer the argument in Lemma 1, where "slide" the values in the range of the functions on a set of measure zero to produce a desired order type everywhere, as a "sliding argument". This type of argument will be used frequently in the following, and we abbreviate the details of the argument.

Lemma 2. *For any $\mathcal{V} \in R_{2n+1}$, and any ordinal $\boldsymbol{\delta}$, and $H : \theta_{\mathcal{V}} \to \boldsymbol{\delta}$, either there is a $\boldsymbol{\delta}' < \boldsymbol{\delta}$ such that $H(\alpha) \leq \boldsymbol{\delta}'$ for almost all α w.r.t. $\mathcal{V}$, or there is a measure one set A w.r.t. $\mathcal{V}$ such that $H \restriction A$ is dominated by a monotonically increasing function into $\boldsymbol{\delta}$.*

Proof. We consider first the case $\mathcal{V} = S^{\ell,m}_{2n+1}$, with $\ell > 1$. We fix such a $\boldsymbol{\delta}$ and H. We assume the first clause fails and show the second. In particular, for all $\alpha < \theta_{\mathcal{V}}$, we have that for cofinally many $\beta < \theta_{\mathcal{V}}$, $H(\alpha) < H(\beta)$. We then consider the partition $\mathcal{P}$: we partition $F_1, F_2 : \boldsymbol{\delta}^1_{2n+1} \to \boldsymbol{\delta}^1_{2n+1}$ of the correct type ordered as follows: if α_1, α_2 are represented w.r.t. $v = v(\mathcal{V})$ by $f_1, f_2 : \theta_v \to \boldsymbol{\delta}^1_{2n+1}$, and $\theta_1, \theta_2 = \sup_{\text{a.e.}} f_1, f_2$ respectively, then if $\theta_1 < \theta_2$, $F_{1,2}(\alpha_1) < F_{1,2}(\alpha_2)$; if $\theta_1 > \theta_2$, $F_{1,2}(\alpha_1) > F_{1,2}(\alpha_2)$, and if $\theta_1 = \theta_2$, $F_1(\alpha_1) < F_2(\alpha_2)$. We partition such F_1, F_2 according to whether or not $H([F_1]_{\mu^v}) < H([F_2]_{\mu^v})$. From our above assumption, and a sliding argument as in Lemma 1, it follows that on the homogeneous side of the partition, the property stated in $\mathcal{P}$ holds. We fix a c.u.b. subset $C \subseteq \boldsymbol{\delta}^1_{2n+1}$ homogeneous for $\mathcal{P}$, and take A to be the measure one set determined by C. For $\eta \in A$ represented by $F : \boldsymbol{\delta}^1_{2n+1} \to C$ of the correct type, we define $H'(\eta) = \sup_{\eta'}(H(\eta'))$, where the sup ranges over η' represented by $F' : \boldsymbol{\delta}^1_{2n+1} \to C$ of the correct type, and $\eta' < [F_0]_{\mu^v}$, where $F_0 : \boldsymbol{\delta}^1_{2n+1} \to \boldsymbol{\delta}^1_{2n+1}$ is defined by: for α represented by $f : \theta_v \to \boldsymbol{\delta}^1_{2n+1}$ w.r.t. v, $F_0(\alpha) = \sup_{\alpha'} F(\alpha')$, where the sup ranges over α' represented w.r.t. v by f' such that $\sup_{\text{a.e.}} f' \leq \sup_{\text{a.e.}} f$. This is well defined.

It is clear that $H' \restriction A$ is monotonic increasing and dominates H. To show that it has range in $\boldsymbol{\delta}$, and finish the proof of the lemma, it is enough to establish the following claim:

If $F_1, F_2 : \boldsymbol{\delta}^1_{2n+1} \to C$ are of the correct type and $(F_1)_0 < (F_2)_0$ a.e. w.r.t. μ^v (where F_0 is defined above for $F : \boldsymbol{\delta}^1_{2n+1} \to \boldsymbol{\delta}^1_{2n+1}$), then $H([F_1]) < H([F_2])$.

To establish the claim, it is enough to show, for such F_1, F_2, that there are F_1', F_2' satisfying:

1) $F_1' = F_1$, $F_2' = F_2$ a.e. w.r.t. μ^v.

2) F_1', F_2' have range a subset of C and are of the correct type everywhere.

3) F_1', F_2' are ordered as in $\mathcal{P}$.

To establish this we use a sliding argument similar, but somewhat different to that of Lemma 1: We consider that partition $\bar{\mathcal{P}}$: we partition $f : \theta_v \to \delta^1_{2n+1}$ of the correct type according to whether or not $F_2([f]) > (F_1)_0([f])$.

It follows readily that on the homogeneous side fo the partition, the property stated in $\mathcal{P}'$ holds. We let C_2 be a c.u.b. subset of δ^1_{2n+1} homogeneous for $\mathcal{P}'$, where C_2 consists of limit ordinals and is closed under $\mathcal{J}_v$ and $(F_1)_0, (F_2)_0$. We let C_2' be its set of closure points. We define F_1', F_2' as follows:

1) for α represented by $f : \theta_v \to C_2'$ of the correct type, we set $F_1'(\alpha) = F_1,\ F_2'(\alpha) = F_2$.

2) for α not as in 1, if α is represented by $f : \theta_v \to \delta^1_{2n+1}$, and $\theta = \sup_{\text{a.e.}} f \in C'$, and $f < \theta$ a.e.w.r.t. v, we set $F_1'(\alpha) = F_1(\alpha_2),\ F_2'(\alpha) = F_2(\alpha_2)$, where α_2 is represented by $f_2 : \theta_v \to \delta^1_{2n+1}$ defined as follows: we let α' be the least ordinal greater than α and represented by some $h : \theta_v \to \theta$ monotonic increasing (it follows readily from Lemma 2 and induction that such an α' exists). We fix a monotonic $h : \theta_v \to \theta$ representing α'. For $\beta < \theta_v$, we then set $f_2(\beta) =$ the $2\omega \cdot ((\beta - 1) + f(\beta) + 1)^{st}$-element of C after $h(\beta)$, where $\beta - 1 = \beta$ if β is a limit ordinal. It follows that α_2 is well-defined.

3) for α not as in 1, $\theta \in C'$, and $f = \theta$ a.e.w.r.t. v, we set $F_1'(\alpha) = N_C(\sup_{\alpha'} F_1(\alpha'))$, where the sup ranges over α' represented by f' with $\theta' = \sup_{\text{a.e.}} f' \leq \theta$, and $f' < \theta'$ almost everywhere. Here $N_C(\beta) =$ the next element in $C > \beta$. We also set $F_2'(\alpha) = N_c(\sup_{\alpha'} F_2(\alpha'))$, α' as above.

4) for α not as in 1, and $\theta \notin C'$, we proceed inductively: We set $F_1'(\alpha) = N_c(\max(\sup_{\alpha'<\alpha} F_1'(\alpha'), \sup_{\beta} F_2'(\beta)))$, where β ranges over ordinals represented by $g : \theta_v \to \delta^1_{2n+1}$ with $\sup_{\text{a.e.}} g < \theta$. Also, $F_2'(\alpha) = N_c(\max(\sup_{\alpha'<\alpha} F_2'(\alpha), \sup_{\beta} F_1'(\beta)))$, where now β ranges over ordinals represented by g with $\sup_{\text{a.e.}} g \leq \theta$. It is immediate that $F_1' = F_1,\ F_2' = F_2$ almost everywhere, that F_1', F_2' have range a subset of C, and are non-normal of uniform cofinality ω. The fact that $F_1',\ F_0'$ are strictly increasing and ordered as in $\mathcal{P}$ follows from the definitions of C', F_1', F_2' upon consideration of several cases. The claim now follows. This establishes Lemma 2 in this case. The other cases are similar.

Lemma 3. *If $\mathcal{V}_1, \mathcal{V}_2 \in \cup_{m \leq n} R_{2m+1}$, and $\mathcal{V}_1 \sim \mathcal{V}_2$ then $\mathcal{V}_1 \times \mathcal{V}_2 = \mathcal{V}_2 \times \mathcal{V}_1$, that is, if $A \subseteq \theta_{\mathcal{V}_1} \times \theta_{\mathcal{V}_2}$, and A has measure one with respect to $\mathcal{V}_1 \times \mathcal{V}_2$ (that is, for almost all $\alpha \in \theta_{\mathcal{V}_1}$, for almost all $\beta \in \theta_{\mathcal{V}_2}$, $(\alpha, \beta) \in A$), then A has measure one w.r.t. $\mathcal{V}_2 \times \mathcal{V}_2$ (for almost all $\beta \in \theta_{\mathcal{V}_2}$, for almost all $\alpha \in \theta_{\mathcal{V}_1}$, $(\alpha, \beta) \in A$).*

Proof. We again consider the case $\mathcal{V}_1 = S^{\ell_1,m_1}_{2n+1}$, $\mathcal{V}_2 = S^{\ell_2,m_2}_{2n+1}$, $\ell_1, \ell_2 > 1$, $\ell_1 \neq \ell_2$, the other cases being similar. We fix A having measure one w.r.t. $\mathcal{V}_1 \times \mathcal{V}_2$, and having measure zero w.r.t. $\mathcal{V}_2 \times \mathcal{V}_1$ towards a contradiction. We consider the partition $\mathcal{P}$: we partition pairs of function $F_1, F_2 : \boldsymbol{\delta}^1_{2n+1} \to \boldsymbol{\delta}^1_{2n+1}$ of the correct type, and ordered as in the partition of Lemma 1, according to whether or not $([F_1],[F_2]) \in A$.

We similarly consider the partition $\mathcal{P}'$: we partition $F_2, F_1 : \boldsymbol{\delta}^1_{2n+1} \to \boldsymbol{\delta}^1_{2n+1}$ as above, ordered as in the partition of Lemma 1 (with the roles of F_1, F_2 reversed), according to whether or not $([F_1],[F_2]) \in \neg A$.

It then follows from a sliding argument similar to that of Lemma 1 that on the homogeneous side of these partitions, the properties stated in $\mathcal{P}$, $\mathcal{P}'$ hold.

We fix a c.u.b. $C \subseteq \boldsymbol{\delta}^1_{2n+1}$ homogeneous for $\mathcal{P}$, $\mathcal{P}'$, and fix $F_1, F_2 : \boldsymbol{\delta}^1_{2n+1} \to C$ of the correct type, and ordered as in $\mathcal{P}$. Hence $([F_1],[F_2]) \in A$. We then let C' be the set of closure points of a c.u.b. subset of C consisting of limit ordinals, closed under J_{v_1}, J_{v_2} (where $v_1 = v(\mathcal{V}_1)$, $v_2 = v(\mathcal{V}_2)$), and closed under $(F_1)_0$, $(F_2)_0$ (as in Lemma 2). It then follows from a sliding agreement as in Lemma 1 (using the fact that cof $\theta_{v_1} \neq$ cof θ_{v_2}) that there are F_2', F_1' such that

1) $F_2' = F_2$, $F_1' = F_1$ almost everywhere.

2) F_2', F_1' have range a subset of C and are of the correct type.

3) F_2', F_1' are ordered as in $\mathcal{P}'$.

Hence it follows that $([F_1'],[F_2']) = ([F_1],[F_2]) \notin A$, a contradiction, which established Lemma 3.

A simple but frequently used lemma is the following:

Lemma 4. *For any c.u.b. subset $C \subseteq \boldsymbol{\delta}^1_{2n+1}$, there is a c.u.b. $C' \subseteq C$ such that for any measure V on Ξ_{2n+1} = the predecessor of $\boldsymbol{\delta}^1_{2n+1}$, and any $f : \theta_v \to C'$, if f is of the correct type almost everywhere with respect to v (i.e. there is a measure one set A w.r.t. v such that $f \restriction A$ is of the correct type), then there is an $f_2 : \theta_v \to C$ such that $f_2 = f_1$ almost everywhere w.r.t. v, and such that f_2 is of the correct type everywhere.*

Proof. Given C, take C' to be the set of closure points of a c.u.b. subset of C consisting of limit ordinals. Then, for $f : \theta_v \to C'$, and $A \subseteq \theta_v$ of measure one such that $f \restriction A$ is of the correct type, define f_2 by:

1) for $\alpha \in A$, set $f_2(\alpha) = f(\alpha)$.

2) for $\alpha \notin A$, set $f_2(\alpha) = N_c(\sup_{\alpha' < \alpha} F_2'(\alpha))$. It follows readily that f_2 has the required properties.

III. A Global Embedding Theorem. We let τ denote an $n - *$ tuple with respect to the measures in $\bigcup_{m \leq n} S_{2m+1} \cup \bigcup_{m \leq n} W_{2m+1}$, and let $<^\tau$ denote the corresponding ordering on tuples of integers and ordinals $< \Xi_{2n+3}$. Since we are assuming I_{2n+1}, $\mu_{2n+3}^{<^\tau}$ is defined.

We state the main theorem of this section:

Theorem. *For each $n - *$ tuple τ and corresponding measure $\mu_{2n+3}^{<^\tau}$, there is an integer m such that the ultrapower of $\boldsymbol{\delta}_{2n+3}^1$ by $\mu_{2n+3}^{<^\tau}$ is $\leq$ the ultrapower of $\boldsymbol{\delta}_{2n+3}^1$ by W_{2n+3}^m.*

In order to prove this, we require an inductive hypothesis:

$\mathcal{B}_{2n+1}$: We let T be on $n - *$ tuple and $<^T$ the corresponding ordering on tuples $< \Xi_{2n+1}$. Then there is an integer m and measure $S_{2n-1}^{\ell',m}$, where $\ell' = 2^n - 1$ is maximal, such that $K_{2n-1}^{\ell',m} = \theta_{S_{2n-1}^{\ell',m}} >$ the order type of $<^T$, and there is a measure u on Ξ_{2n+1} satisfying the following: there is a family of maps f_α, for $\alpha \in \theta_u$, $f_\alpha : <^T \rightarrow K_{2n-1}^{\ell',m}$ defined almost everywhere with respect to the measures $v^{(i_1,\ldots,i_k)}$ (so actually f_α is an equivalence class w.r.t. $\mu^{<^T}$), and satisfying

1) f_α is order-preserving almost everywhere w.r.t. the $v^{(i_1,\ldots,i_k)}$.

2) If $A \subseteq K_{2n-1}^{\ell',m}$ has measure one, then for almost all α w.r.t. u, and all indices $(i_1,\ldots,i_k)$ $\in$ dom $<^T$, for almost all $\beta < \theta^{(i_1,\ldots,i_k)}$ w.r.t. $v^{(i_1,\ldots,i_k)}$ we have that $f_\alpha^{(i_1,\ldots,i_k)}(\beta) \in A$, where $f_\alpha^{(i_1,\ldots,i_k)}$ is the induced function defined by $f_\alpha^{(i_1,\ldots,i_k)}(\beta) = f_\alpha(\beta^{(i_1,\ldots,i_k)})$. This is well defined.

2) for almost all α, $\sup_{\text{a.e.}} f_\alpha < K_{2n-1}^{\ell',m}$. Here, of course $\sup_{\text{a.e.}} f_\alpha$ denotes $\max_{(i_1,\ldots,i_k)} (\sup_{\text{a.e.w.r.t. } v^{(i_1,\ldots,i_k)}} f_\alpha^{(i_1,\ldots,i_k)})$.

3) if $S_{2n-1}^{\ell',m}$ satisfies the above, then so does $S_{2n-1}^{\ell',m'}$ for any $m' > m$.

We have defined $\mathcal{B}_{2n+3}$ as well.

We first show that the theorem follows from $\mathcal{B}_{2n+3}$. We produce an embedding M from the ultrapower of $\boldsymbol{\delta}_{2n+3}^1$ by $\mu_{2n+3}^{<^\tau}$ into the ultrapower by W_{2n+3}^m. We let $(I_1,\ldots,I_k,\ldots)$ denote the indices in the domain of τ. We let F_1 from tuples $(\ldots,\alpha_{(I_1,\ldots,i_k)},\ldots) \rightarrow \boldsymbol{\delta}_{2n+3}^1$ be given with respect to $\mu_{2n+3}^{<^T}$, and define $M([F_1])$ to be the ordinal represented with respect to W_{2n+3}^m by F_2 defined as follows: given $f : K_{2n+1}^{\ell,m} \rightarrow \boldsymbol{\delta}_{2n+3}^1$ (where $\ell = 2^{n+1} - 1$ is maximal),

$F_2([f])$ is represented w.r.t. the measure $\mathcal{U}$ (corresponding to u as in $\mathcal{B}_{2n+1}$) by the function which assigns to $\gamma \in \theta_{\mathcal{U}}$ the ordinal $F_1(\ldots,\beta_{(I_1,\ldots,I_k)},\ldots)$, where $\beta_{(I_1,\ldots,I_k)}$ is represented w.r.t. $V^{(I_1,\ldots,I_k)}$ (as in the $n-*$ suple τ) by the function $(f \circ \mathcal{F}_\gamma)^{(I_1,\ldots,I_k)}$, where F_γ is the map corresponding to γ as in $\mathcal{B}_{2n+3}$, and the superscript denote the induced map. It follows readily from $\mathcal{B}_{2n+3}$ that this is well defined.

We assume $\mathcal{B}_{2n+1}$ throughout the remainder of this section, and proceed to establish $\mathcal{B}_{2n+3}$. We fix an $n-*$ tuple τ, and corresponding ordering $<^\tau$ on Ξ_{2m+3}. We let $(I_1,\ldots,I_k)$, $V^{(I_1,\ldots,I_k)}$ denote induces and measures corresponding to τ, for $k \leq n$, $k \leq n-1$ respectively. We have that $V^{(I_1,\ldots,I_k)} = \prod_r {}^rV^{(I_1,\ldots,I_k)}$, where ${}^rV^{(I_1,\ldots,I_k)}$ is of the form $S_{2n+1^rC(I_1,\ldots,I_k)}^{{}^rT((I_1,\ldots,I_k)}$, for some $T(I_1,\ldots,I_k)$, $C(I_1,\ldots,I_k)$, or $\mu_{2n+1}^{<^rT(I_1,\ldots,I_k)}$, or a measure ${}^rv^{(I_1,\ldots,I_k)} \in \bigcup_{m\leq n} \mathcal{R}_{2m-1}$.

From $\mathcal{B}_{2n+1}$, there is an m and measures ${}^ru^{(I_1,\ldots,I_k)}$, corresponding to each ${}^rT(I_1,\ldots,I_k)$ in the first case above, such that the statement of $\mathcal{B}_{2n+1}$ is satisfied by $<^{{}^rT(I_1,\ldots,I_k)}$, $S_{2n-1}^{\ell',m}$ (here $\ell' = 2^n - 1$), and ${}^ru^{(I_1,\ldots,I_k)}$. We fix such an m.

We consider $S_{2n+1}^{\ell,m}$, where $\ell = 2^{n+1} - 1$ is maximal, and claim that $\mathcal{B}_{2n+3}$ is satisfied by $<^\tau$ and $S_{2n+1}^{\ell,m}$. We may also take any $m' > m$ in the following.

We define the measure $\mathcal{U}$ as in the statement of $\mathcal{B}_{2n+3}$:

We let Ω be the set of tuples of ordinals and integers of the form:

$$\langle \alpha_0, I_1, \ldots, {}^r\alpha_1^{(I_1,\ldots,I_k)}, \ldots, I_2, \ldots {}^r\alpha_2^{(I_1,\ldots,I_k)}, \ldots, I_{n-1}, \ldots, {}^r\alpha_{\alpha_n}^{(I_1,\ldots,I_k)}, \ldots, I_n\rangle,$$

or an initial segment of such a sequence, where:

1) $\alpha_0 < \boldsymbol{\delta}_{2n+1}^1$

2) the indices $(I_1,\ldots,I_k)$ correspond to indices in the domain of τ.

3) the indices $(I_1,\ldots,I_k)$ associated with ${}^r\alpha_J$ correspond to the indices $(I_1,\ldots,I_k)$ in ${}^rC(I_1,\ldots,I_J)$ if ${}^rV^{(I_1,\ldots,I_J)} \in \mathcal{S}_{2n+1}$, and to the indices $(i_1,\ldots,i_k) \in \operatorname{dom} {}^rT(I_1,\ldots,I_J)$ if ${}^rV^{(I_1,\ldots,I_J)} \in W_{2n+1}$. If ${}^rV^{(I_1,\ldots,I_J)} \in \bigcup_{m<n} \mathcal{R}_{2m+1}$, we omit these superscripts.

4) a) for fixed r, j, the indices $(I_1,\ldots,I_k)$ associated with ${}^r\alpha_J$ are ordered as in the fixed (but arbitrary) ordering on them which we use to identify the measure space of $S_{2n+1^rC(I_1,\ldots,I_J)}^{{}^rT(I_1,\ldots,I_J)}$ with an ordinal.

b) for $r \neq r'$, ${}^r\alpha_J^{(I_1,\ldots,I_k)}$ precedes ${}^{r'}\alpha_J^{(I_1',\ldots,I_k')}$ if in identifying $V^{(I_1,\ldots,I_J)} = \prod_r {}^rV^{(I_1,\ldots,I_J)}$ with an ordinal, we order by the rth component before the r'th component.

5) ${}^r\alpha_J^{(I_1,\ldots,I_k)} < J_{{}^ru^{(I_1,\ldots,I_J)}}(\alpha_0)$ for ${}^rV^{(I_1,\ldots,I_k)} \in \mathcal{S}_{2n+1}$, and otherwise ${}^r\alpha_J^{(I_1,\ldots,I_k)} < \alpha_0$.

We let $<^{\Omega}$ denote the Kleene-Brouwer ordering on these tuples.

For indices $(I_1, \ldots, I_k) \in \operatorname{dom} \tau$, with $k > 1$, we define a set $M(I_1, \ldots, I_k)$ of indexed ordinals and an ordering $<^{M(I_1,\ldots,I_k)}$ on them as follows:

$M(I_1, \ldots, I_k)$ consists of indexed ordinals of the form $r_{\theta(I_1,\ldots,I_\ell)}^{(I_1,\ldots,I_m)}$ where $(I_1, \ldots, I_\ell)$ is an initial segment of $(I_1, \ldots, I_k)$, $(I_1, \ldots, I_m)$ is an index in ${}^{r}\mathcal{C}(I_1, \ldots, I_\ell)$ and $\theta \in {}^{r}u^{(I_1,\ldots,I_\ell)}$, for ${}^{r}V^{(I_1,\ldots,I_k)} \in \mathcal{S}_{2n+1}$; and of the form $r_{\theta(I_1,\ldots,I_\ell)}^{(i_1,\ldots,i_k)}$, where $(I_1, \ldots, I_\ell)$ is an initial segment of $(I_1, \ldots, I_k)$, $(i_1, \ldots, i_k) \in \operatorname{dom} {}^{r}T(I_1, \ldots, I_\ell)$ and $\theta \in r_V(i_1, \ldots, i_k)$ if ${}^{r}V^{(I_1,\ldots,I_k)} \in W_{2n+1}$, together with ordinals $\theta < \theta(K_{2n-1}^{\ell',m})$.

We define ${}^{r_1}\theta_{1(I_1^1,\ldots,I_\ell^1)}^{(I_1^1,\ldots,I_{m_1}^1)} <^{M(I_1,\ldots,I_k)} {}^{r_2}\theta_{2(I_1^2,\ldots,I_\ell^2)}^{(I_1^2,\ldots,I_{m_2}^2)}$ to hold provided one of the following is satisfied:

1) $r_1 \neq r_2$, ${}^{r_1}V^{(I_1^1,\ldots,I_{\ell_1}^1)}$, ${}^{r_2}\theta^{(I_1^2,\ldots,I_{\ell_2}^2)} \in \mathcal{S}_{2n+1}$. In this case, we require that there is a c.u.b. $C \subseteq \boldsymbol{\delta}_{2n+1}^1$ such that for $\mathcal{F} : <_{{}^{r}C(I_1,\ldots,I_\ell)}^{{}^{r}T(I_1,\ldots,I_\ell)} \rightarrow C$ of the correct type with $r = r_1$ or r_2, representing a pair $([{}^{r_1}F], [{}^{r_2}F])$ in the product measure ${}^{r_1}V^{(I_1,\ldots,I_\ell)} \times {}^{r_2}V^{(I_1,\ldots,I_\ell)}$ (for $r_1 < r_2$, and otherwise representing $([{}^{r_2}F], [{}^{r_1}F])$ w.r.t. ${}^{r_2}V \times {}^{r_1}V$), we have that for almost all $f : K_{2n-1}^{\ell',m} \rightarrow \boldsymbol{\delta}_{2n+1}^1$ of the correct type w.r.t. $\mu_{2n+1}^{\ell',m}$ that if $f_1 : <^{{}^{r_1}T(I_1,\ldots,I_{\ell_1})} \rightarrow K_{2n-1}^{\ell',m}$, $f_2 : <^{{}^{r_2}T(I_1,\ldots,I_{\ell_2})} \rightarrow K_{2n-1}^{\ell',m}$, denote the embeddings (determined a.e.) from $\mathcal{B}_{2n+1}$ corresponding to ${}^{r_1}u^{(I_1,\ldots,I_{\ell_1})}$, ${}^{r_2}u^{(I_1,\ldots,I_{\ell_2})}$, and if $g_1 = f \circ f_1$, $g_2 = f \circ f_2$, then ${}^{r_1}F(\langle \mathcal{S}_1 \rangle) < {}^{r_2}F(\langle \mathcal{S}_2 \rangle)$, where $\mathcal{S}_1$ is the tuple corresponding to $g_1 : <^{{}^{r}T(I_1,\ldots,I_{\ell_1})} \rightarrow \boldsymbol{\delta}_{2n+1}^1$, and $(I_1^1, \ldots, I_{m_1}^1)$ as in the definition of ${}^{r_1}V$, and similarly for $\mathcal{S}_2$.

2) $r_1 = r_2$ and ${}^{r_1}V^{(I_1,\ldots,I_k)} \in \mathcal{S}_{2n+1}$ (hence so are ${}^{r_1}V^{(I_1,\ldots,I_{\ell_1})}$ and ${}^{r_1}V^{(I_1,\ldots,I_{\ell_2})}$). We then require that there is a c.u.b. $C \subseteq \boldsymbol{\delta}_{2n+1}^1$ such that for $F : <_{{}^{r}C(I_1,\ldots,I_k)}^{{}^{r}T(I_1,\ldots,I_k)} \rightarrow C$ of the correct type, with induced functions, $F_1 : <_{{}^{r}C(I_1,\ldots,I_{\ell_1})}^{{}^{r}T(I_1,I_{\ell_1})} \rightarrow C$ and $F_2 : <_{{}^{r}C(I_1,\ldots,I_{\ell_2})}^{{}^{r}T(I_1,\ldots,I_{\ell_2})} \rightarrow C$, that for almost all $f : K_{2n-1}^{\ell',m} \rightarrow \boldsymbol{\delta}_{2n+1}^1$, and f_1, f_2, g_1, g_2 as above, $F_1(\langle \mathcal{S}_1 \rangle) < F_2(\langle \mathcal{S}_2 \rangle)$, where $\mathcal{S}_1$ is the sequence corresponding to g_1, $(I_1', \ldots, I_{m_1}')$, and similarly for $\mathcal{S}_2$.

3) ${}^{r_1}V \in W_{2n+1}$ and ${}^{r_2}V \in \mathcal{S}_{2n+1}$.

4) $r_1 = r_2$, ${}^{r_1}V$, ${}^{r_2}V \in \mathcal{W}_{2n+1}$. We then require that for almost all $f : <^{{}^{r}T(I_1,\ldots,I_k)} \rightarrow \boldsymbol{\delta}_{2n+1}^1$ w.r.t. $\mu_{2n+1}^{{}^{r}T(I_1,\ldots,I_k)}$ with induced functions $f_1 : <^{{}^{r}T(I_1,\ldots,I_{\ell_1})} \rightarrow \boldsymbol{\delta}_{2n+1}^1$, $f_2 : <^{{}^{r}T(I_1,\ldots,I_{\ell_2})} \rightarrow \boldsymbol{\delta}_{2n+1}^1$, and induced functions $f_1^{(i_1^1,\ldots,i_{k_1}^1)} : \theta_{v(i_1,\ldots,i_{k_1-1})} \rightarrow \boldsymbol{\delta}_{2n+1}^1$, and $f_2^{(i_1^2,\ldots,i_{k_2}^2)}$, that $f_1^{(i_1^1,\ldots,i_{k_1}^1)}(\theta_1) < f_2^{(i_1^2,\ldots,i_{k_2}^2)}(\theta_2)$.

5) $r_1 \neq r_2$, ${}^{r_1}V$, ${}^{r_2}V \in W_{2n+1}$. In this case, we require $r_1 < r_2$.

6) ${}^{r_1}V \in W_{2n+1}$ and $\theta_2 \in K_{2n-1}^{\ell',m}$.

7) $\theta_1, \theta_2 \in K^{\ell',m}_{2n-1}$ and $\theta_1 < \theta_2$.

8) ${}^{r_1}V \in \mathcal{S}_{2n+1}$ and $\theta_2 \in K^{\ell',m}_{2n-1}$. We require that there is a c.u.b. $C \subseteq \delta^1_{2n+1}$ such that for $F : <^{r_T(I_1,\dots,I_{\ell_1})}_{r_C(I_1,\dots,I_{\ell_1})} \to C$ of the correct type we have that for almost all $f : K^{\ell',m}_{2n-1} \to \delta^1_{2n+1}$ of the correct type, with f_1, g_1 as above, that $F(\langle S \rangle) < f(\theta_2)$, where S corresponds to g, $(I^1_1, \dots, I^1_{m_1})$ as in the definition of ${}^{r_1}V$.

9) $\theta_1 \in K^{\ell',m}_{2n-1}$ ${}^{r_2}V \in \mathcal{S}_{2n+1}$. Similar to 8) where we require that $f(\theta_1) < F(\langle S \rangle)$.

We identify $\theta^{()}_1$, $\theta^{()}_2$ in $<^{M(I_1,\dots,I_k)}$ if neither $\theta_1 <^M \theta_2$ or $\theta_2 <^M \theta_1$ holds.

The measure $\mathcal{U}$ is a measure on tuples $(\dots, \eta_{(I_1,\dots,I_k)}, \dots)$ indexed by indices in τ, induced by the strong partition relation on δ^1_{2n+1} functions $H : <^\Omega \to \delta^1_{2n+1}$ of the correct type, and the measures $\prod_r {}^r v \times \mu^{<M(I_1,\dots,I_k)}_{2n+1}$ for $k > 1$, and $\mu^{K^{\ell',m}_{2n-1}}_{2n+1}$ for $k = 1$. Here $\Pi_r\, {}^r v$ denotes the subproduct of V consisting of those measures in $\cup_{m<n} \mathcal{R}_{2m+1}$.

That is, A has measure one if there is a c.u.b. subset C of δ^1_{2n+1} such that for $H : <^\Omega \to C$ of the correct type, $(\dots, \eta_{(I_1,\dots,I_k)}, \dots) \in A$, where for fixed $(I_1, \dots, I_k)$, $\eta_{(I_1,\dots,I_k)}$ is represented w.r.t. $\prod_r {}^r v \times \mu^{<M(I_1,\dots,I_k)}_{2n+1}$ by the function H_2 which assigns to the pair of tuples $(\theta_{r_1}, \theta_{r_2}, \dots, \theta_{r_p})$, $(\gamma_0, \dots, {}^r\gamma^{(I_1,\dots,I_m)}_{(I_1,\dots,I_\ell)}, \dots)$ (where an $h : <^{M(I_1,\dots,I_k)} \to \delta^1_{2n+1}$ represents the γ's w.r.t. the measures $r^{(I_1,\dots,I_\ell)}_u$ if ${}^rV^{(I_1,\dots,I_\ell)} \in \mathcal{S}_{2n+1}$, w.r.t. $v^{(i_1,\dots,i_k)}$ if ${}^rV \in \mathcal{W}_{2n+1}$, and w.r.t. $S^{\ell',m}_{2n-1}$ for γ_0) the ordinal $H(\langle T \rangle)$, where T denotes the sequence

$$T = \langle \alpha_0, I_1, \dots, {}^r\alpha^{(I_1,\dots,I_m)}_1, \dots, I_J, \dots, {}^r\alpha^{(I_1,\dots,I_m)}_J, \dots, \rangle$$

as in the definition of Ω, where:

1) $\alpha_0 = \gamma_0$

2) ${}^r\alpha^{(I_1,\dots,I_m)}_J = {}^r\gamma^{(I_1,\dots,I_m)}_{(I_1,\dots,I_J)}$ if ${}^rV \in \mathcal{S}_{2n+1}$

3) ${}^r\alpha^{(i_1,\dots,i_k)}_J = {}^r\gamma^{(i_1,\dots,i_k)}_{(I_1,\dots,i_J)}$ if ${}^rV \in W_{2n+1}$

4) ${}^r\alpha_J = \Pi^{(I_1,\dots,I_k)}_{(I_1,\dots,I_J)}(\theta_r)$ if ${}^rV^{(I_1,\dots,I_k)} \in \bigcup_{m<n} \mathcal{R}_{2m+1}$.

This completes the definition of the measure $\mathcal{U}$.

We proceed to define the family of embeddings $\mathcal{F}_\gamma$ corresponding to $\mathcal{U}$ as in the statement of $\mathcal{B}_{2n+3}$.

We fix $\gamma \in \theta_\mathcal{U}$ which is represented w.r.t. the measures $\prod_r {}^r v \times \mu^{M(I_1,\dots,I_k)}_{2n+1}$ by a function $H : <^\Omega \to \delta^1_{2n+1}$ of the correct type (which happens for almost all γ w.r.t. $\mathcal{U}$). We proceed to define $\mathcal{F}_\gamma : <^\tau \to K^{\ell,m}_{2n+1}$. We fix an element $\alpha^{(I_1,\dots,I_k)}$ in the domain of $<^\tau$, where

$(I_1,\ldots,I_k) \in \operatorname{dom} \tau$, and define $\mathcal{F}_\gamma(\alpha^{(I_1,\ldots,I_k)})$ (for almost all α w.r.t. $V^{(I_1,\ldots,I_{k-1})}$ for $k > 1$). We see that $\alpha = (\ldots, {}^r\xi, \ldots, {}^r\xi_{(i_1,\ldots,i_\ell)}, \ldots, {}^r\xi_{(I_1,\ldots,I_m)}, \ldots)$, the components ${}^r\xi$ corresponding to ${}^rV \in \bigcup_{m<n} \mathcal{R}_{2m+1}$, ${}^r\xi_{(i_1,\ldots,i_k)}$ to ${}^rV \in \mathcal{W}_{2n+1}$, and ${}^r\xi_{(I_1,\ldots,I_m)}$ to $V \in \mathcal{S}_{2n+1}$. In the latter two cases, we have functions ${}^rF : <^{{}^rT(I_1,\ldots,I_k)} \to \boldsymbol{\delta}^1_{2n+1}$, and ${}^rF : <^{{}^rT(I_1,\ldots,I_\ell)}_{{}^rC(I_1,\ldots,I_k)} \to \boldsymbol{\delta}^1_{2n+1}$ representing, ${}^r\xi_{(i_1,\ldots,i_k)}$, ${}^r\xi_{(I_1,\ldots,I_m)}$ w.r.t. the measures $v^{(i_1,\ldots,i_k)}$ and the measure $\mu_{2n+1}^{<^{{}^rT(I_1,\ldots,I_k)}}$ respectively. We define a function $G : \boldsymbol{\delta}^1_{2n+1} \to \boldsymbol{\delta}^1_{2n+1}$ representing $\mathcal{F}_\gamma(\alpha^{(I_1,\ldots,I_k)})$ w.r.t. $\mu_{2n+1}^{S^{\ell',m}_{2n-1}}$. We let $g : K^{\ell',m}_{2n-1} \to \boldsymbol{\delta}^1_{2n+1}$ of the correct type be given, and define $G([g])$. We set $G([g]) = H(\langle T \rangle)$, where $T = \langle \beta_0, I_1, \ldots, {}^r\beta_1^{(I_1,\ldots,I_m)}, \ldots, I_J, \ldots, {}^r\beta_J^{(I_1,\ldots,I_m)}, \ldots, I_k \rangle$ is as in the definition of Ω, and the β's are given by:

1) $\beta_0 = [g]_{S^{\ell',m}_{2n-1}}$

2) If ${}^rV^{(I_1,\ldots,I_J)} \in \bigcup_{m<n} \mathcal{R}_{2m+1}$, then ${}^r\beta_J = \Pi^{(I_1,\ldots,I_k)}_{(I_1,\ldots,I_J)}({}^r\xi)$.

3) If ${}^rV^{(I_1,\ldots,I_J)} \in \mathcal{W}_{2n+1}$, then ${}^r\beta_J^{(i_1,\ldots,i_k)} = {}^r\xi^{(i_1,\ldots,i_k)}$.

4) If ${}^rV^{(I_1,\ldots,I_J)} \in \mathcal{S}_{2n+1}$, we represent ${}^r\beta_J^{(I_1,\ldots,I_m)}$ (where we recall $(I_1,\ldots,I_m)$ is an index in ${}^r\mathcal{C}(I_1,\ldots,I_J)$), w.r.t. ${}^ru^{(I_1,\ldots,I_J)}$ by a function ${}^r\mathcal{M}_J^{(I_1,\ldots,I_m)}$ defined by: for θ in the measure space ${}^ru^{(I_1,\ldots,I_J)}$, and $f_\theta : <^{{}^rT(I_1,\ldots,I_J)} \to K^{\ell',m}_{2n-1}$ the corresponding embedding from $\mathcal{B}_{2n+1}$ (defined almost everywhere), and $\mathcal{Q} = g \circ f_\theta : <^{{}^rT(I_1,\ldots,I_J)} \to \boldsymbol{\delta}^1_{2n+1}$, we set ${}^r\mathcal{M}_J^{(I_1,\ldots,I_m)}(\theta) = {}^rF(\langle \mathcal{S}(\mathcal{Q}, (I_1,\ldots,I_J), (I_1,\ldots,I_m)) \rangle)$, where $\mathcal{S}(\,)$ denotes the corresponding sequence as in the definition of ${}^rV$. This is well-defined for almost all g. We use here the fact that the $\theta \in K^{\ell',m}_{2n-1}$ are cofinal in the ordering $<^{M(I_1,\ldots,I_k)}$.

This completes the definition of $\mathcal{F}_\gamma$. To show that this is well-defined, we verify the following:

1) For fixed H, $({}^{r_1}\xi, \ldots, {}^{r_p}\xi)$, ${}^rF$, g; ${}^r\mathcal{M}_J^{(I_1,\ldots,I_m)}(\theta)$ is well defined. This follows since f_θ is determined almost everywhere w.r.t. the $v^{(i_1,\ldots,i_k)}$, hence so is $\mathcal{Q}$ above, hence the sequence $\mathcal{S}$ above is well-defined.

2) For fixed H, $({}^{r_1}\xi, \ldots, {}^{r_p}\xi)$, ${}^rF$, $\mathcal{F}_\gamma(\alpha^{(I_1,\ldots,I_k)})$ is well defined with respect to $\mu_{2n+1}^{S^{\ell',m}_{2n-1}}$. If $g_1, g_2 : K^{\ell',m}_{2n-1} \to \boldsymbol{\delta}^1_{2n+1}$ agree on a measure one set A w.r.t. $S^{\ell',m}_{2n-1}$ then for any $(I_1,\ldots,I_J)$ as above, it follows from $\mathcal{B}_{2n+1}$ that for almost all θ w.r.t. ${}^ru^{(I_1,\ldots,I_J)}$ that $f_\theta : <^{{}^rT(I_1,\ldots,I_J)} \to K^{\ell',m}_{2n-1}$ has range a.e. in A w.r.t. the measures $v^{(i_1,\ldots,i_k)}$, where $(i_1,\ldots,i_k) \in \operatorname{dom} {}^rT(I_1,\ldots,I_J)$. We are assuming ${}^rV^{(i_1,\ldots,i_k)} \in \mathcal{S}_{2n+1}$, the other cases being immediate. Hence, for $\mathcal{Q}_1 = g_1 \circ f_\theta$, $\mathcal{Q}_2 = g_2 \circ f_\theta$, for each $(I_1,\ldots,I_m)$ in ${}^r\mathcal{C}(I_1,\ldots,I_J)$, it follows that $\mathcal{S}(Q_1, (I_1,\ldots,I_J)(I_1,\ldots,I_m))$ $= \mathcal{S}(Q_2, (I_1,\ldots,I_J), (I_1,\ldots,I_m))$, and hence ${}^r\mathcal{M}_J^{(I_1,\ldots,I_m)}(1)(\theta) = {}^r\mathcal{M}_J^{(I_1,\ldots,I_m)}(2)(\theta)$ for almost all θ w.r.t. ${}^ru^{(I_1,\ldots,I_J)}$, and all $(I_1,\ldots,I_m)$. Hence, ${}^r\beta_J^{(I_1,\ldots,I_m)}(1) = {}^r\beta_J^{(I_1,\ldots,I_m)}(2)$, and hence

the sequences T_1, T_2 corresponding to g_1, g_2 are the same, hence $G([g_1]) = G([g_2])$.

3) For fixed $H : <^{\Omega}\to \delta^1_{2n+1}$, $\mathcal{F}_\gamma$ is well defined w.r.t. the choice of the ${}^rF$ representing $\alpha^{(I_1,\ldots,I_k)}$, for ${}^rV^{(I_1,\ldots,I_k)} \in \mathcal{S}_{2n+1}$ or $\mathcal{W}_{2n+1}$. For each such r, we let ${}^rF_1, {}^rF_2 : <^{{}^rT(I_1,\ldots,I_k)}_{{}^rC(I_1,\ldots,I_k)}\to \delta^1_{2n+1}$ or from $<^{{}^rT(I_1,\ldots,I_k)}\to \delta^1_{2n+1}$ be given, representing ${}^r\alpha^{(I_1,\ldots,I_k)}$. We let C be a c.u.b. subset of δ^1_{2n+1} such that for r's of the first type, $f : <^{{}^rT(I_1,\ldots,I_k)}\to C$ of the correct type, $(\mathcal{I}_1,\ldots,\mathcal{I}_m)$ an index in ${}^r\mathcal{C}(I_1,\ldots,I_k)$, and $\mathcal{S} = \mathcal{S}(f,(I_1,\ldots,I_k),(\mathcal{I}_1,\ldots,\mathcal{I}_m))$ the corresponding sequence, ${}^rF_1(\langle\mathcal{S}\rangle) = {}^rF_2(\langle\mathcal{S}\rangle)$. For $g : K^{\ell',m}_{2n-1} \to C$ of the correct type (where we also assume C is sufficiently closed w.r.t. the F's so the definition of $\mathcal{F}_\gamma$ makes sense), $J \leq k$, $(\mathcal{I}_1,\ldots,\mathcal{I}_m)$ an index in ${}^r\mathcal{C}(I_1,\ldots,I_J)$ (we consider here the case ${}^rV^{(I_1,\ldots,I_J)} \in \mathcal{S}_{2n+1}$) it follows that for almost all θ w.r.t. ${}^ru^{(I_1,\ldots,I_J)}$ that $Q = g \circ f_\theta$ has range almost everywhere in C and is of the correct type almost everywhere, since g is. By Lemma 4 we may assume Q has range in C and is of the correct type everywhere. Hence, for almost all θ, ${}^r\mathcal{M}_J^{(\mathcal{I}_1,\ldots,\mathcal{I}_m)}(1)(\theta) = {}^rF_1(\langle\mathcal{S}(Q,(I_1,\ldots,I_J),(\mathcal{I}_1,\ldots,\mathcal{I}_m))\rangle) = {}^rF_2(\langle\mathcal{S}(Q,(I_1,\ldots,I_J),(\mathcal{I}_1,\ldots,\mathcal{I}_m))\rangle) = {}^r\mathcal{M}_J^{(\mathcal{I}_1,\ldots,\mathcal{I}_m)}(2)(\theta)$. Hence, the ${}^r\beta$'s appearing in the sequence T are the same for ${}^rF_1, {}^rF_2$ in the case ${}^rV \in \mathcal{S}_{2n+1}$, and the other cases are immediate. Hence $G([g])_{F_1} = G([g])_{F_2}$.

4) If $H_1, H_2 : <^{\Omega}\to \delta^1_{2n+1}$ of the correct type agree a.e. w.r.t. $\mu^{<M(I_1,\ldots,I_k)}_{2n+1}$, then the embeddings $\mathcal{F}_1, \mathcal{F}_2 : <^{\tau}\to K^{\ell,m}_{2n+1}$ corresponding to H_1, H_2 agree, for each fixed $(I_1,\ldots,I_k) \in$ dom τ, almost everywhere w.r.t. $V^{(I_1,\ldots,I_{k-1})} = \prod_r {}^rV^{(I_1,\ldots,I_{k-1})}$, if $k > 1$, and agree if $k = 1$.

We fix a c.u.b. $C \subseteq \delta^1_{2n+1}$ such that for $(I_1,\ldots,I_k) \in$ dom τ, and the corresponding ordering $<^{M(I_1,\ldots,I_k)}$, if $h : <^{M(I_1,\ldots,I_k)}\to C$ is of the correct type, then for all $({}^{r_1}\xi,\ldots,{}^{r_p}\xi) \in A$, a measure one set w.r.t. $\Pi_r {}^rv$, $H_1(\langle T\rangle) = H_2(\langle T\rangle)$, where $T = \langle\beta_0, I_1,\ldots,I_J,\ldots, {}^r\beta_J^{(\mathcal{I}_1,\ldots,\mathcal{I}_m)},\ldots, I_k\rangle$ is the sequence corresponding to h as in the definition of $\mathcal{U}$.

For $k=1$ we require that for $h{:}K^{\ell',m}_{2n-1} \to C$, of the correct type $H_1(\langle[g], I_1\rangle) = H_2(\langle[g], I_1\rangle)$.

For each $\theta_1^{()}, \theta_2^{()} \in$ dom $<^{M(I_1,\ldots,I_k)}$, hence $\theta_1^{()} = {}^r\theta^{(\mathcal{I}_1,\ldots,\mathcal{I}_{m_1})}_{1(I_1,\ldots,I_{J_1})}$ or ${}^r\theta^{(i_1,\ldots,i_k)}_{1(I_1,\ldots,I_{J_1})}$ or $\theta_1 < K^{\ell',m}_{2n-1}$, it follows that there is a c.u.b. $C_{\theta_1,\theta_2} \subseteq \delta^1_{2n+1}$ such that for ${}^rF_1 : <^{{}^rT(I_1,\ldots,I_k)}_{{}^rC(I_1,\ldots,I_k)}\to C_{\theta_1,\theta_2}$ or : $<^{{}^rT(I_1,\ldots,I_k)}\to C_{\theta_1,\theta_2}$ of the correct type representing components of $\alpha^{(I_1,\ldots,I_k)}$ in the measure space $\Pi_r\ {}^rV^{(I_1,\ldots,I_k)}$ for ${}^rV \in \mathcal{W}_{2n+1} \cup \mathcal{S}_{2n+1}$, the following are satisfied:

1) a) If ${}^{r_1}V, {}^{r_2}V \in \mathcal{S}_{2n+1}$, and ${}^{r_1}\theta^{(\mathcal{I}^1_1,\ldots,\mathcal{I}^1_{m_1})}_{1(I^1_1,\ldots,I^1_{\ell_1})} <^{M(I_1,\ldots,I_k)} {}^{r_2}\theta^{(\mathcal{I}^2_1,\ldots,\mathcal{I}^2_{m_2})}_{2(I^2_1,\ldots,I^2_{\ell_2})}$ then there is a c.u.b. $D_{\theta_1\theta_2} \subseteq \delta^1_{2n+1}$ (which depends on the ${}^rF$ as well) such that for $g : K^{\ell',m}_{2n-1} \to D$ of the correct type,

$$f_{\theta_1} : <^{{}^{r_1}T(I^1_1,\ldots,I^1_{\ell_1})} \to K^{\ell',m}_{2n-1},\quad f_{\theta_2} : <^{{}^{r_2}T(I^2_1,\ldots,I^2_{\ell_2})} \to K^{\ell',m}_{2n-1},$$

and the corresponding embeddings, $Q_1 = g \circ f_{\theta_1}$, $Q_2 = g \circ f_{\theta_2}$, then ${}^{r_1}F(\langle\mathcal{S}(Q_1,(I^1_1,\ldots,I^1_{\ell_1}),$

$(I_1^1,\ldots,I_{m_1}^1)\rangle)) < {}^{r_2}F(S\langle Q_2,(I_1^2,\ldots,I_{\ell_2}^2),(I_1^2,\ldots,I_{m_2}^2)\rangle)))$.

b) similar to a) with $\theta_1 >^M \theta_2$.

c) if neither $\theta_1 <^M \theta_2$ or $\theta_2 <^M \theta_1$ then we require that ${}^{r_1}F(\langle\,\rangle) = {}^{r_2}F(\langle\,\rangle)$.

2) a) If ${}^{r_1}V \in \mathcal{S}_{2n+1}$, $\theta_2 \in K_{2n-1}^{\ell',m}$, and ${}^{r_1}\theta_{1(\,)}^{(\,)} <^M \theta_2$, we proved as in 1a) where we require that ${}^{r_1}F(\langle\,\rangle) < g(\theta_2)$.

b) and c) similar to b, c above.

3a) similar to 2a), with $\theta_1 \in K_{2n-1}^{\ell',m}$, ${}^{r_2}V \in \mathcal{S}_{2n+1}$ and $\theta_1 <^M \theta_{2(\,)}^{(\,)}$.

b), c) as above.

4a) If ${}^{r_1}V$, ${}^{r_2}V \in \mathcal{W}_{2n+1}$ and

$${}^{r_1}\theta_{1(I_1^1,\ldots,I_{m_1}^1)}^{(i_1^1,\ldots,i_{k_1}^1)} <^M {}^{r_2}\theta_{2(I_1^2,\ldots,I_{m_2}^2)}^{(i_1^2,\ldots,i_{k_2}^2)}, \text{ then } {}^{r_1}F^{(i_1^1,\ldots,i_{k_1}^1)} < {}^{r_2}F^{(i_1^2,\ldots,i_{k_2}^2)}.$$

b), c) similar to above.

Since the $\theta_{r_u(I_1,\ldots,I_k)}$, $\theta_{r_v(I_1,\ldots,I_k)}$, $K_{2n-1}^{\ell',m} < \delta_{2n+1}^1$, it follows readily that there is a c.u.b. $C_2 \subseteq C$ such that for each $\theta_1 <^M \theta_2$, if the ${}^rF$'s are of the correct type into C_2 representing a tuple in the product space $V^{(I_1,\ldots,I_k)}$, then 1-4 above hold.

For

$$\alpha^{(I_1,\ldots,I_k)} = (\ldots,{}^r\xi,\ldots,{}^r\xi^{(i_1,\ldots,i_k)},\ldots,{}^r\xi_{(I_1,\ldots,I_\ell)}^{(I_1,\ldots,I_m)},\ldots)$$

where $({}^{r_1}\xi,\ldots,{}^{r_p}\xi) \in A$, and for r such that ${}^rV \in \mathcal{S}_{2n+1} \cup \mathcal{W}_{2n+1}$, the ${}^r\xi$'s are represented by ${}^rF : <_{rc(I_1,\ldots,I_k)}^{r_T(I_1,\ldots,I_k)} \to C_2$, or ${}^rF : <^{r_T(I_1,\ldots,I_k)} \to C_2$ of the correct type (and ordered as in the product measure), we claim that $\mathcal{F}_{H_1}(\alpha^{(I_1,\ldots,I_k)}) = \mathcal{F}_{H_2}(\alpha^{(I_1,\ldots,I_k)})$.

For such fixed $({}^{r_1}\xi,\ldots,{}^{r_p}\xi)$, ${}^rF$, it follows from the δ_{2n+1}^1-additivity of $\mu_{2n+1}^{K_{2n-1}^{\ell',m}}$ that there is a c.u.b. $D \subseteq \delta_{2n+1}^1$ such that for $g : K_{2n-1}^{\ell',m} \to D$ of the correct type, 1-4 above hold. We may also assume $D \subseteq C_2$, and that the least element of $D > \sup {}^rF$ for ${}^rV \in \mathcal{W}_{2n+1}$. It then follows that for $g : K_{2n-1}^{\ell',m} \to D$ of the correct type that the corresponding $h : <^{M(I_1,\ldots,I_k)} \to \delta_{2n+1}^1$ is order-preserving. Since the ${}^rF$ have range in C_2, and the ${}^rF$ are of the correct type, it also follows that h has range in C_2, and is of the correct type. Hence $\mathcal{F}_{H_1}(\alpha^{(I_1,\ldots,I_k)}) = H_1(\langle T\rangle) = H_2(\langle T\rangle) = \mathcal{F}_{H_2}(\alpha^{(I_1,\ldots,I_k)})$, where T is the sequence corresponding to h, $(I_1,\ldots,I_k)$.

For $k = 1$, this is immediate.

We now establish that for almost all γ w.r.t. $\mathcal{U}$, that $\mathcal{F}_\gamma$ is order-preserving almost everywhere from $<^r$ into $K_{2n+1}^{\ell,m}$. We fix $H : <^\Omega \to \delta_{2n+1}^1$ of the correct type representing γ

(which happens for almost all γ). We show that $\mathcal{F}_\gamma$ is order-preserving almost everywhere. We let $\alpha_1^{(I_1^1,\dots,I_{k_1}^1)}$, $\alpha_2^{(I_1^2,\dots,I_{k_2}^2)}$ be given, and fix $({}^{r_1}\xi_1,\dots,{}^{r_p}\xi_1)$, ${}^rF_1 : <_{{}^rC(I_1^1,\dots,I_{k_1}^1)}^{{}^rT(I_1^1,\dots,I_{k_1}^1)} \to \boldsymbol{\delta}_{2n+1}^1$ or : $<^{{}^rT(I_1^1,\dots,I_{k_1}^1)} \to \boldsymbol{\delta}_{2n+1}^1$ of the correct type representing the components of α_1 for ${}^rV \in \mathcal{S}_{2n+1} \cup \mathcal{W}_{2n+1}$, and similarly for α_2, which happens for almost all α_1, α_2. We assume that $\alpha_1^{(\,)} <^r \alpha_2^{(\,)}$ and show that $\mathcal{F}_\gamma(\alpha_1^{(\,)}) < \mathcal{F}_\gamma(\alpha_2^{(\,)})$. We let G_1, G_2 represent $\mathcal{F}_\gamma(\alpha_1^{(\,)})$, $\mathcal{F}_\gamma(\alpha_2^{(\,)})$ w.r.t. $\mu_{2n+1}^{K_{2n-1}^{\ell',m}}$ as in the definition of $\mathcal{F}_\gamma$. We show that for amost all $g : K_{2n-1}^{\ell',m} \to \boldsymbol{\delta}_{2n+1}^1$ of the correct type, $G_1([g]) < G_2([g])$. We consider the following cases:

1) $k_1 = 1$ and $k_2 \geq 1$. In this case (from the definition of $<^r$) $I_1^1 < I_1^2$. From the definition of $\mathcal{F}$ and the ordering $<^\Omega$, it follows that for almost all g, $G_1([g]) = H(\langle [g], I_1^1\rangle) < H(\langle [g], I_1^2, \dots, \beta\text{'s}, \dots, \rangle) = G_2([g])$, where the β's appear if $k_2 > 1$.

2) $k_1 > 1$ and $k_2 = 1$. Hence $I_1^1 \leq I_1^2$. If $I_1^1 < I_1^2$, we proceed as above. If $I_1^1 = I_1^2$, the result follows since for almost all g,

$$G_1([g]) = H(\langle T_1\rangle) = H(\langle [g], I_1^1, \dots, {}^r\beta_J^{(\,)}\rangle) < H(\langle [g], I_1^2\rangle) = H(\langle T_2\rangle),$$

since T_1 extends T_2.

3) $k_1, k_2 > 1$.

We let $\bar{\alpha}_1$ denote the sequence $\bar{\alpha}_1 = \langle I_1^1, \Pi_{(I_1)}^{(I_1,\dots,I_{k_1-1})}(\alpha_1), \dots, I_{k_1-1}^1, \alpha_1, I_{k_1}^1\rangle$, and similarly for $\bar{\alpha}_2$.

a) There is a least position where $\bar{\alpha}_1$, $\bar{\alpha}_2$ disagree, which of the form I_{q+1}. Hence, $I_1^1 = I_1^2 = I_1, \dots, I_q^1 = I_q^2 = I_q$, $I_{q+1}^1 < I_{q+1}^2$. For fixed g, we consider the functions $h_1 :<^{M(I_1,\dots,I_{k_1})} \to \boldsymbol{\delta}_{2n+1}^1$, $h_2 : <^{M(I_1,\dots,I_{k_2})} \to \boldsymbol{\delta}_{2n+1}^1$, which we may assume are order-preserving of the correct type. For $\bar{q} \leq q$ since $\Pi_{(I_1,\dots,I_q)}^{(I_1,\dots,I_{k_1-1})}(\alpha_1) = \Pi_{(I_1,\dots,I_q)}^{(I_1,\dots,I_{k_2-1})}(\alpha_2)$, it follows that:

1) that if $({}^{r_1}\xi(1),\dots,{}^{r_p}\xi(1))$, $({}^{r_1}\xi(2),\dots,{}^{r_p}\xi(2))$ enumerate the components of $\Pi_{(I_1,\dots,I_q)}^{(I_1,\dots,I_{k_1-1})}(\alpha_1)$ and $\Pi_{(I_1,\dots,I_q)}^{(I_1,\dots,I_{k_2-1})}(\alpha_2)$ corresponding to ${}^rV \in \cup_{m<n} \mathcal{R}_{2m+1}$, then ${}^{r_1}\xi(1) = {}^{r_1}\xi(2), \dots, {}^{r_p}\xi(1) = {}^{r_p}\xi(2)$.

2) For r with ${}^rV \in \mathcal{W}_{2n+1}$, the induced functions ${}^rF_1^{(i_1,\dots,i_k)} = {}^rF_2^{(i_1,\dots,i_k)}$ agree almost everywhere w.r.t. $\nu^{(i_1,\dots,i_k)}$ for all indices $(i_1,\dots,i_k) \in \operatorname{dom} {}^rT(I_1,\dots,I_{\bar{q}})$. Hence, the ${}^r\beta_{(I_1,\dots,I_q)}^{(i_1,\dots,i_k)}$) (as in T_1, T_2) agree for the two sequences T_1, T_2 corresponding to ${}^rF_1$, ${}^rF_2$.

3) For r with ${}^rV \in \mathcal{S}_{2n+1}$, there is a c.u.b. $C \subseteq \boldsymbol{\delta}_{2n+1}^1$ such that for $g : K_{2n-1}^{\ell',m} \to C$ of the correct type, for all indices $(I_1,\dots,I_m)$ in ${}^r\mathcal{C}(I_1,\dots,I_{\bar{q}})$ we have that for almost all θ w.r.t. ${}^ru^{(I_1,\dots,I_q)}$ that ${}^r\mathcal{M}_{\bar{q}}^{(I_1,\dots,I_m)}(1)(\theta) = {}^r\mathcal{M}_{\bar{q}}^{(I_1,\dots,I_m)}(2)(\theta)$, where 1, 2 refer to the definitions relative to ${}^rF_1$, ${}^rF_2$. Hence it follows that there is a c.u.b. $C \subseteq \boldsymbol{\delta}_{2n+1}^1$ such that for $g : K_{2n-1}^{\ell',m} \to C$,

if T_1, T_2 denote the sequences as in the definition of $\mathcal{F}_{\gamma_1}$, $\mathcal{F}_{\gamma_2}$, then T_1, T_2 agree up to I_{q+1}, and also $I^1_{q+1} < I^2_{q+1}$. Since H is order-preserving, it follows that $G([q_1]) < G([q_2])$.

b) There is a least position where $\bar{\alpha}_1, \bar{\alpha}_2$ disagree which of the form

$$\Pi^{(I_1^1,\dots,I^1_{k_1-1})}_{(I_1,\dots,I_q)}(\alpha_1) < \Pi^{(I_1^2,\dots,I^2_{k_2-1})}_{(I_1,\dots,I_q)}(\alpha_2).$$

Hence $I^1_1 = I^2_1, \dots, I^1_q = I^2_q$, and if $\bar{r}$ is the least component with respect to the product $V^{(I_1,\dots,I_q)} = \prod_r {}^rV^{(I_1,\dots,I_q)}$ such that the above tuples disagree, then $\Pi^{(I_1^1,\dots,I^1_{k_1-1})}_{(I_1,\dots,I_q)}(\alpha_1)(r) < \Pi^{(I_1^2,\dots,I^2_{k_2-1})}_{(I_1,\dots,I_q)}(\alpha_2)(r)$. It then follows that there is a c.u.b. $C \subseteq \boldsymbol{\delta}^1_{2n+1}$ such that for $g : K^{\ell',m}_{2n-1} \to C$ of the correct type, we have

1) For $\bar{q} < q$, ${}^r\beta_{\bar{q}}(1) = {}^r\beta_{\bar{q}}(2)$, for all r, where the β's are elements of the sequences T_1, T_2.

2) For $r' < \bar{r}$, ${}^{r'}\beta_q(1) = {}^{r'}\beta_q(2)$.

3) If ${}^{\bar{r}}V^{(I_1,\dots,I_q)} \in \bigcup_{m<n} \mathcal{R}_{2m+1}$, then ${}^{\bar{r}}\beta_q(1) < {}^{\bar{r}}\beta_q(2)$, since ${}^{\bar{r}}\beta_q(1) = \Pi^{(I_1^1,\dots,I^1_{k_1-1})}_{(I_1,\dots,I_q)}(\alpha_1)(r)$ in this case, and similarly for ${}^{\bar{r}}\beta_q(2)$.

4) If ${}^{\bar{r}}V^{(I_1,\dots,I_q)} \in \mathcal{W}_{2n+1}$, then in the fixed ordering on indices $(i_1,\dots,i_k) \in \operatorname{dom} {}^rT(I_1,\dots,I_q)$ which we use to identify $\theta_{{}^rV^{(I_1,\dots,I_q)}}$ with an ordinal, if $(\bar{i}_1,\dots,\bar{i}_k)$ denotes the least index s.t. $\left[{}^{\bar{r}}F_1^{(\bar{i}_1,\dots,\bar{i}_k)}\right] \neq \left[{}^{\bar{r}}F_2^{(\bar{i}_1,\dots,\bar{i}_k)}\right]$, then the first is smaller. Hence ${}^{\bar{r}}\beta_q^{(\bar{i}_1,\dots,\bar{i}_k)}(1) < {}^{\bar{r}}\beta_1^{(\bar{i}_1,\dots,\bar{i}_k)}(2)$, for the least index $(\bar{i}_1,\dots,\bar{i}_k)$ where ${}^{\bar{r}}\beta_1^{(\bar{i}_1,\dots,\bar{i}_k)}(1)$ and ${}^{\bar{r}}\beta_q^{(\bar{i}_1,\dots,\bar{i}_k)}(2)$ disagree.

5) If ${}^{\bar{r}}V^{(I_1,\dots,I_q)} \in \mathcal{S}_{2n+1}$, then for the least index $(I_1,\dots,I_m) \in \operatorname{dom} {}^{\bar{r}}\mathcal{C}(I_1,\dots,I_q)$, such that $\left[{}^{\bar{r}}\mathcal{M}_q^{(I_1,\dots,I_m)}(1)\right] \neq \left[{}^{\bar{r}}\mathcal{M}_q^{(I_1,\dots,I_m)}(2)\right]$, the first is smaller. This follows since for the least

$$(I_1,\dots,I_m) \text{ s.t. } \Pi^{(I_1^1,\dots,I^1_{k_1-1})}_{(I_1,\dots,I_q)}(\alpha_1)(I_1,\dots,I_m) \neq \Pi^{(I_1^2,\dots,I^2_{k_2-1})}_{(I_1,\dots,I_q)}(\alpha_2)(I_1,\dots,I_m),$$

the first is smaller. Hence for the least $(I_1,\dots,I_m)$ s.t. ${}^{\bar{r}}\beta_q^{(I_1,\dots,I_m)}(1) \neq {}^{\bar{r}}\beta_q^{(I_1,\dots,I_m)}(2)$, the first is smaller.

Hence, for such g, it follows that for the least position where T_1, T_2 disagree, the first is smaller. Since H is order-preserving, $G_1([g]) < G_2([g])$.

c) $\bar{\alpha}_1$ is an extension of $\bar{\alpha}_2$. Proceeding as above, it follows that for almost all g, the sequence T_1 extends T_2, so $G_1([g]) < G_2([g])$.

Hence, for almost all γ w.r.t. $\mathcal{U}$, $\mathcal{F}_\gamma$ is order-preserving almost everywhere.

We now establish $\mathcal{B}_{2n+3}$: We let $A \subseteq K_{2n+1}^{\ell,m}$ have measure one w.r.t. $S_{2n+1}^{\ell,m}$. We let C be a c.u.b. subset of δ_{2n+1}^1 such that if $G : \delta_{2n+1}^1 \to C$ of the correct type, then the ordinal β represented by G w.r.t. $\mu_{2n+1}^{S_{2n-1}^{\ell',m}} \in A$.

We let B be the measure one set w.r.t. $\mathcal{U}$ determined by C. We fixed $\gamma \in B$, and a function $H : <^{\Omega} \to C$ of the correct type representing γ with respect to the measures $\prod_r {}^r\nu^{(I_1,\ldots,I_k)} \times \mu_{2n+1}^{<M(I_1,\ldots,I_k)}$ for $(I_1, \ldots, I_k) \in \operatorname{dom} \tau$. For $\alpha^{(I_1,\ldots,I_k)} \in \operatorname{dom} <^{\tau}$ and ${}^rF$ representing the components of α for the ${}^r\mathcal{V} \in \mathcal{W}_{2n+1} \cup \mathcal{S}_{2n+1}$, (which happens for almost all α), and G representing $\mathcal{F}_\gamma(\alpha^{(I_1,\ldots,I_k)})$ w.r.t. $\mu_{2n+1}^{S_{2n-1}^{\ell',m}}$, it follows from the definition of $\mathcal{F}_\gamma$ that for almost all $g : K_{2n-1}^{\ell',m} \to \delta_{2n+1}^1$, $G([g]) = H(\langle T \rangle)$, for the sequence T as in the definition of $\mathcal{F}_\gamma$, and hence $\in C$.

Since H is of the correct type, it follows that G is non-normal on a measure one set, and has uniform cofinality ω.

Also, since β_O (as in the definition of T) $= [g]$, it follows that G is strictly increasing on a measure one set.

By a sliding argument (as in Lemma 3), we may assume G is of the correct type and has range in C everywhere. Hence $\mathcal{F}_\gamma(\alpha^{(I_1,\ldots,I_k)}) \in B$.

This establishes $\mathcal{B}_{2n+3}$.

IV. A Local Embedding Theorem.

Theorem. *For any measure $\mathcal{V} \in \cup_{m \leq n} \mathcal{S}_{2m+1} \cup_{m \leq n} \mathcal{W}_{2m+1}$, and regular cardinal $\kappa < \delta_{2n+3}^1$, there is a measure $R \in R_{2n+1}$ and a c.u.b. $C \subseteq \delta_{2n+3}^1$ such that for $\alpha \in C$,* cof $(\alpha) = \kappa$, *$J_{\mathcal{V}}(\alpha) \leq J_R(\alpha)$, where $J_{\mathcal{V}}$, J_R denote the embeddings from the ultrapowers by the measures $\mathcal{V}, R$.*

We use the notation $\mathcal{V} = \prod_r {}^r\mathcal{V}$ where each ${}^r\mathcal{V}$ is basic.

We require the following inductive hypothesis:

H_{2n+1}: For any basic measures ${}^r\mathcal{V}_1 \in \cup_{m \leq n} \mathcal{S}_{2m-1} \cup \cup_{m \leq n} \mathcal{W}_{2m-1}$, for $1 \leq r \leq r_0$, and corresponding product measure $\mathcal{V}_1 = \prod_r {}^r\mathcal{V}_1$, there are measures ${}^r\mathcal{V}_2$, $l \leq r \leq r_0$, in $\cup_{m \leq n} R_{2m-1}$, a measure u on Ξ_{2n-1}, and a map $\alpha \to [f_\alpha]_u$ such that for almost all $\alpha = (\alpha_1, \ldots, \alpha_r)$ w.r.t. $\mathcal{V}_2$, the ordinal $[f_\alpha]_u$ is represented by an $f_\alpha : \theta_u \to \theta_{\mathcal{V}_1}$ defined a.e. w.r.t. u satisfying:

1) For any $A_1 \in \theta_{\mathcal{V}_1}$ of measure one w.r.t. $\mathcal{V}_1$, there is an $A_2 \subseteq \theta_{\mathcal{V}_2}$ of measure one w.r.t. $\mathcal{V}_2$ such that for all $\alpha \in A_2$, f_α has range almost everywhere in A_1.

2) There is a measure one set $A_2 \subseteq \theta_{\mathcal{V}_2}$ such that for $\alpha \in A_2$, $\sup f_\alpha \overset{\text{a.e.}}{<} \theta_{\mathcal{V}_1}$ (that is there are $\beta_1 < \theta_{1_{\mathcal{V}_1}}, \ldots, \beta_{r_0} < \theta_{r_{0_{\mathcal{V}_1}}}$ s.t. for almost all η w.r.t. u, if $f_\alpha(\eta) = (\gamma_1, \ldots, \gamma_{r_0})$, then $\gamma_1 < \beta_1, \ldots, \gamma_{r_0} < \beta_{r_0}$).

3) If ${}^r\mathcal{V}'_1$, $1 \leq r \leq r_0$ are basic measures in $\cup_{m \leq n} \mathcal{S}_{2m-1} \cup \cup_{m \leq n} \mathcal{W}_{2m-1}$ such that the embeddings $\Pi^{{}^r\mathcal{V}'_1}_{{}^r\mathcal{V}_1}$ are defined for $1 \leq r \leq r_0$, then there are ${}^r\mathcal{V}'_2 \sim^r \mathcal{V}_2$ such that 1) and 2) hold for $\mathcal{V}'_1$, $\mathcal{V}'_2$.

We also require that if ${}^r\mathcal{V}'_2 < {}^r\mathcal{V}_2$ for $1 \leq r \leq r_0$, then H_{2n+1} holds with $\mathcal{V}'_2 = \Pi_r\, {}^r\mathcal{V}'_2$ replacing $\mathcal{V}_2$ (with a possibly different u, etc.).

4) If $\mathcal{V}_1 = \Pi\, {}^r\mathcal{V}_1$, where each ${}^r\mathcal{V}_1 \in R_{2n-1}$ and ${}^{r_1}\mathcal{V}_1 \sim {}^{r_2}\mathcal{V}_1$ for $1 \leq r_1, r_2 \leq r_0$, then there is a $\mathcal{V}_2 \in R_{2n-1}$ with $\mathcal{V}_2 \sim \mathcal{V}_1$, and a measure $\mathcal{U}$ on Ξ_{2n-1} such that H_{2n+1} 1), 2), 3) above are satisfied.

We require the following additional hypothesis:

D_{2n+1}: 1) We let ${}^r\mathcal{V}_1$, ${}^r\mathcal{V}_2$, u be as in H_{2n+1} 1), 2), 3). Then for any $1 \leq r \leq r_0$, there is an $r' \leq r_0$ and a measure one set A w.r.t. $\mathcal{V}_2$ such that if $(\beta_1, \ldots, \beta_r, \ldots, \beta_{r_0})$, $(\bar\beta_1, \ldots, \bar\beta_r, \ldots, \bar\beta_{r_0})$ are in A and $\beta'_r < \bar\beta'_r$ then for almost all γ w.r.t. u we have that $f_{(\vec\beta)}(\gamma)(r) < f_{(\vec{\bar\beta})}(\gamma)(r)$, these denoting the rth components of these tuples.

We further require that if ${}^r\mathcal{V}'_1 \leq {}^r\mathcal{V}_1$ for $1 \leq r \leq r_0$, are such that H_{2n+1} 1), 2), 3) are also satisfied by ${}^r\mathcal{V}'_1$, ${}^r\mathcal{V}_2$ (for some u') then the r' for the measures ${}^r\mathcal{V}'_1$ is the same as for the ${}^r\mathcal{V}_1$, and in fact, for $\vec\beta$, $\vec{\bar\beta}$ as above, for almost all γ w.r.t. u', $(\Pi^{\mathcal{V}'_1}_{\mathcal{V}_1} f_{(\vec\beta)}(\gamma))(r) < (\Pi^{\mathcal{V}'_1}_{\mathcal{V}_1} f_{(\vec{\bar\beta})}(\gamma))(r)$.

2) If ${}^r\mathcal{V}_1 \in R_{2n+1}$, ${}^{r_1}\mathcal{V}_1 \sim {}^{r_2}\mathcal{V}_1$, $\mathcal{V}_2 \in R_{2n-1}$ and u are as in $H_{2n+1}(4)$, then there is a measure one set A w.r.t. $\mathcal{V}_2$ such that for $\beta_1 < \beta_2$ in A, for almost all γ w.r.t. $\mathcal{U}$, $f_{\beta_1}(\gamma)(r) < f_{\beta_2}(\gamma)(r)$ for all $1 \leq r \leq r_0$.

We first establish the theorem from H_{2n+3} and K_{2n+3}:

We let ${}^r\mathcal{V}_1$, $1 \leq r \leq r_0$ be basic measures in $\cup_{m \leq n} \mathcal{S}_{2m+1} \cup \cup_{m \leq n} \mathcal{W}_{2n+1}$ and $\mathcal{V}_1 = \Pi_r\, {}^r\mathcal{V}_1$. We let C be a c.u.b. subset of δ^1_{2n+3} closed under $\mathcal{J}_{\mathcal{V}}$ for all measures $\mathcal{V}$ in $\cup_{m \leq n} \mathcal{S}_{2m+1} \cup \cup_{m \leq n} \mathcal{W}_{2m+1}$. We first show that there are measures ${}^r\mathcal{V}_2 \in R_{2n+1}$ s.t. for all $\alpha \in C$, $\mathcal{J}_{\mathcal{V}_1}(\alpha) \leq \mathcal{J}_{\mathcal{V}_2}(\alpha)$, where $\mathcal{V}_2 = \Pi_r\, {}^r\mathcal{V}_2$. We let the ${}^r\mathcal{V}_2$ be as in H_{2n+3}, and fix $\alpha \in C$. We define an embedding $\mathcal{E}$ from $\mathcal{J}_{\mathcal{V}_1}(\alpha)$ into $\mathcal{J}_{\mathcal{V}_2}(\alpha)$ as follows: If $F : \theta_{\mathcal{V}_1} \to \alpha$, then we represent $\mathcal{E}([F]_{\mathcal{V}_1})$ by $G : \theta_{\mathcal{V}_2} \to \alpha$ defined as follows, if $\beta = ({}^1\beta, \ldots, {}^{r_0}\beta) \in \theta_{\mathcal{V}_2}$, we represent $G(\beta)$ w.r.t. $\mathcal{U}$ (as in H_{2n+3}) by the function $F \circ f_\beta : \theta_u \to \alpha$. It follows from H_{2n+3} and the definition of C that $\mathcal{E}$ is well-defined.

We now assume without loss of generality that $\mathcal{V}_1 = \Pi_r\ {}^r\mathcal{V}_1$ where ${}^r\mathcal{V}_1 \in \cup_{m\leq n} R_{2m+1}$. We fix a regular cardinal $\kappa < \delta^1_{2n+3}$ and proceed to show that for some $\mathcal{V}_2 \in \cup_{m\leq n} R_{2m+1}$, $\mathcal{J}_{\mathcal{V}_1}(\alpha) \leq \mathcal{J}_{\mathcal{V}_2}(\alpha)$ for all $\alpha \in C$, cof$(\alpha) = K$. By Lemma 3, we may assume that $\mathcal{V}_1$ is of the form $\mathcal{V}_1 = \Pi^{q_0}_{r=1}\ {}^r\mathcal{V}_1 \times \Pi^{r_0}_{r=q_0+1}\ {}^r\mathcal{V}_1$, where ${}^{r_1}\mathcal{V}_1 \sim {}^{r_2}\mathcal{V}_1$ for $1 \leq r_1,\ r_2 \leq q_0$, and ${}^{r_1}\mathcal{V}_1 \not\sim {}^{r_2}\mathcal{V}_1$ for $r_1 \leq q_0,\ r_2 > q_0$, and cof $\theta({}^{r_1}\mathcal{V}_1) = \kappa$. It now follows from Lemmas 1, 2 and the definition of C that if $\mathcal{V}_1' = \Pi^{r_0}_{r=q_0+1}\ {}^r\mathcal{V}_1$, then $\mathcal{J}_{\mathcal{V}_1'}$, then $\mathcal{J}_{\mathcal{V}_1'}(\alpha) = \alpha$. Hence, we may assume without loss of generality that $\mathcal{V}_1 = \Pi_r\ {}^r\mathcal{V}_1$ where ${}^{r_1}\mathcal{V}_1 \sim {}^{r_2}\mathcal{V}_1$ for all $1 \leq r_1,\ r_2 \leq r_0$. We then elect $\mathcal{V}_2$ as in $H_{2n+3}(4)$, and proceed as above to define an embedding from $\mathcal{J}_{\mathcal{V}_1}(\alpha)$ into $\mathcal{J}_{\mathcal{V}_2}(\alpha)$.

We now assume H_{2n+1}, D_{2n+1}, and proceed to establish H_{2n+3}, D_{2n+3}.

We first consider $H_{2n+3}(1) - (3)$.

We let ${}^r\mathcal{V}_1$, for each $1 \leq r \leq r_0$ be basic measures in $\cup_{m\leq n} \mathcal{S}_{2m+1} \cup \cup_{m\leq n} \mathcal{W}_{2m+1}$. For each r, we will define ${}^r\mathcal{V}_2$, and ${}^r\mathcal{U}_1$ and will have $\mathcal{V}_2 = \Pi_r\ {}^r\mathcal{V}_2$, $\mathcal{U} = \Pi_r\ {}^r\mathcal{U}$. Hence, in defining $\mathcal{V}_2, \mathcal{U}$, we consider a fixed ${}^r\mathcal{V}_1$, and we suppress the superscript r thoughout the definition.

We consider the following cases:

I) $\mathcal{V}_1\ (= {}^r\mathcal{V}_1) \in W_{2n+1}$. We fix an $n - *$ tuple T and corresponding ordering on Ξ_{2n+1} such that $\mathcal{V}_1 = \mu^{<^T}_{2n+1}$. We fix a measure $S^{\ell',m}_{2n-1}$ s.t. the hypothesis B_{2n+1} from the global embedding theorem is satisfied by $<^T$ and $S^{\ell',m}_{2n-1}$, with measure u. We then let $\mathcal{V}_2 = \mu^{S^{\ell',m}_{2n-1}}_{2n+1} = W^m_{2n+1}$, and let the $\mathcal{U}$ in H_{2n+3} be the u from B_{2n+1}.

II) $\mathcal{V}_1 \in \mathcal{S}_{2n+1}$, and $n_1 \notin$ dom $<^T$ (where, we recall, $n_1 =$ the largest integer s.t. $n_1 \in$ dom T). In this case, V^{n_1} is defined. For all indices $(n_1, i_2, ..., i_k) \in$ dom T with $v^{(n_1,i_2,...,i_k)}$ defined, $v^{(n_1,i_2,...,i_k)} < V^{(n_1)}$, that is $\Pi^{(n_1,i_2,...,i_k)}_{(n_1)}$ is defined. From H_{2n+1} (1) and (3) it follows that there is a measure $\tilde{v}_2$ s.t. for all $v^{(n_1,i_2,...,i_k)}$, H_{2n+1} holds for $v^{(n_1,...,i_k)}$, $\tilde{v}_2$ and a measure $u^{(n_1,...,i_k)}$, and $\tilde{v}_2 = w_1 \times \cdots \times w_n$ where $w_1, ..., w_n$ are basic measures in $\cup_{m\leq n} R_{2m-1} = \cup_{m\leq n} \mathcal{S}_{2m-1} \cup \cup_{m\leq n} \mathcal{W}_{2m-1}$. We also let $v^{(n_1)} = \Pi_s\ {}^s v^{(n_1)}$, a produce of basic measures. We let $\bar{s}$ be the s such that in identifying $\theta_{v^{(n_1)}}$ with $\Pi_s \theta_{{}^s_v{}^{(n_1)}}$, we order by $\theta_{{}^{\bar{s}}_v{}^{(n_1)}}$ most significantly. We let $\bar{\bar{s}} \leq n$ be, from D_{2n+1}, the integer s.t. there is a measure one set A w.r.t. $\tilde{v}_2$ such that for $(\beta_1, ..., \beta_n)$, $(\beta_1', ..., \beta_n') \in A$ and $\beta_{\bar{\bar{s}}} < \beta'_{\bar{\bar{s}}}$, we have that for all $(n_1, i_2, ..., i_k)$ that for almost all γ w.r.t. $u^{(n_1,...,i_k)}$ that $(\Pi^{(n_1,...,i_k)}_{(n_1)} f_{\vec{\beta}}(\gamma))(\bar{s}) < \Pi^{(n_1,...,i_k)}_{(n_1)} f_{(\vec{\beta}')}(\gamma))(\bar{s})$.

We let $\tilde{v}_2^2 = \tilde{v}_2 \times \tilde{v}_2 = (w_1 \times \cdots \times w_n) \times (w_1 \times \cdots \times w_n)$, and let $E_1, ..., E_p$ enumerate the components of $\tilde{v}_2^2$ equivalent (w.r.t. $\sim$) to $w_{\bar{\bar{s}}}$. By H_{2n+1} (4), we let $v_2 \in \cup_{m\leq n} R_{2n+1}$ be s.t. H_{2n+1} holds for $E_1 \times \cdots \times E_{p'}\ v_{2'}$ and some measure u_2. We then let $\mathcal{V}_2 = S^{\ell,m}_{2n+1}$, where $v(S^{\ell,m}_{2n+1}) = v_2$ (here ℓ is not necessarily maximal). We may also take any $m' \geq m$ in the following.

We proceed to define the measure $\mathcal{U}$.

We define two orderings $<^{\Omega_1}$, and $<^{\Omega_2}$ on tuples of ordinals and integers.

The domain of $<^{\Omega_1}$ consists of tuples of the form

$$T = \langle \alpha_0, I_1, ..., \alpha^{(i_1)}, ..., I_2, ..., \alpha^{(i_1,i_2)}, ..., I_k, ..., \alpha^{(i_1,...,I_k)},\ I_n, ..., \alpha^{(i_1,...,i_n)} \rangle$$

or an initial segment of such, satisfying:

1) $\alpha_0 \leq \boldsymbol{\delta}^1_{2n+1}$.

2) $(I_1, ..., I_k)$ is an index in $\mathcal{C}$.

3) the ordinals $\alpha^{(i_1,...,i_k)}$ following I_k are indexed by indices $(i_1, ..., i_k)$ in $C_{(I_1,...,I_k)}$, and occur in the same order as in $C_{(I_1,...,I_k)}$.

4) $\alpha^{(i_1,...,i_k)} < \alpha_0$.

We let $<^{\Omega_1}$ be the Kleene-Brouwer ordering on the tuples.

We let $<^{\Omega_2}$ denote the usual ordering on $\boldsymbol{\delta}^1_{2n+1}$.

We define the measure $\mathcal{U}$ as follows: We define A to have measure one if there is a c.u.b. $C_2 \subseteq \boldsymbol{\delta}^1_{2n+1}$ s.t. for all $H_2 : <^{\Omega_2} \to C_2$ of the correct type, there is a c.u.b. $C_1 \subseteq \boldsymbol{\delta}^1_{2n+1}$ s.t. for all $H_1 : <^{\Omega_1} \to C_1$ of the correct type $\alpha = (\ldots, \alpha^{(I_1,...,I_k)}, \ldots) \in A$, where for fixed $(I_1, ..., I_k)$, an index in $\mathcal{C}$, $\alpha^{(I_1,...,I_k)}$ is defined through the following sequence of definitions:

a) $\alpha^{(I_1,...,I_k)}$ is represented w.r.t. $\mu^{<^T}_{2n+1}$ by a function $\bar{F}$,

b) for fixed $f : <^T \to \boldsymbol{\delta}^1_{2n+1}$ of the correct type, $\bar{F}([f])$ is represented w.r.t. v_2 (as above) by a function g.

c) for $\theta < \theta_{v_2}$, $g(\theta) = H_2(\beta)$, where β is represented w.r.t. u_2 (as above) by a function h_2.

d) for $\gamma \in \theta_{u_2}$, we represent $h_2(\gamma)$ w.r.t. $D_1 \times \cdots \times D_q$, where $D_1, \ldots, D_q$ enumerate the components of $\tilde{v}^2_2$ not equivalent to the $E_1, \ldots, E_p$, by the function $\bar{g}_1$, where if $i_1, \ldots, i_q$ enumerate the integers $a \leq 2n$ where $W_{i_a} \sim D_i$ for some i, then $\bar{g}_1(\xi_{i_1}, \ldots, \xi_{i_q}) = g_1(\xi_1, \ldots, \xi_n, \xi'_1, \ldots, \xi'_n)$ where $\{\xi_1, \ldots, \xi_n, \xi'_1, \ldots, \xi'_k\} = \{\xi_{i_1}, \ldots, \xi_{i_q}\} \cup \{\xi_{J_1}, \ldots, \xi_{J_p}\}$, where $J_1, \ldots, J_p$ enumerate the integers $a \leq 2n$ where $w_a \sim E_1$, and $(\xi_{J_1}, \ldots, \xi_{J_p}) = f_\theta(\gamma)$, where f is as in H_{2n+1} for $\tilde{v}^2_2$ and v_2.

e) g_1 is defined by $g_1(\xi_1, \ldots, \xi_n, \xi'_1, \ldots, \xi'_n) = H_1(\langle T \rangle)$, where $T = \langle \alpha_0, I_1, \ldots,$

$\alpha^{(i_1)}, \ldots, I_k, \ldots, \alpha_k^{(i_1,\ldots,i_k)}, \ldots\rangle$ is defined by:

i) α_0 is represented w.r.t. $u^{(n_1)}$ by a function h_1^0 defined as follows: We let $f_\rho : \theta_{V^{(n_1)}} \to \delta_{2n+1}^1$ represent $\sup_{\text{a.e.}} f^{(n_1)}$ w.r.t. $V^{(n_1)}$. For $\eta \in \theta_{u^{(n_2)}}$, we set $h_1^0(n) = f_p(f_{(\xi_1',\ldots,\xi_n')}(\eta))$.

ii) $\alpha^{(n_1,i_2,\ldots,i_k)}$ is represented w.r.t. $u^{(n_1,i_2,\ldots,i_k)}$ by a function $h_1^{(n_1,i_2,\ldots,i_k)}$ defined by: for $\eta \in \theta_{u^{(n_1,i_2,\ldots,i_k)}}$, we set $h_1^{(n_1,i_2,\ldots,i_k)}(\eta) = f(\beta^{(n_1,i_2,\ldots,i_k)})$ where $\beta = f_{(\xi_1,\ldots,\xi_n)}(\eta)$, where $f_{(\,)}$ is as in H_{2n+1} for $v^{(n_1,i_2,\ldots,i_k)}$ and $\tilde{v}^2$.

iii) $\alpha^{(i_1,\ldots,i_k)}$, for $i_1 < n_1$ is represented w.r.t. $v^{(i_1,\ldots,i_k)}$ by the induced function $f^{(i_1,\ldots,i_k)}$.

We show that $\mathcal{U}$ is well-defined. For fixed H_2, H_1, $f : <^T\to \delta_{2n+1}^1$, we first claim that $\bar{F}([f])$ is well-defined. By H_{2n+1}, the functions $f_\theta, f_{(\xi_1,\ldots,\xi_n)}, f_{(\xi_1',\ldots,\xi_n')}$ are well-defined almost everywhere. We also note the following facts:

1) For almost all f there is a measure one set A w.r.t. $\tilde{v}_2^2$ s.t. for $(\xi_1, \ldots, \xi_n, \xi_1', \ldots, \xi_n') \in A$ and T the corresponding sequence as above, (for any fixed $(I_1, \ldots, I_k)$), we have that $\alpha_0 > \alpha^{(i_1)}, \ldots, \alpha^{(i_1,\ldots,i_k)}$. This follows from H_{2n+1} (i) and (ii).

2) By Lemma 3, we have that there is a measure one set E w.r.t. $E_1 \times \cdots \times E_p$ s.t. for $(\xi_{J_1}, \ldots, \xi_{J_p}) \in E$, for almost all $(\xi_{i_1}, \ldots, \xi_{i_q})$ w.r.t. $D_1 \times \cdots \times D_q$, $(\xi_1, \ldots, \xi_n, \xi_1', \ldots, \xi_n') \in A$, where, as above, $(\xi_1, \ldots, \xi_n, \xi_1', \ldots, \xi_n')$ is the enumeration of $(\xi_{i_1}, \ldots, \xi_{i_q}) \cup (\xi_{J_1}, \ldots, \xi_{J_p})$ by the subscripts.

We next claim that for fixed H_2, H_1, for almost all $f : <^T\to \delta_{2n+1}^1$, $\bar{F}([f])$ depends only on the equivalence class of f. We fix H_2, H_1, $f : <^T\to \delta_{2n+1}^1$ of the correct type, and let f_2 agree with f a.e. w.r.t. the measures $v^{(i_1,\ldots,i_k)}$ (we also assume f is into a sufficiently closed set so that $\bar{F}([f])$ is defined). We also fix $f_p, f_p(2)$ representing $\sup_{\text{a.e.}} f$ w.r.t. $V^{(n_1)}$. We then have that there are measure one sets $A^{(i_1,\ldots,i_k)}$ s.t. $f_1 f_2$ agree for $\theta^{(i_1,\ldots,i_k)} \in A^{(i_1,\ldots,i_k)}$. It also follows readily that there is a measure one set A^0 w.r.t. $V^{(n_1)}$ s.t. for $\alpha \in A^0$, $f_p(\alpha) = f_p(2)(\alpha)$. (This follows since $\sup_\delta$ a.e. $f(\langle n_1, \delta\rangle) = \sup_\delta$ a.e. $f_2(\langle n_1, \delta\rangle))$. It then follows that for almost all $(\xi_1, \ldots, \xi_n, \xi_1', \ldots, \xi_n')$ w.r.t. $\tilde{v}_2^2$ that (for any fixed $(I_1, \ldots, I_k)$), $T = T_2$, using $H_{2n+1}(1)$, where T, T_2 denote the sequences as in the definition of $\mathcal{U}$ corresponding to f, f_2. It then follows readily from $H_{2n+1}(1)$ and Lemma 3 that for almost all θ w.r.t. v_2 that $g(\theta) = g_2(\theta)$, g, g_2 as in the definition of $\mathcal{U}$. Hence $\bar{F}(f) = \bar{F}(f_2)$.

Hence, $\mathcal{U}$ is well defined.

III. $\mathcal{V}_1 \in S_{2n+1}$, and $n_1 \in \operatorname{dom} <^T$. We set $\mathcal{V}_2 = S_{2n+1}^{1,m}$ where $m = 2[$(the number of indices $(i_1, \ldots, i_\ell)$ appearing in all the $C_{(I_1,\ldots,I_k)}$ for $(I_1, \ldots, I_k)$ an index in $\mathcal{C}) + n]$.

We proceed to define $\mathcal{U}(= {}^{\eta}\mathcal{U})$ in this case.

We define measures $\mathcal{U}^{(I_1,\dots,I_k)}$ and $\mathcal{U}^{(I_1,\dots,I_k)}_{(i_1,\dots,i_\ell)}$ for $(I_1,\dots,I_k)$ an index in C, and $(i_1,\dots,i_\ell)$ on index in $C_{(I_1,\dots,I_k)}$.

We let $b =$ the number of indices of the form (I_1) in C, and we set $\mathcal{U}^{(I_1)} =$ the b-fold product of the ω-cofinal normal measure on $\boldsymbol{\delta}^1_{2n+1}$.

For fixed I_1, we let $(i^1),\dots,(i^p)$ denote the integers in C_{I_1} less than n_1. For $1 \leq q \leq p$ and the index $i_1 = i^q$ in C_{I_1}, we define $\mathcal{U}^{(I_1)}_{(i_1)}$ as follows: We let $<^{(I_1)}_{(i_1)}$ be the lexicographic ordering on tuples $(\alpha_1,\dots,\alpha_q)$ of ordinals $< \boldsymbol{\delta}^1_{2n+1}$, where $\alpha_2,\dots,\alpha_q < \alpha_1$. We then define A to have measure one if there is a c.u.b. subset of $\boldsymbol{\delta}^1_{2n+1}$, C, such that for $H : <^{\Omega}\ {}^{(I_1)}_{(i_1)} \to C$ of the correct type, $\alpha \in A$, where α is represented w.r.t. $\mu^{<T}_{2n+1}$ by the function which assigns to $f : <^T \to \boldsymbol{\delta}^1_{2n+1}$ of the correct type, the ordinal $H(\langle\alpha_1,\dots,\alpha_q\rangle)$, where if f represents the ordinals $\alpha^{(i_1,\dots,i_k)}$, then $\alpha_1 =$ the largest of $\{\alpha^{i^1},\dots,\alpha^{i^q}\}$, $\alpha_2 =$ the second largest of these, etc.

In general, we let $\mathcal{U}^{(I_1,\dots,I_k)}$ be the b-fold product, where $b =$ the number of indices of the form $(I_1,\dots,I'_k)$ extending $(I_1,\dots,I_{k-1})$ in C, of the measure $\mathcal{U}^{(I_1,\dots,I_{k-1})}_{(i_1,\dots,i_\ell)}$, where $(i_1,\dots,i_\ell)$ is the index in $C_{(I_1,\dots,I_{k-1})}$ occuring last.

For $(i_1,\dots,i_\ell)$ an index in $C_{(I_1,\dots,I_k)}$, we define $\mathcal{U}^{(I_1,\dots,I_k)}_{(i_1,\dots,i_\ell)}$ as follows: We let $(\vec{i}^1),\dots,(\vec{i}^q)$ denote the elements of $C_{(I_1,\dots,I_k)}$ preceding (and including) $(i_1,\dots,i_\ell)$, as well as the elements of $C_{(I_1)},\dots,C_{(I_1,\dots,I_{k-1})}$, except that we exclude (n_1) if it occurs. We let $<^{\Omega}\ {}^{(I_1,\dots,I_k)}_{(i_1,\dots,i_\ell)}$ be the ordering on tuples $(\alpha_1,\dots,\alpha_q)$ as above. We then define A to have measure one if there is a c.u.b. $C \subseteq \boldsymbol{\delta}^1_{2n+1}$ such that for all $H : <^{\Omega}\ {}^{(I_1,\dots,I_k)}_{(i_1,\dots,i_\ell)} \to C$ of the correct type, $\alpha \in A$, where α is represented w.r.t. $\mu^{<T}_{2n+1}$ by the function which assigns to $f : <^T \to \delta^1_{2n+1}$ of the correct type the ordinal $H(\langle\alpha_1,\dots,\alpha_q\rangle)$, where $\alpha_1 =$ the largest of $\{\alpha^{(\vec{i}^1)},\dots,\alpha^{(\vec{i}^q)}\}$, $\alpha_2 =$ the second largest of these, etc., where f represents the ordinals $\alpha^{(i_1,\dots,i_k)}$.

We then let $\mathcal{U}$ be the product

$$\mathcal{U} = \left[\mathcal{U}^{(I_1)}\right]^{m+2} \times \prod_{I_1} \prod_{i_1 \in C_{I_1}} \left[\mathcal{U}^{(I_1)}_{(i_1)}\right]^{m+2} \times \cdots \times$$

$$\Big[\prod_{(I_1,\dots,I_{n-1}) \in C} \mathcal{U}^{(I_1,\dots,I_n)}\Big]^{m+2} \times$$

$$\prod_{(I_1,\dots,I_n) \in C} \quad \prod_{(i_1,\dots,i_\ell) \in C_{(I_1,\dots,I_n)}} \left[\mathcal{U}^{(I_1,\dots,I_n)}_{(i_1,\dots,i_k)}\right]^{m+2},$$

where we order the factors in each product according to the ordering of indices in each $(I_1,\dots,I_k)$.

This completes the definition of the measure $\mathcal{U}(=\mathcal{U}^r)$ in this case, and hence completes the definition of $\mathcal{U}$.

We now proceed to define the family of embeddings $\mathcal{F}$ as in H_{2n+3}: We fix $\beta = (\beta_1, ..., \beta_r) \in \mathcal{V}_2 = \Pi_r\ \mathcal{V}_2^r$, and define $\mathcal{F}_\beta$. We fix functions $G_1, ..., G_r$ representing $\beta_1, ..., \beta_r$ as elements in $\mathcal{V}_2^r$. We fix a tuple $\theta = (\theta_1, ..., \theta_r) \in \Pi_r\ \mathcal{U}^r = \mathcal{U}$. We set $\mathcal{F}_\beta(\theta) = (\mathcal{F}_1(G_1, \theta_1), ..., \mathcal{F}_{r_0}(G_r, \theta_r)) = (\alpha_1, ..., \alpha_r)$, say, where it remains to define $\alpha_r = \mathcal{F}_r(G_r, \theta_r)$, $1 \le r \le r_0$. We again suppress the subscript r.

Definition of $\mathcal{F}_r(G_r, \theta_r) = \mathcal{F}(G, \theta)$.

We consider the following cases:

1) $\mathcal{V}_1 \in \mathcal{W}_{2n+1}$. In this case, $G : S_{2n-1}^{\ell', m} \to \boldsymbol{\delta}_{2n+1}^1$, and θ is in the measure space $\mathcal{U}$ as in B_{2n+1} for T corresponding to $\mathcal{V}_1 = \mu_{2n+1}^{<T}$. We represent α by $f : <^T \to \boldsymbol{\delta}_{2n+1}^1$ given by $f : G \circ f_\theta$, where $f_\theta : <^T \to S_{2n-1}^{\ell', m}$ is as in B_{2n+1}. Here α is a tuple $\alpha = (\cdots, \alpha^{(i_1, ..., i_k)}, \cdots)$. This is well defined, and f is of the correct type almost everywhere.

2) $\mathcal{V}_1 \in \mathcal{S}_{2n+1}$, and $n_1 \notin \text{dom} <^T$. We fix functions $\bar{F}^{(I_1, ..., I_k)}$ representing $\theta = (\cdots, \theta^{(I_1, ..., I_k)}, \cdots)$ with respect to $\mu_{2n+1}^{<T}$. We then represent $\alpha = (\cdots, \alpha^{(I_1, ..., I_k)}, \cdots)$ w.r.t. $\mu_{2n+1}^{<T}$ by the function $F^{(\vec{I})}([f]) = G \circ \bar{F}^{(\vec{I})}([f])$ for $f : <^T \to \boldsymbol{\delta}_{2n+1}^1$ of the correct type. This is well defined.

3) $\mathcal{V}_1 \in \mathcal{S}_{2n+1}$ and $n_1 \in \text{dom} <^T$. We represent $\alpha^{(I_1, ..., I_k)}$, where $\alpha = (\cdots, \alpha^{(I_1, ..., I_k)}, \cdots)$, with respect to $\mu_{2n+1}^{<T}$ by a function $F^{(I_1, ..., I_k)}$ defined as follows: We fix functions

$$H_{0(i_1, ..., i_\ell)}^{(I_1, ..., I_k)},\ H_{1(i_1, ..., i_\ell)}^{(I_1, ..., I_k)},\ H_{p(i_1, ..., i_\ell)}^{(I_1, ..., I_k)},\ 1 \le p \le m$$

representing the components of θ corresponding to the factor $\left[\mathcal{U}_{(i_1, ..., i_\ell)}^{(I_1, ..., I_k)}\right]^{m+2}$ in $\mathcal{U}$, where the index $(i_1, ..., i_\ell)$ may not appear.

For a fixed $f : <^T \to \boldsymbol{\delta}_{2n+1}^1$, we set $F^{(I_1, ..., I_k)}([f]) = G(\alpha_2, ..., \alpha_m, \alpha_1)$, where $\alpha_1, ..., \alpha_m$ are defined as follows:

We set $\alpha_1 = f(\langle n_1 \rangle)$.

We set $\alpha_2 = H_0^{(I_1)}(I_1)$ (we recall $H_0^{(I_1)}$ is a sequence of ordinals $< \boldsymbol{\delta}_{2n+1}^1$). If $k = 1$ and C_{I_1} is empty, we set $\alpha_3 = H_2^{(I_1)}(I_1), ..., \alpha_m = H_{m-1}^{(I_1)}(I_1)$, and otherwise, $\alpha_3 = H_1^{(I_1)}(I_1)$.

We let $r_i =$ the number of indices in $C_{(I_1, ..., I_i)}$ for $i \le 1 \le k$.

In general, we assume that α_j has been defined for $j \le u = \sum_{q \le i}(2r_q + 2)$, and that α_u is of the form $H_{1(i_1,...,i_\ell)}^{(I_1,...,I_i)}(...,\theta^{(\vec{j})},...)$, where $(i_1,...,i_\ell)$ denotes the last index in $C_{(I_1,...,I_i)}$, and $(\vec{j})$ denotes the indices in $C_{(I_1,...,I_i)}$ and $C_{(I_1,...,I_{i'}')}$ for $i' < i$, and the $\theta^{(\vec{j})}$ are represented w.r.t. $V^{(\vec{j})}$ by the functions induced from $f : <^T \rightarrow \boldsymbol{\delta}_{2n+1}^1$, and the $\theta^{(\vec{j})}$ are ordered as in the definition of $<^\Omega \, {}_{(i_1,...,i_\ell)}^{(I_1,...,I_i)}$.

We then set $\alpha_{u+1} = H_O^{(I_1,...,I_{i+1})}(I_{i+1})(...,\theta^{(j_1,...,j_\ell)},...)$, which makes sense since $<^\Omega$ ${}^{(I_1,...,I_{i+1})} = <^\Omega \, {}_{(i_1,...,i_\ell)}^{(I_1,...,I_i)}$ for $(i_1,...,i_\ell)$ the last index in $C_{(I_1,...,I_i)}$. We also set $\alpha_{u+2} = H_1^{(I_1,...,I_{i+1})}(I_{i+1})(...,\theta^{(\vec{j})},...)$.

For $2 \leq v \leq r_{i+1}$, we set $\alpha_{u+2\cdot v-1}$ equal to $H_{O(u_1,...,i_\ell)}^{(I_1,...,I_{i+1})}(...,\theta^{(\vec{j})},...)$, where $(i_1,...,i_\ell)$ is the $v-1$st index in $C_{(I_1,...,I_{i+1})}$, and the $(\vec{j})$'s enumerate the indices in $C_{(I_1,...,I_{i+1})}$ occurring before $(i_1,...,i_\ell)$ in $C_{(I_1,...,I_{i+1})}$ along with the indices in $C_{(I_1,...,I_{i'})}$, $i' < i+1$, and occurring in the same order as in $<^\Omega \, {}_{(i_1,...,i_\ell)}^{(I_1,...,I_{i+1})}$. We also set $\alpha_{u+2\cdot v} = H_{1(i_1,...,i_\ell)}^{(I_1,...,I_{i+1})}(...,\theta^{(\vec{j})},...)$.

If $i+1 < k$, we set $\alpha_{u+2\cdot r_{i+1}+1} = H_{O(i_1,...,i_\ell)}^{(I_1,...,I_{i+1})}(...,\theta^{(\vec{j})},...)$ where $(i_1,...,i_\ell)$ is the last index in $C_{(I_1,...,I_{i+1})}$, and the $(\vec{j})$ are as above. In this case, we also set $\alpha_{u+2\cdot r_{i+1}+2} = H_{1(i_1,...,i_\ell)}^{(I_1,...,I_{i+1})}$ $(...,\theta^{(\vec{j})},...)$.

If $i+1 = k$, we set $\alpha_{u+2\cdot r_{i+1}+1} = H_{O(i_1,...,i_\ell)}^{(I_1,...,I_{i+1})}(...,\theta^{(\vec{j})},...)$ as above. We then set $\alpha_{u+2\cdot r_{i+1}+c}$ $= H_{c(i_1,...,i_\ell)}^{(I_1,...,I_{i+1})}(...,\theta^{(\vec{j})},...)$, out to $\alpha_m = \alpha_{u+2\cdot r_{i+1}+(m-u-2\cdot r_{i+1})}$.

We show that this is well defined (for the fixed $G : <^m \rightarrow \boldsymbol{\delta}_{2n+1}^1$). We let $H_1,...,H_p$ and $H_1',...,H_p'$ (for the appropriate integer p) be given representing the same θ as in the definition of $\mathcal{U}$. Hence for all $q \leq p$, if $H_q = H_{(i_1,...,i_\ell)}^{(I_1,...,I_k)}$ then $H_q = H_q'$ almost everywhere with respect to $<^\Omega \, {}_{(i_1,...,i_\ell)}^{(I_1,...,I_k)}$. We fix a c.u.b. subset of $\boldsymbol{\delta}_{2n+1}^1$, C, such that for $f : <^T \rightarrow C$ of the correct type, for all $1 \leq q \leq p$, $H_q(\langle T \rangle) = H_q'(\langle T \rangle)$, where T is the sequence corresponding to f as in the definition of the measure $\mathcal{U}_{(i_1,...,i_k)}^{(I_1,...,I_k)}$ corresponding to H_q.

It then follows that for $f : <^T \rightarrow C$ of the correct type, that if $(\alpha_1, \alpha_2, ..., \alpha_m)$, $(\alpha_1', \alpha_2', ..., \alpha_m')$ denote the sequences as in the definition of $\mathcal{F}$, then $\alpha_1 = \alpha_1', ..., \alpha_m = \alpha_m'$, and hence $\mathcal{F}(G; \vec{H}) = \mathcal{F}(G; \vec{H}')$, and hence $\mathcal{F}$ is well defined with respect to the choice of the H's representing θ.

We now show that $\mathcal{F}$ is well defined with respect to the choice of the functions $G_1, ..., G_{r_0}$ representing β. It is again enough to show this componentwise, and we again suppress the subscript r. We suppose, then, that $[G_1] = [G_2]$, and establish that for almost all θ w.r.t. $\mathcal{U}$ that $\mathcal{F}(G_1; \theta) = \mathcal{F}(G_2; \theta)$. We consider the following cases:

1) $\mathcal{V}_1 \in W_{2n+1}$. If $G_1, G_2 : S_{2n-1}^{\ell',m} \rightarrow \boldsymbol{\delta}_{2n+1}^1$ agree on a measure one set A w.r.t $S_{2n-1}^{\ell',m}$, then

from $\mathcal{B}_{2n+1}$, for almost all θ, f_θ has range in A almost everywhere, and hence $G_1 \circ f_\theta = G_2 \circ f_\theta$ agree almost everywhere w.r.t. the $v^{(i_1,\dots,i_k)}$ (and agree if $k=1$).

2) $\mathcal{V}_1 \in \mathcal{S}_{2n+1}$, and $n_1 \notin \text{dom} <^T$. We fix $G_1 : \boldsymbol{\delta}^1_{2n+1} \to \boldsymbol{\delta}^1_{2n+1}$ of the correct type, and fix G_2 s.t. $G_1 = G_2$ a.e. w.r.t. $\mu_{2n+1}^{<S^{\ell',m}_{2n-1}}$. We let C be a c.u.b. subset of $\boldsymbol{\delta}^1_{2n+1}$ s.t. for $g : S^{\ell',m}_{2n-1} \to C$ of the correct type $G_1([g]) = G_2([g])$. If the above fails, then we get (from the strong partition relation on $\boldsymbol{\delta}^1_{2n+1}$) a c.u.b. $C_2 \subseteq C$ s.t. for $H_2 : <^{\Omega_2} \to C_2$ of the correct type, there is a c.u.b. $C_1 (= C_1(H_2))$ s.t. for $H_1 : <^{\Omega_1} \to C_1$ of the correct type $\mathcal{F}(G_1; H_2, H_1) \neq \mathcal{F}(G_2; H_2, H_1)$. We fix an index $(I_1, \dots, I_k)$. If $\mathcal{F}(G_1; H_2, H_1) = (\dots, \alpha_1^{(I_1,\dots,I_k)}, \dots)$ and similarly for $\mathcal{F}(G_2; H_2, H_1)$, and $\alpha_1^{(I_1,\dots,I_k)}$, $\alpha_2^{(I_1,\dots,I_k)}$ are represented by functions $F_1^{(\vec{I})}$, $F_2^{(\vec{I})}$ as in the definition of $\mathcal{F}$, then we show that for almost all $f : <^T \to \boldsymbol{\delta}^1_{2n+1}$, $F_1^{(\vec{I})}([f]) = F_2^{(\vec{I})}([f])$ to get a contradiction. We have that $F_1^{(\vec{I})}([f]) = G_1 \circ \bar{F}^{(\vec{I})}([f]), F_2^{(\vec{I})} = G_2 \circ \bar{F}^{(\vec{I})}([f])$, where $\bar{F}$ is as in the definition of $\mathcal{U}$ (i.e. represents θ corresponding to H_2, H_1). $\bar{F}^{(\vec{I})}([f])$ is represented with respect to $S^{\ell',m}_{2n-1}$ $(= v(\mathcal{V}_2))$ by a function g, where for $\theta < \theta(S^{\ell',m}_{2n-1})$, $g(\theta) = H_2(\gamma)$, for some γ as in the definition of $\mathcal{U}$, and hence $g(\theta) \subseteq C_2 \subseteq C$. Since H_2 is of the correct type, g is of uniform cofinality ω.

It remains to show that g is almost everywhere strictly increasing.

It follows from D_{2n+1} that there is an r' and a measure one set A w.r.t. $\tilde{v}^2_2$ s.t. if $(\xi_1, \dots, \xi_n, \xi'_1, \dots, \xi'_n)$, $(\eta_1, \dots, \eta_n, \eta'_1, \dots, \eta'_n)$ are in A, then:

a) if $\xi'_r < \eta_r$, then for all indices $(n_1, i_2, \dots, i_k) \in \text{dom } T$, for almost all δ w.r.t. $u^{(n_1,i_2,\dots,i_k)}$, $(\Pi_{(n_1)}^{(n_1,\dots,i_k)} \circ f_{(\xi_1,\dots,\xi_n)}(\delta))(r) < \Pi_{(n_1)}^{(n_1,i_2,\dots,i_k)} \circ f_{(\eta_1,\dots,\eta_n)}(\delta))(r)$, where $f_{(\xi_1,\dots,\xi_n)}$, $f_{(\eta_1,\dots,\eta_n)}$ denote the embeddings as in H_{2n+1}, and r corresponds to the component of the measure $v^{(n_1)}$ such that in identifying $\theta_{v^{(n_1)}}$ with an ordinal, we order first by the rth component of the product measure $v^{(n_1)}$. Hence, we have $f((f_{(\xi_1,\dots,\xi_n)}(\delta))^{(n_1,\dots,i_k)} < f((f_{(\eta_1,\dots,\eta_n)}(\delta))^{(n_1,\dots,i_k)})$.

b) if $\xi'_{r'} < \eta'_{r'}$, then for almost all δ w.r.t. $\mathcal{U}^{(n_1)}$, $\theta_1 = (f_{(\xi'_1,\dots,\xi'_n)}(\delta))(r) < (f_{(\eta'_1,\dots,\eta'_n)}(\delta))(r)$ $= \theta_2$, and hence $\sup_{\beta:\rho(\beta)<\rho(\theta_1)} f(\langle n_1 \beta \langle) \leq \sup_{\beta:\rho(\beta)<\rho(\theta_2)} f(\langle n_1, \beta \rangle)\beta\rangle)$, (where ρ is the minimal pressing down, monotonic, non-constant a.e. function on $\theta_{v^{(n_1)}}$.).

Hence, it follows that if g_1 denotes the function as in the definition of $\mathcal{U}$, then $g_1(\xi_1, \dots, \xi_n, \xi'_1, \dots, \xi'_n) < g_1(\eta_1, \dots, \eta_n, \eta'_1, \dots, \eta'_n)$ for such $\vec{\xi}$, $\vec{\eta}$.

From $H_{2n+1^{(i)}}$ and D_{2n+1} it also follows that there is a measure one set A_2 w.r.t. $S^{\ell',m}_{2n-1}$ such that for $\theta_1 < \theta_2$ in A_2, for almost all δ w.r.t. $\mathcal{U}_2$ (as in the definition of $\mathcal{U}$, $\mathcal{U}_2$ = the measure from H_{2n+1} for $\tilde{v}^2_2$ and $S^{\ell',m}_{2n-1}$), if $f_{\theta_1}(\delta) = (\xi_{J_1}, \dots, \xi_{J_p})$, $f_{\theta_2}(\delta) = (\eta_{J_1}, \dots, \eta_{J_p})$, then for almost all $(\gamma_{i_1}, \dots, \gamma_{i_q})$ w.r.t. $D_1 \times \dots \times D_q$ (as in the definition of $\mathcal{U}$), if we consider the corresponding sequences $(\xi_1, ..., \xi_n, \xi'_1, ..., \xi'_n)$ and $(\eta_1, ..., \eta_n, \eta'_1, ..., \eta'_n)$, we have that $\vec{\xi}$, $\vec{\eta}$ are in A, and $\xi_{r'} < \eta_{r'}$, $\xi'_{r'} < \eta'_{r'}$. Hence, it follows that for $\theta_1 < \theta_2$ in A, $g(\theta_1) < g(\theta_2)$, so g is almost everywhere strictly increasing.

Hence $\mathcal{F}$ is well defined in this case.

3) $\mathcal{V}_1 \in \mathcal{S}_{2n+1}$ and $n_1 \in \operatorname{dom} <^T$. We let $G_1 : <^m \rightarrow \boldsymbol{\delta}^1_{2n+1}$ be of the correct type, and $G_2 : <^m \rightarrow \boldsymbol{\delta}^1_{2n+1}$ be such that $G_1 = G_2$ a.e. w.r.t. the m-fold product of the ω-cofinal normal measure on $\boldsymbol{\delta}^1_{2n+1}$. We let C be a c.u.b. subset of $\boldsymbol{\delta}^1_{2n+1}$ such that for $\alpha_2 < \cdots < \alpha_m < \alpha_1$ in C of cofinality ω, $G_1(\alpha_2, ..., \alpha_m, \alpha_1) = G_2(\alpha_2, ..., \alpha_m, \alpha_1)$. We fix an index $(I_1, ..., I_k)$ and show $\alpha_1^{(I_1,...,I_k)} = \alpha_2^{(I_1,...,I_k)}$ as in the previous case. We fix functions $\vec{H} : <^{\Omega}\ {}^{(I_1,...,I_k)}_{(i_1,...,i_\ell)} \rightarrow C$ of the correct type representing θ (which happens for almost all θ w.r.t. $\mathcal{U}$). For $F_1^{(I_1,...,I_k)}$ representing $\alpha_1^{(I_1,...,I_k)}$ w.r.t. $\mu^{<T}_{2n+1}$ as in the definition of $\mathcal{F}$, we have that for almost all $f : <^T \rightarrow \boldsymbol{\delta}^1_{2n+1}$ that $F_1^{(I_1,...,I_k)}([f]) = G_1(\alpha_2, ..., \alpha_k, \alpha_1)$, and similarly for F_2. Since the range of the H's is in C, it follows that the $\alpha_2, ..., \alpha_m \in C$. We may also assume $f : <^T \rightarrow C$, so that $\alpha_1 \in C$. Also, since the H's are of the correct type, the α_i have cofinality ω. It remains to show that $\alpha_2 < \cdots < \alpha_m < \alpha_1$.

We clearly have that for almost all f $\alpha_1 = f(\langle n_1 \rangle) > \alpha_i = H_i([f])$, since the H_i do not depend on $f(\langle n_1 \rangle)$.

It also follows readily that we may assume (which happens for almost all θ) that the $H_1, ..., H_p$ are chosen s.t. for almost all f, $H_i([f]) < H_{i+1}([f])$. This uses the fact that the indices $\vec{j}$ occurring in the definition of $<^{\Omega}\ {}^{(I_1,...,I_k)}_{(i_1,...,i_\ell)}$ corresponding to H_i are a subset of those corresponding to H_{i+1}. Hence it follows that $\alpha_2 < \cdots < \alpha_m$, since if $\alpha_i = H_a([f])$, $a_{i+1} = H_b([f])$, then $a < b$, from the definition of $\mathcal{F}$.

Hence $\mathcal{F}$ is well defined in this case.

This completes the proof that $\mathcal{F}$ is well defined.

We proceed to establish H_{2n+3}.

We first consider $H_{2n+3}(i)$.

We let A have measure one with respect to $\mathcal{V}_1 = \Pi_r\, \mathcal{V}_1^r$. From the definition of the product measure and the strong partition relation on $\boldsymbol{\delta}^1_{2n+1}$, it follows readily that there is a c.u.b. $C \subseteq \boldsymbol{\delta}^1_{2n+1}$ such that if $\alpha = (\alpha_1, ..., \alpha_r, ..., \alpha_{r_0})$ is represented by functions $F_1, ..., F_r$ (as in the definition of the measures $\mathcal{V}_1^r$, then $\alpha \in A$ provided the following hold:

1) $F_1, ..., F_r$ have range in C and are of the correct type (where $F_i : <^{T_i} \rightarrow \boldsymbol{\delta}^1_{2n+1}$ if $\mathcal{V}_1^i \in$

$\mathcal{W}_{2n+1}$ and $F_i : <^{\Omega_i} \rightarrow \delta^1_{2n+1}$ if $\mathcal{V}^i_1 \in \mathcal{S}_{2n+1}$).

2) a) If $i < j$ and $\mathcal{V}^i_1, \mathcal{V}^j_1 \in \mathcal{W}_{2n+1}$, then $\sup_{\text{a.e.}} F_i < \inf F_j$.

b) If $i < j$, $\mathcal{V}^i_1, \mathcal{V}^j_1 \in \mathcal{S}_{2n+1}$, $n_1 \notin \operatorname{dom} <^{T_i}$, $n_1 \notin \operatorname{dom} <^{T_j}$ (where the n_1 may be different for i, j), and $\operatorname{cof} \theta_{v_i^{n_1}} = \operatorname{cof} \theta_{v_j^{n_1}}$ (where $v_i^{n_1}$ is the measure v^{n_1} corresponding to $<^{T_i}$ and similarly for j), then there is a c.u.b. $C_2 \subseteq \delta^1_{2n+1}$ such that if $f_1 : <^{T_i} \rightarrow C_2$, $f_2 : <^{T_j} \rightarrow C_2$ are of the correct type and $\sup_{\text{a.e.}} f_1 = \sup_{\text{a.e.}} f_2$, then $F_i([f_1]) < F_j([f_2])$.

c) Same as b above where $n_1 \in \operatorname{dom} <^{T_i}$ and $n_1 \in \operatorname{dom} <^{T_j}$, and we remove the restriction on $\operatorname{cof} \theta_{v_i^{n_i}} = \operatorname{cof} \theta_{n_j^{n_1}}$. Here, of course, $\sup_{\text{a.e.}} f_1$ means $f_1(\langle n_1 \rangle)$ and similarly for f_2.

We fix such a c.u.b. set C. We fix $\beta = (\beta_1, ..., \beta_r, ..., \beta_{r_0})$ in the measure space $\mathcal{V}_2 = \Pi_r \mathcal{V}^r_2$ represented by functions $G_1, ..., G_r, ..., G_{r_0}$ satisfying the following (which happens for almost all β w.r.t. $\mathcal{V}_2$):

i) $G_1, ..., G_{r_0}$ have range in C and are of the correct type.

ii) a) If $i < j$, $V^i_2, V^j_2 \in \mathcal{W}_{2n+1}$, then $\sup_{\text{a.e.}} G_i < \inf G_j$.

b) If $i < j$, $V^i_2, V^j_2 \in \mathcal{S}^{\ell,m}_{2n+1}$ with $\ell > 1$ (ℓ, m may depend on i, j) and $\operatorname{cof} \theta(\mathcal{S}^{\ell_i,m_i}_{2n-1}) = \operatorname{cof} \theta(\mathcal{S}^{\ell_j,m_j}_{2n-1})$ (this is in fact equivalent to $\ell_i = \ell_j$), then there is a c.u.b. subset $C_2 \subseteq \delta^1_{2n+1}$ such that for $g_1 : \theta(\mathcal{S}^{\ell_i,m_i}_{2n-1}) \rightarrow C_2$, $g_2 : \theta(\mathcal{S}^{\ell_i,m_i}_{2n-1}) \rightarrow C_2$ of the correct type, if $\sup_{\text{a.e.}} g_1 = \sup_{\text{a.e.}} g_2$, then $G_1([g_1]) < G_j([g_2])$.

c) if $i < j$, $V^i_2, V^j_2 \in \mathcal{S}^{1,m}_{2n+1}$, then there is a c.u.b. $C_2 \subseteq \delta^1_{2n+1}$ such that for $(\gamma_2, ..., \gamma_m, \gamma_1)$, $(\delta_2, ..., \delta_m, \delta_1)$ in C_2, if $\gamma_1 = \delta_1$ then $G_1(\gamma_2, ..., \gamma_m, \gamma_1) < G_2(\delta_2, ..., \delta_m, \delta_1)$.

We fix such $G_1, ..., G_{r_0}$.

It then follows readily that for almost all $\theta = (\theta_1, ..., \theta_{r_0})$ w.r.t. $\mathcal{U} = \Pi_r \mathcal{U}^r$, where we fix $\vec{H}_1, ..., \vec{H}_{r_0}$ representing $\theta_1, ..., \theta_{r_0}$ (where $\vec{H}_r = (H_{r,1}, H_{r,2})$ if $n_1 \notin \operatorname{dom} <^{T_r}$ and $\vec{H}_r = (..., H^{(I_1,...,I_k)}_{r(i_1,...,i_\ell)}, ...)$ if $n_1 \in \operatorname{dom} <^{T_r}$), that if $\alpha = (\alpha_1, ..., \alpha_{r_0}) = \mathcal{F}_\beta(\theta)$, where α_i is represented by F_i as in the definition of $\mathcal{F}$, then the following are satisfied:

a) If $i < j$ and $\mathcal{V}^i_1, \mathcal{V}^j_1 \in \mathcal{W}_{2n+1}$, then $\sup_{\text{a.e.}} F_i < \inf F_j$.

b) If $\mathcal{V}^i_1 \in \mathcal{S}_{2n+1}$ and $n_1 \notin \operatorname{dom} <^{T_i}$, then for almost all $f_i : <^{T_i} \rightarrow \delta^1_{2n+1}$, if $F_i([f_i]) = G_i([g])$ where $g : \theta(\mathcal{S}^{\ell_i,m_i}_{2n-1}) \rightarrow \delta^1_{2n+1}$ is as in the definition of $\mathcal{F}$, then $\sup_{\text{a.e.}} g = \sup_{\text{a.e.}} f_i$.

c) If $\mathcal{V}^i_1 \in \mathcal{S}_{2n+1}$ and $n_1 \in \operatorname{dom} <^{T_i}$, then for almost all $f_i : <^{T_i} \rightarrow \delta^1_{2n+1}$, $F_i([f_i]) = G_i(\alpha_2, ..., \alpha_m, \alpha_1)$ where $\alpha_1 = \sup f_i$.

a) and c) are immediate from the definition of $\mathcal{F}$ and the choice of the G_i. b) follows from H_{2n+1} (i), (ii), the fact that measure one sets are cofinal in v^{n_1}, and the fact that almost all f_i are sufficiently closed with respect to the H_i.

That is, there is a measure one set E w.r.t. $E_1 \times \cdots \times E_p$ (as in the definition of $\mathcal{U}$) s.t. for $(\xi_{i_1}, ..., \xi_{i_p}) \in E$, for almost all $(\xi_{j_1}, ..., \xi_{j_q})$ w.r.t. $D_1 \times \cdots \times D_q$ and $(\xi_1, ..., \xi_n, \xi'_1, ..., \xi'_n)$ the enumeration of these ordinals in the order corresponding to the measures in $\tilde{v}_2^2$, we have that $g_1(\xi_1, ..., \xi_n, \xi'_1, ..., \xi'_n) = H_{i,1}(\langle T \rangle)$, where (for a fixed $(I_1, ..., I_k)$), $T = \langle \alpha_0,\ I_1, ..., I_k,$ $\alpha^{(i_1,...,i_k)}, ...\rangle$ and $\alpha_0, ..., \alpha^{(i_1,...,i_k)} <$ sup range f_i. This follows from H_{2n+1} (ii) and the fact that the range of f_i may be taken as closed under the $j_{u^{(n_1,i_2,...,i_k)}}$.

It further follows from the definition of the measures $E_1, ..., E_p$ and D_{2n+1} that there is a measure one set E' w.r.t. $E_1 \times \cdots \times E_p$ s.t. if $(\xi_i, ..., \xi_{i_p}) \in E'$ then $\sup_{\xi_{j_1}, ..., \xi_{j_q}} g_1(\xi_1, ..., \xi_n,$ $\xi'_1, ..., \xi'_k) <$ sup range f_i.

It then follows from another application of H_{2n+1} (ii) that for almost all γ w.r.t. $S^{\ell',m}_{2n-1}(= v(\mathcal{V}_2))$ that $g(\gamma)$ is represented w.r.t. u_2 (as in the definition of $\mathcal{U}$, u_2 is the measure from H_{2n+1} corresponding to $E_1 \times \cdots \times E_p$ and $S^{\ell',m}_{2n-1}$) by a function $g_\gamma(\delta)$ s.t. $\sup_{\text{a.e.}} g_\gamma(\delta) <$ sup range f_i.

Hence, we may assume that $g(\gamma) <$ sup range f_i almost everywhere. Hence $\sup_{\text{a.e.}} g \leq$ sup range f_i.

Also $\sup_{\text{a.e.}} g \geq$ sup range f_i follows from H_{2n+1} (i) and the fact that measure one sets in v^{n_1} are cofinal in $\theta_{v^{n_1}}$.

This establishes b) above.

The second claim 2) above about the F_i now follows from a), b), c) and the choice of the G_i.

Hence, it remains to establish 1) above, and it is enough to establish this componentwise. We again suppress the subscript r. We consider the following cases:

1) $\mathcal{V}_1 \in \mathcal{W}_{2n+1}$. In this case, $\mathcal{F}_\beta(\theta)$ is represented by $\bar{f} : <^T\rightarrow \delta^1_{2n+1}$ given by $\bar{f} : G \circ f_\theta$, where f_θ is as in $\mathcal{B}_{2n+1} : <^T\rightarrow \theta(S^{\ell',m}_{2n-1})$. Since f_θ is of the correct type almost everywhere, and G is of the correct type, it follows that $\bar{f}$ is of the correct type almost everywhere, and we may assume (by Lemma 4) that it is of the correct type everywhere. Since range $G \subseteq C$, range $\bar{f} \subseteq C$.

2) $\mathcal{V}_1 \in \mathcal{S}_{2n+1}$ and $n_1 \notin \text{dom} <^T$. We fix $\vec{H}$ of the correct type representing θ as in the definition of $\mathcal{U}$ (which happens for almost all θ w.r.t. $\mathcal{U}$). Since for almost all $f : <^T \rightarrow \boldsymbol{\delta}^1_{2n+1}$, $F([f]) = G([g])$, $g : \theta(S^{\ell',m}_{2n-1}) \rightarrow \boldsymbol{\delta}^1_{2n+1}$, it follows that for almost all f, $F([f]) \in C$, and has uniform cofinality ω.

We show that F is order-preserving almost everywhere.

We fix indices $(\mathcal{I}_1, ..., \mathcal{I}_k)$, $(\mathcal{J}_1, ..., \mathcal{J}_\ell)$ in $\mathcal{C}$, and fix $f_1, f_2 : <^T \rightarrow \boldsymbol{\delta}^1_{2n+1}$ of the correct type, and let $S_1 = \langle \alpha_0, \mathcal{I}_1, ..., \alpha^{(i_1)}, ..., \mathcal{I}_k, ..., \alpha^{(i_1, ..., i_k)}, ... \rangle$, $S_2 = \langle \beta_0, \mathcal{J}_1, ..., \beta^{(j_1)}, ..., \mathcal{J}_k, ..., \beta^{(j_1, ..., j_\ell)}, ... \rangle$ denote the corresponding sequences as in the definition of $\mathcal{V}_1$. We assume that $S_1 <^{K.B.} S_2$, and proceed to show that $F(\langle S_1 \rangle) < F(\langle S_2 \rangle)$. We consider the following cases:

a) $\alpha_0 < \beta_0$. That is $\sup_{\text{a.e.}} f_1 < \sup_{\text{a.e.}} f_2$. It then follows from H_{2n+1} (i) that for almost all $(\xi_1, ..., \xi_n, \xi'_1, ..., \xi'_n)$ that if $T_1 = \langle \gamma_0, \mathcal{I}_1, ..., \gamma^{(i_1)}, ..., \mathcal{I}_i, ..., \gamma^{(i_1, ..., i_k)}, ... \rangle$, $T_2 = \langle \delta_0, \mathcal{J}_1, ..., \mathcal{J}_\ell, ..., \delta^{(i_1, ..., i_\ell)} \rangle$ denote the sequences as in the definition of $\mathcal{U}$, then $\gamma < \delta_0$, and hence $T_1 <^{\Omega} T_2$, and the result follows.

b) $\alpha_0 = \beta_0$ and there is a lexicographically least position in which S_1, S_2 disagree which is of the form $\alpha^{(i_1, ..., i_m)} < \beta^{(i_1, ..., i_m)}$, where $\mathcal{I}_1 = \mathcal{J}_1, ..., \mathcal{I}_m = \mathcal{J}_m$. It then follows from H_{2n+1} (i) that for almost all $(\xi_1, ..., \xi_n, \xi'_1, ..., \xi'_k)$ that T_1, T_2 agree up to $\gamma^{(i_1, ..., i_m)}$ and that $\gamma^{(i_1, ..., i_m)} < \delta^{(i_1, ..., i_m)}$.

c) $\alpha_0 = \beta_0$, and the least position where S_1, S_2 disagree is of the form $\mathcal{I}_m < \mathcal{J}_m$. This is similar to the previous case.

d) S_1 extends S_2. As above, by H_{2n+1} (i), we have that for almost all $(\xi_1, ..., \xi_n, \xi'_1, ..., \xi'_k)$, T_1 extends T_2, hence $H_1(\langle T_1 \rangle) < H_1(\langle T_2 \rangle)$, from which the result follows readily.

3) $\mathcal{V}_1 \in \mathcal{S}_{2n+1}$ and $n_1 \in \text{dom} <^T$. We again fix functions $\vec{H}$ representing θ as in the definition of $\mathcal{U}$, and fix indices $(\mathcal{I}_1, ..., \mathcal{I}_k)$, $(\mathcal{J}_1, ..., \mathcal{J}_\ell)$, functions f_1, f_2 of the correct type and suppose $S_1 <^{K.B.} S_2$, where S_1, S_2 denote the corresponding sequences. We let $\alpha_2, ..., \alpha_m, \alpha_1$, $\beta_2, ..., \beta_m, \beta_1$ denote the sequences of ordinals as in the definition of $\mathcal{F}$. We consider the following cases:

a) $\sup f_1 = f_1(\langle n_1 \rangle) < f_2(\langle n_1 \rangle) = \sup f_2$. In the case, $\alpha_1 < \beta_1$, and $G(\alpha_2, \ldots, \alpha_m, \alpha_1) < G(\beta_2, \ldots, \beta_m, \beta_1)$ follows from the choice of G.

b) $\alpha_1 = \beta_1$ and there is a lexicographically least position where S_1, S_2 disagree which is of the form $\alpha^{(i_1,\ldots,i_m)} < \beta^{(i_1,\ldots,i_m)}$, where $I_1 = J_1, \ldots, I_m = J_m$. We then have that for each $H^{(I_1,\ldots,I_v)}_{(i'_1,\ldots,i'_w)}$, where $v < m$ or $v = m$ and $(i'_1, \ldots, i'_w)$ precedes $(i_1, \ldots, i_m)$, that $H^{(I_1,\ldots,I_v)}_{(i'_1,\ldots,i'_w)}([f_1]) = H^{(I_1,\ldots,I_v)}_{(i'_1,\ldots,i'_w)}([f_2])$, since from the definition of $\mathcal{U}^{(I_1,\ldots,I_v)}_{(i'_1,\ldots,i'_w)}$, H depends only on the ordinals $\gamma^{(i''_1,\ldots,i''_{w''})}$ represented by f for indices $(i''_1, \ldots, i''_{w''})$ preceding and including $(i'_1, \ldots, i'_w)$, and hence $\gamma_1^{(i''_1,\ldots,i''_{w''})} = \gamma_2^{(i''_1,\ldots,i''_{w''})}$ for f_1, f_2. Hence $\alpha_2 = \beta_2, \ldots, \alpha_k = \beta_k$ for k corresponding to the H preceding $H^{(I_1,\ldots,I_m)}_{0(i_1,\ldots,i_m)}$. Since $H^{(I_1,\ldots,I_m)}_{0(i_1,\ldots,i_m)}$ is order-preserving and $\gamma_1^{(i_1,\ldots,i_m)} < \gamma_2^{(i_1,\ldots,i_m)}$, it then follows that $\alpha_{k+1} < \beta_{k+1}$. Since G is order-preserving, it then follows that $G(\alpha_2, \ldots, \alpha_m, \alpha_1) < G(\beta_2, \ldots, \beta_m, \beta_1)$.

c) $\alpha_1 = \beta_1$, and the least position where S_1, S_2 disagree is of the form $I_m < J_m$. Proceeding as above, we have that $\alpha_2 = \beta_2, \ldots, \alpha_k = \beta_k$ corresponding to the functions H preceding $H^{(I_1,\ldots,I_m)}(1)$ used in the definition of $\mathcal{F}$. It then follows that $\alpha_{k+1} < \beta_{k+1}$ since $\alpha_{k+1} = H^{(I_1,\ldots,I_m)}(I_m)([f]) < H^{(I_1,\ldots,I_m)}(J_m)([f]) = \beta_{k+1}$, since in the b-fold product measure $\mathcal{U}^{(I_1,\ldots,I_m)}$, we may assume the functions $H^{(I_1,\ldots,I_m)}(1), \ldots, H^{(I_1,\ldots,I_m)}(b)$ are such that if $i < j$ then $H^{(\vec{I})}(i)([f]) < H^{(\vec{I})}(j)([f])$. Hence, $G(\alpha_2, \ldots, \alpha_m, \alpha_1) < G(\beta_2, \ldots, \beta_m, \beta_1)$.

d) $\alpha_1 = \beta_1$, and S_1 extends S_2. We again have that $\alpha_2 = \beta_2, \ldots, \alpha_k = \beta_k$ corresponding to H's preceding $H_0(i_1, \ldots, i_m)^{(I_1,\ldots,I_m)}$, where $(i_1, \ldots i_m)$ denotes the last index in $C_{(I_1,\ldots,I_\ell)}$, and $(I_1, \ldots, I_\ell) = (J_1, \ldots J_\ell)$ here. We then have from the definition of $\mathcal{F}$ that $\alpha_{k+1} = H^{(I_1,\ldots I_\ell)}_{0(i_1,\ldots,i_m)}([f_1]) = H^{(I_1,\ldots,I_\ell)}_{0(i_1,\ldots i_m)}([f_2]) = \beta_{k+1}$ as above. Also, $\alpha_{k+2} = H^{(I_1,\ldots,I_\ell)}_{1(i_1,\ldots,i_m)}([f_1]) < H^{(I_1,\ldots,I_\ell)}_{2(i_1,\ldots,i_m)}([f_2]) = \beta_{k+2}$ without loss of generality. Hence $G(\alpha_2, \ldots, \alpha_m, \alpha_1) < G(\beta_2, \ldots, \beta_m, \beta_1)$.

This establishes 1) above in all cases.

This completes the proof of $H_{2n+3}(i)$.

We now consider $H_{2n+3}(ii)$: It is enough to establish this componentwise. We consider the following cases:

1) $\mathcal{V}_1 \in \mathcal{W}_{2n+1}$. In this case, for all β w.r.t. $\mathcal{V}_2$ represented by a $G : \theta(S^{\ell',m}_{2n-1}) \to \boldsymbol{\delta}^1_{2n+1}$ of the correct type we have from the definition of $\mathcal{F}$ that for almost all θ w.r.t. $\mathcal{U}$ that $\mathcal{F}_\beta(\theta) = [G \circ f_\theta]$, and $\sup_{\text{a.e.}} G \circ f_\theta \leq \sup_{\text{a.e.}} G$ (in fact strictly $<$), which establishes $H_{2n+3}(ii)$.

2) $\mathcal{V}_1 \in \mathcal{S}_{2n+1}$ and $n_1 \notin \operatorname{dom} T$. We fix β represented by a $G : \boldsymbol{\delta}^1_{2n+1} \to \boldsymbol{\delta}^1_{2n+1}$ of the correct type. It then follows as in the proof of $H_{2n+3}(i)$ that for almost all θ w.r.t. $\mathcal{U}$ that if $\bar{F}$ represents θ w.r.t. $\mu^{<T}_{2n+1}$ that for almost all $f : <^T \to \boldsymbol{\delta}^1_{2n+1}$ that $\bar{F}([f]) = [g]_{S^{\ell',m}_{2n-1}}$, where $\sup_{\text{a.e.}} g = \sup_{\text{a.e.}} f$. Hence, if F represents $\mathcal{F}_\beta(\theta)$ w.r.t. $\mu^{<T}_{2n+1}$ as in the definition of $\mathcal{F}$, then for

almost all f, $F([f]) = G([g]) < G(J_{S^{\ell',m}_{2n-1}}(\sup_{\text{a.e.}} f))$. This establishes $H_{2n+3}(ii)$.

3) $\mathcal{V}_1 \in \mathcal{S}_{2n+1}$ and $n_1 \in \text{dom } T$. We again fix β in the measure space $\mathcal{V}_2$ represented by a $G : <^m \rightarrow \delta^1_{2n+1}$ of the correct type. For any θ in the measure space $\mathcal{U}$, it follows from the definition of $\mathcal{F}$ that if F represents $\mathcal{F}_\beta(\theta)$ w.r.t. $\mu^{<^T}_{2n+1}$, then for almost all $f : <^T \rightarrow \delta^1_{2n+1}$, $F([f]) = G(\alpha_2, \ldots, \alpha_m, \alpha_1)$, where $\alpha_1 = \sup f = f(\langle n_1 \rangle)$, and $\alpha_2, \ldots, \alpha_m < \alpha_1$. Hence $F([f]) < G'(\alpha_1) \equiv \sup_{\alpha_2, \ldots, \alpha_{m'} < \alpha_1} G(\alpha_2, ..., \alpha_m, \alpha_1) < \delta^1_{2n+1}$. This establishes $H_{2n+3}(ii)$.

We now consider $H_{2n+3}(iii)$. It is enough to establish this componentwise. We again consider the following cases:

1) $\mathcal{V}_1 \in \mathcal{W}_{2n+1}$. For $\mathcal{V}'_1$ as in $H_{2n+3}(iii)$ and corresponding measures $\mathcal{V}_2, \mathcal{V}'_2$, we have that $\mathcal{V}_2 = \mu^{S^{\ell',m}_{2n-1}}_{2n+1}$, $\mathcal{V}'_2 = \mu^{S^{\ell',m'}_{2n-1}}_{2n+1}$ for some m'; here ($\ell' = 2^n - 1$ is maximal) . From the definition of $\sim$, we have that $\mathcal{V}_2 \sim \mathcal{V}'_2$.

2) $\mathcal{V}_1 \in \mathcal{S}_{2n+1}$ and $n_1 \notin \text{dom } <^T$. We let $\mathcal{V}'_1$ be as in H_{2n+3} (iii). It follows from the definition of Π that since $\Pi^{\mathcal{V}'_1}_{\mathcal{V}_1}$ is defined, $v^{n_1} = v'^{n_1}$, a measure on $\theta(v^{n_1}) = \theta_1 \times \cdots \times \theta_b$ for some b, where in identifying $\theta(v^{n_1})$ with an ordinal, we order first by θ_c, say, for some $1 \leq c \leq b$ (which is the same for $\mathcal{V}_1, \mathcal{V}'_1$). Hence, it follows from $H_{2n+1}(iii)$ that $\tilde{v}^2_2 = (E_1 \times \cdots \times E_p)^2$ where $E_i \sim E_j$ and $E_i \in R_{2n-1}$ for $1 \leq i,\ j \leq p$, and $v'^2_2 = (E'_1 \times \cdots \times E'_p)^2$, where $E'_i \sim E_i$. Hence, it follows from $H_{2n+1}(iv)$ that $v_2 = v(\mathcal{V}_2) \sim E_i \sim E'_i \sim v'_2$, and hence $\mathcal{V}_2 \sim \mathcal{V}'_2$ from the definition of $\sim$.

3) $\mathcal{V}_1 \in \mathcal{S}_{2n+1}$ and $n_1 \in \text{dom } <^T$.

We let $\mathcal{V}'_1$ be as in $H_{2n+3}(iii)$. Since $\Pi^{\mathcal{V}'_1}_{\mathcal{V}_1}$ is defined, it follows that $n_1 \in \text{dom } <^{T'}$, (since $<^{T'}$ extends $<^T$) and hence $\mathcal{V}_1 = S^{1,m}_{2n+1}$, $\mathcal{V}'_1 = S^{1,m'}_{2n+1}$ for some m;, and from the definition of $\sim$ it follows that $\mathcal{V}_1 \sim \mathcal{V}'_1$.

This establishes $H_{2n+3}(iii)$.

We now consider $D_{2n+3}(i)$.

For $\mathcal{V}_1 = \Pi_r\ \mathcal{V}^r_1$, $\mathcal{V}_2 = \Pi_r\ \mathcal{V}^r_2$, $\mathcal{U} = \Pi_r\ \mathcal{U}^r$ as in D_{2n+3}, we recall that $\mathcal{F}$ is defined componentwise, that is, for $\beta = (\beta_1, \ldots, \beta_{r_0})$ in the measure space $\mathcal{V}_2$, $\theta = (\theta_1, \ldots, \theta_{r_0})$ in the measure space $\mathcal{U}$, $\mathcal{F}_\beta(\theta) = (\mathcal{F}_{\beta_1}(\theta_1), \ldots, \mathcal{F}_{\beta_r}(\theta_r)) \in$ the measure space $\mathcal{V}_1$. For any fixed $r \leq r_0$, we take $r' = r$, and consider $\mathcal{V}^r_2$, $\mathcal{U}^r$.

1) If $\mathcal{V}^r_1 \in \mathcal{W}_{2n+1}$ and $\beta^r_1 < \beta^r_2$ are represented by functions $G_1,\ G_2 : \theta(S^{\ell',m}_{2n-1}) \rightarrow \delta^1_{2n+1}$ of the correct type, so $G_1 < G_2$ almost everywhere, then for almost all γ w.r.t. u (as in B_{2n+1})

$G_1 \circ f_\gamma < G_2 \circ f_\gamma$, using B_{2n+1}. This establishes $D_{2n+3}(i)$ in this case.

2) If $\mathcal{V}_1^r \in \mathcal{S}_{2n+1}$ and $n_1 \notin \text{dom} <^T$, from the definition of $\mathcal{F}$, for $\beta_1^r < \beta_2^r$ represented by $G_1, G_2 : \boldsymbol{\delta}_{2n+1}^1 \to \boldsymbol{\delta}_{2n+1}^1$ of the correct type, and $\gamma \in \theta(\mathcal{U})$ represented by $\bar{F}$ w.r.t. $\mu_{2n+1}^{<^T}$, if F_1, F_2 represent $\mathcal{F}_{\beta_1^r}(\gamma)$, $\mathcal{F}_{\beta_2^r}(\gamma)$ w.r.t. $\mu_{2n+1}^{<^T}$, then for almost all f w.r.t. $\mu_{2n+1}^{<^T}$, $F_1([f]) = G_1(\bar{F}([f])) < G_2(\bar{F}([f])) = F_2([f])$, since we may assume $\bar{F}$ (which happens for almost all γ w.r.t. $\mathcal{U}$) has the property that for almost all f, $\bar{F}([f])$ is in a measure one set A w.r.t. $\mu_{2n+1}^{S_{2n-1}^{\ell',m}}$ on which $G_1 < G_2$.

3) If $\mathcal{V}_1^r \in \mathcal{S}_{2n+1}$ and $n_1 \in \text{dom} <^T$, $\beta_1^r < \beta_2^r$ are represented by $G_1, G_2 : <^m \to \boldsymbol{\delta}_{2n+1}^1$ of the correct type, then it follows from the definition of $\mathcal{F}$ that if $\gamma \in \theta(\mathcal{U})$ is represented by functions $\vec{H}$ of the correct type (as in the definition of $\mathcal{U}$), and F_1, F_2 represent $\mathcal{F}_{\beta_1^r}(\gamma)$, $\mathcal{F}_{\beta_2^r}(\gamma)$ w.r.t. $\mu_{2n+1}^{<^T}$, then for almost all $f : <^T \to \boldsymbol{\delta}_{2n+1}^1$, $F_1([f]) = G_1(\alpha_2, \ldots, \alpha_m, \alpha_1) < G_2(\alpha_2, \ldots, \alpha_m, \alpha_1) = F_2([f])$, since for almost all γ we have that for almost all f, the α's above are in a measure one set A one which $G_1 < G_2$.

This establishes $D_{2n+1}(i)$.

We now consider $H_{2n+3}(iv)$ and $D_{2n+3}(ii)$. We consider the following cases:

1) $\mathcal{V}_1^r \in \mathcal{W}_{2n+1}$, hence is of the form $\mathcal{V}_1^r = \mu_{2n+1}^{S_{2n-1}^{\ell',m_r}}$ ($\ell' = 2^n - 1$ is maximal). We let m be such tht B_{2n+1} holds for the ordering corresponding to T defined by $v^r = S_{2n-1}^{\ell',m_r}$ for $1 \le r \le r_{0'}$ and $S_{2n-1}^{\ell',m}$. We let $\mathcal{V}_2 = \mu_{2n+1}^{S_{2n-1}^{\ell',m}}$, and let $\mathcal{U}$ be the measure as in B_{2n+1}. We construct $\mathcal{F}$ as in $H_{2n+3}(i - iii)$ in this case. In fact, our previous consideration of $H_{2n+3}(i - iii)$ for the case $\mathcal{V}_1 \in \mathcal{W}_{2n+1}$ included this case.

$D_{2n+3}(ii)$ follows readily form B_{2n+1}.

2) $\mathcal{V}_1^r \in \mathcal{S}_{2n+1}$ and is of the form $S_{2n+1}^{\bar{\ell}_r,m_r}$ where $\bar{\ell}_r > 1$, = the measure induced from the strong partition relation on $\boldsymbol{\delta}_{2n+1}^1$ and $\mu_{2n+1}^{\ell_r,m_r}$, say. Since $\mathcal{V}^{r_1} \sim \mathcal{V}^{r_2}$ it follows that $\ell_1 = \ell_2 = \cdots = \ell_{r_0} = \ell$, say.

Since $S_{2n-1}^{\ell,m_r} \sim S_{2n-1}^{\ell,m_r'}$, it follows from $H_{2n+1}(iv)$ that there is an m s.t. $H_{2n+1}(iv)$ holds from $v_1 = \Pi_r\, S_{2n-1}^{\ell,m_r}$ and $v_2 = S_{2n-1}^{\ell,m}$. We let $\mathcal{V}_2 = S_{2n+1}^{\bar{\ell},m}$.

We let $\mathcal{U}$ be the measure on r_0 tuples of ordinals $(\theta_1, \ldots, \theta_{r_0})$ defined as follows: A has measure one w.r.t. $\mathcal{U}$ if there is a c.u.b. subset of $\boldsymbol{\delta}_{2n+1}^1$, C, such that for $H : \boldsymbol{\delta}_{2n+1}^1 \to C$ of the correct type, $(\theta_1, \ldots, \theta_{r_0}) \in A$ where θ_r is represented with respect to $\mu_{2n+1}^{S_{2n-1}^{\ell,m_r}}$ by a function $\bar{F}_r$, where for $f : \theta(S_{2n-1}^{\ell,m_r}) \to \boldsymbol{\delta}_{2n+1}^1$ of the correct type, $\bar{F}_r([f]) = [g_r]_{S_{2n-1}^{\ell,m}}$, where g_r is defined by: for $\alpha < \theta(S^{\ell,m}_{2n-1})$, $g_r(\alpha) = H([h])$, where $h : \theta(u) \to \boldsymbol{\delta}^1_{2n+1}$ is given by (where u as in $H_{2n+1}(iv)$ for v_1, v_2), for $\beta < \theta(u)$, $h(\beta) = f(f_\alpha(\beta)(r))$, and f_α is as in $H_{2n+1}(iv)$. This is well defined by $H_{2n+1}(iv)$.

We define $\mathcal{F}$ as follows: If β is represented w.r.t. $\mu_{2n+1}^{S^{\ell,m}_{2n-1}}$ by $G : \boldsymbol{\delta}^1_{2n+1} \to \boldsymbol{\delta}^1_{2n+1}$ of the correct type, and $\theta = (\theta_1, \ldots, \theta_{r_0})$ is in the measure space $\mathcal{U}$ represented by $\bar{F}_1, \ldots, \bar{F}_{r_0}$, then we set $\mathcal{F}_\beta(\ldots, \theta_r, \ldots) = (\alpha_1, \ldots, \alpha_{r_0})$, where α_r is represented w.r.t. $\mu_{2n+1}^{S^{\ell,m_r}_{2n-1}}$ by F_r, where for $f : \theta(S^{\ell,m_r}_{2n-1}) \to \boldsymbol{\delta}^1_{2n+1}$ of the connect type, $F_r([f]) = G(\bar{F}_r([f]))$. This is well defined.

The proof of $H_{2n+3}(iv)$ and $D_{2n+3}(ii)$ now follows as in $H_{2n+3}(i - iii)$ and $D_{2n+3}(i)$ above.

3) $\mathcal{V}_1^r \in S_{2n+1}$, and is of the form S^{1,m_r}_{2n+1}. We let $m = (\max_{1 \le r \le r_0} m_r) + 1$, and we set $\mathcal{V}_2 = S^{1,m}_{2n+1}$. We let $\mathcal{U}$ be the $p \equiv r_0 + m$-fold product of the ω-cofinal normal measure on $\boldsymbol{\delta}^1_{2n+1}$.

We define $\mathcal{F}$ as follows: For β represented by $G : <^m\to \boldsymbol{\delta}_{2n+1}$ of the correct type, and $(\gamma_1, \ldots, \gamma_{r_0}, \bar{\gamma}_1, \ldots, \bar{\gamma}_{p-r_0}) \in \theta(\mathcal{U})$, where $\gamma_i, \bar{\gamma}_i < \boldsymbol{\delta}^1_{2n+1}$, we set $\mathcal{F}_\beta(\gamma_1, \ldots, \bar{\gamma}_{p-r_0}) = (\alpha_1, \ldots, \alpha_{r_0})$, where α_r for $1 \le r \le r_0$ is represented by $F_r :<^{m_r}\to \boldsymbol{\delta}^1_{2n+1}$ defined as follows: for $\delta_2 < \cdots < \delta_{m_r} < \delta_1$, we set $F_r(\delta_2, \ldots, \delta_{m_r}, \delta_1) = G(\gamma_r, \bar{\gamma}_1, \ldots, \bar{\gamma}_{m-\bar{m}_r-1}, \delta_2, \ldots, \delta_{m_r}, \delta_1)$. This is well defined for almost all $\delta_2, \ldots, \delta_{m_r}, \delta_1$.

Since $G(\gamma_{r_1}, \ldots, \vec{\gamma}, \delta_1) < G(\gamma_{r_2}, \vec{\gamma}_2, \delta_1)$ for $r_1 < r_2$ (so $\gamma_{r_1} < \gamma_{r_2}$) and any $\delta_1, \vec{\gamma}_1, \vec{\gamma}_2$, it follows readily that $H_{2n+3}(i)$ is satisfied. The remaining parts of H_{2n+3} and D_{2n+3} follow readily as before.

This completes the proof of H_{2n+3} and D_{2n+3} and hence of the local embedding theorem.

V. The Main Lemma. The purpose of this section is to prove a main lemma analyzing functions defined with respect (in a sense to be made precise) to the canonical measures $\cup_{m \le n} R_{2m+1} = \cup_{m \le n} W^m_{2m+1} \cup \cup_{m \le n} S^{\ell,m}_{2n+1}$. We recall that we are still assuming I_{2n+1} and K_{2n+3} as in Section II.

We outline the methods of this section. We introduce a set $\mathcal{D}$ of "descriptions", finitary objects which "describe" functions $F : \boldsymbol{\delta}^1_{2n+3} \to \boldsymbol{\delta}^1_{2n+3}$ with respect to the canonical measures, and a lowering operation $\mathcal{L}$ on them. We then introduce a main inductive hypothesis H_{2n+1}, and a main auxiliary lemma $\bar{H}_{2n+1}$ analyzing functions F from $\boldsymbol{\delta}^1_{2n+3}$ to $\boldsymbol{\delta}^1_{2n+3}$ in terms of $\mathcal{D}$ and $\mathcal{L}$. We will assume H_{2n+1} and $\bar{H}_{2n+1}$ and establish them for $2n+3$. We will also require several auxiliary definitions and conditions.

Definition of $\mathcal{D}$. We proceed to define the set $\mathcal{D} = \cup_n \mathcal{D}_{2n+1} = \cup_n \cup_{\ell,m} \mathcal{D}^{\ell,m}_{2n+1}$, for $+1 \le$

$\ell \leq 2^{n+1} - 1$ or $\ell = -1$. We assume $\mathcal{D}_{2m+1}$ is defined for $m < n$ and define $\mathcal{D}_{2n+1}$. Our definition will be by a simultaneous induction in which we also define an ordering $<$ on $\mathcal{D}_{2n+1}$, two functions h and H associated with descriptions, conditions C, D, A, and a numerical function k. We assume these notions are defined for $\mathcal{D}_{2m+1}$, $m < n$.

Throughout this section, we let $K_1, \ldots, K_t$ denote a sequence of canonical measures of the form R_{2m+1} where $m < n$ or W^m_{2n+1} or $S^{\ell,m}_{2n+1}$ for some ℓ, m. Given such a measure K_j , we let $\mathcal{K}_j$ denote the corresponding function space measure (e.g. in the case $K_j = S^{\ell,m}_{2n+1}$, $\mathcal{K}_j$ will be a measure on functions $h_j : \boldsymbol{\delta}^1_{2n+1} \to \boldsymbol{\delta}^1_{2n+1}$ of the correct type induced from the strong partition relation on $\boldsymbol{\delta}^1_{2n+1}$). We let $h_1, \ldots, h_t$ denote a sequence of functions in the space $\mathcal{K}_1, \ldots, \mathcal{K}_t$.

We define $\mathcal{D}_{2n+1}$ relative to a fixed sequence $K_1, \ldots, K_t$.

We define basic and non-basic descriptions and subdivide these into type -1, 0, and 1. We will denote a general element of $\mathcal{D}^{\ell,m}_{2n+1}$ by $d^{\ell,m}_{2n+1}$. It will be an indexed tuple of the form $d^{(\mathcal{I}_a)}$, where the index $\mathcal{I}_a$ is of the form $\mathcal{I}_a = (f(\bar{K}_1); \bar{K}_2, \ldots, \bar{K}_a)$, $\mathcal{I}_a = (\,; \bar{K}_2, \ldots, \bar{K}_a)$ or $\mathcal{I}_a = (f_k; \bar{K}_2, \ldots, \bar{K}_a)$ where $\bar{K}_2, \ldots, \bar{K}_a$ are measures of the form $v(K_j)$ for some $1 \leq j \leq t$, where $K_j \in S_{2n+1}$ (and we recall that $v(S^{\bar{\ell},\bar{m}}_{2n+1})$ is the measure u such that $S^{\bar{\ell},\bar{m}}_{2n+1}$ is induced from the strong partition relation on $\boldsymbol{\delta}^1_{2n+1}$, functions $F : \boldsymbol{\delta}^1_{2n+1} \to \boldsymbol{\delta}^1_{2n+1}$ and the measure μ^u_{2n+1} on $\boldsymbol{\delta}^1_{2n+1}$), and which are regarded as formal symbols here, and similarly f_k, $f(\bar{K}_1)$ are formal symbols, where $\bar{K}_1$ is the measure $v(S^{\ell,m}_{2n+1})$ and k is an integer. With a slight abuse of notation, we may also think of the symbol $\bar{K}_j$ as coding a particular integer, which we denote by $\tau(\bar{K}_j)$, from 1 to t, such that $\bar{K}_j = v(K_{\tau(\bar{K}_j)})$. Of course, we may have $\bar{K}_j = v(K_{\tau_1}) = v(K_{\tau_2})$ for different integers $\tau_1 \neq \tau_2$, but will assume that such a particular $\tau(\bar{K}_j)$ is coded in the symbol $\bar{K}_j$.

Our induction in the following definitions is reverse induction on a function k, which although defined along with $\mathcal{D}$ below, is actually defined outright.

To define $\mathcal{D}_{2n+1}$, we consider the following cases:

Basic type -1:

a) For $n > 0$, we allow descriptions of the form $d^{-1,m}_{2n+1} = d^{(;\bar{K}_2,\ldots,\bar{K}_a)}$, where $d = (k;)$, and k is an integer $1 \leq k \leq t$, such that $K_k \in W^m_{2n+1}$. $k(d) = k$ in this case.

b) For $n > 0$, we allow $d^{(\mathcal{I}_a)} = d^{(\bar{K}_2,\ldots,\bar{K}_a)} = (k; \bar{d}_2)^{s(\mathcal{I}_a)}$, where $\mathcal{I}_a = (\bar{K}_2, \ldots, \bar{K}_a)$, s is a formal symbol which may or may not appear, $\bar{d}_2 \in \cup_{m<n} \mathcal{D}_{2m+1}$ is defined relative to the sequence of measures $K_{b_1}, \ldots, K_{b_m}$, $\bar{K}_2, \ldots, \bar{K}_a$, where $K_{b_1}, \ldots, K_{b_m}$ enumerate the subsequence of $K_{k+1}, \ldots, K_t$ consisting of those measures in $\cup_{m<n} R_{2m+1}$. We also require that $k_k = W^m_{2n+1}$,

and $\bar{d}_2 \in \mathcal{D}^{\bar{\ell},\bar{m}}_{2n-1}$, where $v(K_k) = S^{\bar{\ell},\bar{m}}_{2n-1}$. Also $k(d) = k$ in this case. For $n = 0$, we allow $D = (k;i)$, where $1 \leq i \leq m$, where $K_k = W^m_{2n+1}$.

Basic type 0: We allow $d = d^{1,m}_{2n+1} = (r)$, where r is an integer $1 \leq r \leq m$. We set $k(d) = \infty$ in this case.

Basic type 1: We allow $d^{\ell,m}_{2n+1} = d^{(f(\bar{K}_1);\bar{K}_2,\ldots,\bar{K}_a)}$ where $d = (\bar{d}_2)^2$, where $1 \leq k \leq t$, the symbol s may or may not appear, $\bar{d}_2$ is defined relative to $\bar{K}_2, \ldots, \bar{K}_a$, $\bar{d}_2 \in \mathcal{D}^{\bar{\ell},\bar{m}}_{2s+1}$, where $s < n$, $v(S^{\ell,m}_{2n+1}) = \bar{K}_1$, and $\bar{K}_1 = S^{\bar{\ell},\bar{m}}_{2s+1}$ or $W^{\bar{m}}_{2s+1}$ depending on whether $\bar{\ell} \geq 1$ or $\bar{\ell} = -1$. We set $k(d) = \infty$.

Non-Basic Descriptions:

a) K_k (for some integer $1 \leq k \leq t$) $= S^{1,m_k}_{2n+1}$, in which case we allow $d = d^{(I_a)}$ where $d = (k; d^{(I_a)}_1, \ldots, d^{(I_a)}_r)^s$ where s may or may not appear, I_a is an index of one of the above forms, $r \leq m$, $r > 1$ if s appears, $d^{(I_a)}_1, \ldots, d^{(I_a)}_r \in \mathcal{D}^{\ell,m}_{2n+1}$ are defined w.r.t. $K_1, \ldots, K_t$ with $k(d^{(I_a)}_1), \ldots, k(d^{(I_a)}_r) > k$, and $d^{(I_a)}_1 > d^{(I_a)}_2, \ldots, d^{(I_a)}_r$ w.r.t. the ordering $<$ to be defined below (being defined simultaneously). We set $k(d) = k$.

b) $K_k = S^{\ell_k,m_k}_{2n+1}$, where $\ell_k > 1$, and we allow $d = d^{(I_a)}$ where $d = (k; d^{(I_a;\bar{K}_{a+1})}_2)^s$, where the symbol s may or may not appear, $\bar{K}_{a+1} = v(K_k)$, d_2 (with the index $I_{a+1} = (I_a; \bar{K}_{a+1})$) is defined relative to $K_1, \ldots, K_t$ and $d_2 \in \mathcal{D}^{\ell,m}_{2n+1}$, with $k(d_2) > k$. We set $k(d) = k$.

The functions h, H we will be defining in our induction will have the following properties:

If $d \in \mathcal{D}^{\ell,m}_{2n+1}$ and satisfies condition C relative to $K_1, \ldots, K_t$ (where C is to be defined) then for almost all $h_1, \ldots, h_t$ w.r.t. $\mathcal{K}_1 \times \cdots \times \mathcal{K}_t$, $h(d; h_1, \ldots, h_t)$ will be defined, and be an ordinal.

The function H will have the properties:

a) If d is as above with $\ell > 1$ then for almost all $h_1, \ldots, h_t$ as above, we have for almost all $f : \theta(\bar{K}_1) \to \boldsymbol{\delta}^1_{2n+1}$ of the correct type (where $I_a = (f(\bar{K}_1); \bar{K}_2, \ldots, \bar{K}_a)$ in this case) that for almost all $\bar{h}_2, \ldots, \bar{h}_a$ w.r.t the product of the function space measures corresponding to $\bar{K}_2, \ldots, \bar{K}_a$ that $H(d; h_1, \ldots, h_t; f; \bar{h}_2, \ldots, \bar{h}_t)$, an ordinal, is defined.

b) Same as above except $\ell = 1$, in which case $I_a = (f_k; \bar{K}_2, \ldots, \bar{K}_a)$, and we require $f : k \to \boldsymbol{\delta}^1_{2n+1}$ (k an integer).

c) For $\ell = -1$, we have that for almost all $h_1, \ldots, h_t, \bar{h}_2, \ldots, \bar{h}_a$, $H(d; h_1, \ldots h_t, \bar{h}_2, \ldots, \bar{h}_a)$ is defined.

We introduce, in our simultaneous induction, two hypotheses concerning the functions h and H.

F_{2n+1}: If d is defined and satisfies condition C relative to $K_1, \ldots, K_t$ then for almost all h_1 if $h_1(1) = h_1(2) = h_1$ almost everywhere (representing elements of K_1), for almost all h_2 if $h_2(1) = h_2(2) = h_2$ almost everywhere,..., for almost all h_t and $h_t(1) = h_t(2) = h_t$ almost everywhere we have that $h(d; h_1(1), \ldots, h_t(1)) = h(d; h_1(2), \ldots, h_t(2))$.

F^2_{2n+1}: If $d^{(I_a)}$ is defined and satisfies condition C relative to $K_1, \ldots, K_t$, then for almost all h_1 and $h_1(1) = h_(2) = h_1$ almost everywhere, ..., for almost all h_t and $h_t(1) = h_t(2) = h_t$ almost everywhere, we have:

i) If $I_a = (f(\bar{K}_1); \bar{K}_2, \ldots, \bar{K}_a)$, then for almost all $f : \theta(\bar{K}_1) \to \boldsymbol{\delta}^1_{2n+1}$, if $[f_1] = [f_2] = [f]$ (w.r.t. $\bar{K}_1$) then for almost all $\bar{h}_2$ and $\bar{h}_2(1) = \bar{h}_2(2) = \bar{h}_2, \ldots$, for almost all $\bar{h}_1$ with $\bar{h}_a(1) = \bar{h}_a(2) = \bar{h}_a$, $H(d; h_1(1), \ldots, h_t(1);\ f_1; \bar{h}_2(1), \ldots, \bar{h}_a(1)) = H(d; h_1(2), \ldots, h_t(2);\ f_2; \bar{h}_2(2), \ldots, \bar{h}_a(2))$.

ii) If $I_a = (f_k; \bar{K}_2, \ldots, \bar{K}_a)$, then same as above with $f : k \to \boldsymbol{\delta}^1_{2n+1}$.

iii) If $I_a = (\bar{K}_2, \ldots, \bar{K}_a)$, same as above except we omit the f.

In particular, we are assuming F_{2m+1}, F^2_{2m+1} for $m < n$.

We now define conditions C and D. We first consider D. We assume $d \in \mathcal{D}_{2n+1}$ is defined relative to $K_1, \ldots, K_t$, and we define when condition D holds for objects of the form d or $(d)^s$ (for $d \in \mathcal{D}_{2n+1}$, $(d)^s$ is not a description, but this does not affect the definitions).

Condition D: Given d or $(d)^s$, where $d = d^{\ell,m}_{2n+1} \in \mathcal{D}_{2n+1}$ and where d is defined relative to $K_1, \ldots, K_t$, we require that d satisfy condition C and further;

a) If $\ell > 1$ (so $I_a = (f(\bar{K}_1); \bar{K}_2, \ldots, \bar{K}_a)$), then, if s does not appear, we require that for almost all $h_1, \ldots, h_t$, if θ represents $h(d; h_1, \ldots, h_t)$ w.r.t. $\mu^{<v}_{2n+1}$ (where $v = v(S^{\ell,m}_{2n+1})$), that there is a measure one set A w.r.t. $\mu^{<v}_{2n+1}$ restricted to which θ is strictly increasing of uniform cofinality ω, and further that if $C \subseteq \boldsymbol{\delta}^1_{2n+1}$ is c.u.b. then for almost all $h_1, \ldots, h_t$, θ has range almost everywhere in C. If s appears, then we require that for almost all $h_1, \ldots, h_t$ $[\theta]$ is the sup of $[\theta']$, for θ' of the correct type with range a.e. in C.

b) If $\ell = 1$ (so $I_a = (f_k; \bar{K}_2, \ldots, \bar{K}_a)$), then for almost all $h_1, \ldots, h_t$, if θ represents $h(d; h_1, \ldots, h_t)$ w.r.t. the m-fold product of the ω-cofinal normal measure on $\boldsymbol{\delta}^1_{2n+1}$, then we require that there is a measure one set A restricted to which θ is strictly-orderpreserving (w.r.t. $<^m$) and of uniform cofinality ω, if s does not appear, and in the case where s appears, we require that $[\theta]$ is a sup of ordinals $[\bar{\theta}]$ represented by such functions.

c) If $\ell = -1$ (so $I_a = (\bar{K}_2, \ldots, \bar{K}_a)$), then for almost all $h_1, \ldots, h_t$, if θ represents $h(d; h_1, \ldots, h_t)$ w.r.t. $\nu = \nu(W^m_{2n+1})$, then there is a measure one set A w.r.t. ν restricted to which θ is of the correct type, if s does not appear, and in the case where s appears, we required $[\theta]$ to be the sup of ordinals $[\bar{\theta}]$ represented by functions of the correct type.

We now define condition C: We again let $d^{(I_a)}$, where $d = d^{\ell,m}_{2n+1} \in \mathcal{D}^{\ell,m}_{2n+1}$ be given and defined relative to $K_1, \ldots, K_t$. We let $k = k(d)$. We consider the following cases:

d basic.

1) $\ell = -1$.

a) $d^{-1,m}_{2n+1} = (k :)$. In this case d satisfies condition C.

b) $d = (k : \bar{d}_2)^s$, where s may or may not appear. We require that $\bar{d}_2$ or $(d_2)^s$ satisfy condition D relative to $\bar{K}_2, \ldots, \bar{K}_a$ depending on whether s does not or does appear in d.

2) $\ell = 1$, $d = d^{1,m}_{2n+1} = (r)$ where $1 \leq r \leq m$. d satisfies D if $m = r = 1$.

3) $\ell > 1$, so $d = d^{\ell,m}_{2n+1} = (\bar{d}_2)^s$, where s may or may not appear. We require $\bar{d}_2$ or $(\bar{d}_2)^s$ to satisfy D w.r.t. $\bar{K}_2, \ldots, \bar{K}_a$, depending on whether s does not or does appear in d. If $d = (\,)$, then we define d to satisfy C.

d non-basic.

1) $K_k = S^{\ell_k,m_k}_{2n+1}$ and $\ell_k > 1$, where $d = d^{\ell,m}_{2n+1} = (k; d_2^{(I_a,K_{a+1})})^s$, where s may or may not appear. We require d_2 to satisfy C w.r.t. $\bar{K}_2, \ldots, \bar{K}_a, \bar{K}_{a+1}$ (defined by induction), and if the symbol s does not appear, we require that for almost all $h_1, \ldots, h_t, f, \bar{h}_2, \ldots, \bar{h}_a$ that the function g defined almost everywhere w.r.t $\bar{K}_{a+1}$ by $g([h_{a+1}]) = H(d_2; h_1, \ldots, h_t; f; \bar{h}_2, \ldots, \bar{h}_a, \bar{h}_{a+1})$ is almost everywhere increasing of uniform cofinality ω. This makes sense since H is defined and satisfies F^2_{2n+1}, by induction, for d_2. If s appears, we require that for almost all $h_1, \ldots, h_t$, $f, \bar{h}_2, \ldots, \bar{h}_a$, that $[g] = \sup[g']$ for g' of the correct type with $\sup\limits_{\text{a.e.}} g = \sup\limits_{\text{a.e.}} g'$.

b) $K_k = S^{1,m_k}_{2n+1}$, where $d = (k; d_1^{(I_a)}, \ldots, d_r^{(I_a)})^s$, $r \leq m_k$, and s may or may not appear. We then require that $d_2^{(I_a)} < d_3^{(I_a)} < \cdots < d_r^{(I_a)} < d_1^{(I_a)}$, and that for almost all $h_1, \ldots, h_t; f; \bar{h}_2, \ldots, \bar{h}_a$, $H(d_1; h_1, \ldots, \bar{h}_a), \ldots, H(d_{r-1}^{(I_a)}; h_1, \ldots, \bar{h}_a)$ have cofinality ω, and if s does not appear, we also require $H(d_{r-1}^{(I_a)}; h_1, \ldots, \bar{h}_a)$ to have cofinality ω.

We now define the functions h and H. We first consider H. We assume d is defined and satisfies Condition C relative to $K_1, \ldots, K_t$.

We consider the following cases; where we introduce the notation: we say a description $d = d_{2n+1}^{\ell,m}$ is of type -1 if $\ell = -1$, of type 0 if $\ell = 1$ and of type 1 if $\ell > 1$:

1) d basic of type -1.

a) $d = (k;\ )$. For fixed $h_1, \ldots, h_t, \bar{h}_2, \ldots, \bar{h}_a$, where $I_a = (\bar{K}_2, \ldots, \bar{K}_a)$ in this case, we set $H(d; h_a, \ldots, h_t; \bar{h}_2, \ldots, \bar{h}_a) = \sup_{\text{a.e.}} h_k$, where $K_k = W_{2n+1}^{m_k}$, $\kappa = \kappa(W_{2n+1}^{m_k})$, and $h_k : \kappa \to \boldsymbol{\delta}_{2n+1}^1$.

b) $d = (k; \bar{d}_2)^s$, where s may or may not appear, and $\bar{d}_2$ or $(\bar{d}_2)^s$ satisfies D relative to $K_{b_1}, \ldots, K_{b_u}$, $\bar{K}_2, \ldots \bar{K}_a$. Here $\bar{d}_2 \in \mathcal{D}_{2s+1}^{\bar{\ell},\bar{m}}$ for $s < n$, where $v(W_{2n+1}^{m_k}) = R_{2s+1}^{\bar{\ell},\bar{m}}$ ($= S_{2s+1}^{\bar{\ell},\bar{m}}$ if $\bar{\ell} > 0$, and $= W_{2s+1}^{\bar{m}_k}$ if $\bar{\ell} = -1$). Since $\bar{d}_2$ satisfies C, we have by induction that for almost all $h_1, \ldots, h_t$, for almost all $\bar{h}_2, \ldots, \bar{h}_a$ that $h(\bar{d}_2; h_{b_1}, \ldots h_{b_u}, \bar{h}_2, \ldots, \bar{h}_a)$ is defined and is an ordinal in the measure space $v(W_{2n+1}^{m_k})$. We set $H(d; h_1, \ldots, h_t, \bar{h}_2, \ldots, \bar{h}_a) = h_k(h(\bar{d}_2;$ $h_{b_u}, \bar{h}_2, \ldots, \bar{h}_a))$ or $\sup_{\beta < h(\bar{d}_2; h_{b_1} \cdots, h_{b_u}, \bar{h}_2, \ldots, \bar{h}_a)} h_k(\beta)$ depending on whether s does not or does appear.

c) For $n = 0$ and $d = (k; i)$, we set $H(d; h_1, \ldots, h_t, \bar{h}_2, \ldots, \bar{h}_a) = h_k(i)$.

2) Basic type 0.

Here $d = d_{2n+1}^{1,m} = (r)$, where $r \leq m$. For fixed $h_1, \ldots, h_t, f, \bar{h}_2, \ldots, \bar{h}_a$, where $f : m \to \boldsymbol{\delta}_{2n+1}^1$, we set $H(d; h_1, \ldots, h_t, f, \bar{h}_2, \ldots, \bar{h}_a) = f(r)$.

3) Basic type 1.

Here $d^{(I_a)} = d^{(f(\bar{K}_1); \bar{K}_2, \ldots, \bar{K}_a)}$, where $d = (\bar{d}_2)^s$, where $\bar{d}_2$ is defined and satisfies Condition C relative to $\bar{K}_2, \ldots, \bar{K}_a$. Hence, for almost all $h_1, \ldots, h_t, f; \bar{h}_2, \ldots, \bar{h}_a$, $h(\bar{d}_2; \bar{h}_2, \ldots, \bar{h}_a)$ is defined, by induction. We set $H(d; h_1, \ldots, h_t, f, \bar{h}_2, \ldots, \bar{h}_a) = f(h(\bar{d}_2; \bar{h}_2, \ldots, \bar{h}_a))$ or $\sup_{\beta < h(\bar{d}_2; \bar{h}_2, \ldots, \bar{h}_a)} f(\beta)$ depending on whether s does not or does appear. This makes sense since $h(\bar{d}_2; \bar{h}_2, \ldots, \bar{h}_a)$ is an element of the measure space $\bar{K}_1$.

If $d = (\ )$, we set $H(d; h_1, \ldots, h_t; f, \bar{h}_2, \ldots, \bar{h}_a) = \sup_{\text{a.e.}} f$.

4) d non-basic with $K_k = S_{2n+1}^{\ell_k, m_k}$ and $\ell_k > 1$. Here $d = (k; d_2^{(I_a+1)})^s$, where s may or may not appear. Since d_2 satisfies C, for almost all $h_1, \ldots, h_t, f; \bar{h}_2, \ldots, \bar{h}_a, \bar{h}_{a+1}$, $H(d_2; h_1, \ldots, h_t,$ $f, \bar{h}_2, \ldots, \bar{h}_a, \bar{h}_{a+1})$ is defined. For fixed $h_1, \ldots, h_t, f\ \bar{h}_2, \ldots, \bar{h}_a$, we let $g : \theta(\bar{K}_{a+1}) \to \boldsymbol{\delta}_{2n+1}^1$ be defined by $g([\bar{h}_{a+1}]) = H(d_2; h_1, \ldots, h_t, f, \bar{h}_2, \ldots, \bar{h}_a, \bar{h}_{a+1})$, which is well defined almost everywhere for almost all $h_1, \ldots, h_t, f, \bar{h}_2, \ldots, \bar{h}_a$ by F_{2n+1}^2 and induction. We then set $H(d; h_1, \ldots, h_t,$

$f, \bar{h}_2, \ldots, \bar{h}_a) = h_k([g])$ or $\sup_{\beta<[g]} h_k(\beta)$ depending on whether s does not or does appear.

5) d non-basic with $K_k = S^{1,m_k}_{2n+1}$. Here $d = (k; d_1^{(I_a)}, \ldots, d_r^{(I_a)})^s$ with $r \leq m$, and s may or may not appear. Since $d_1^{(I_a)}, \ldots, d_r^{(I_a)}$ satisfy C, for almost all $h_1, \ldots, h_t, f, \bar{h}_2, \ldots, \bar{h}_a$, $H(d_i^{(I_a)}; h_1, \ldots, h_t, f, \bar{h}_2, \ldots, \bar{h}_a)$ is defined, and by induction and the definition of $<$ (given below), $H(d_2^{(I_a)}; \cdots) < \cdots < H(d_r^{(I_a)}; \cdots) < H(d_1^{(I_a)}; \cdots)$. We set $H(d; h_1, \ldots, h_t, f, \bar{h}_2, \ldots, \bar{h}_a) = h_k(H(d_2^{(I_a)}; \cdots), \cdots, H(d_r^{(I_a)}; \cdots), H(d_1^{(I_a)}; \cdots))$ if s does not appear and $r = m_k$; $= \sup_{\beta_{r+1}, \ldots, \beta_{m_k} < H(d_1^{(I_a)}; \cdots)} h_k(H(d_2^{(I_a)}; \cdots), \cdots, H(d_r^{(I_a)}; \cdots), \beta_{r+1}, \ldots, \beta_{m_k}, H(d_1^{(I_a)}; \cdots))$ if $r < m_k$ and s does not appear; and $= \sup_{\beta_r < H(d_r^{(I_a)}; \cdots), \beta_{r+1}, \ldots, \beta_{m_k} < H(d_1^{(I_a)}; , \ldots,)} h_k(H(d_2^{(I_a)}, \ldots), \ldots, H(d_{r-1}^{(I_a)}; \cdots), \beta_r, \ldots, \beta_{m_k}, H(d_1^{(I_a)}; \cdots))$ if s appears.

We now consider h.

If $I_a = (f(\bar{K}_1); \bar{K}_2, \ldots, \bar{K}_a)$, we represent $h(d; h_1, \ldots, h_t)$ with respect to $\mu_{2n+1}^{\bar{K}_1}$ by a function $h(d; h_1, \ldots, h_t, f)$, for $f : \theta(\bar{K}_1) \to \boldsymbol{\delta}^1_{2n+1}$ of the correct type. Similarly, if $I_a = (f_k; \bar{K}_2, \ldots, \bar{K}_a)$, we represent $h(d; h_1, \ldots, h_t)$ w.r.t. the k-fold product of the ω-cofinal normal measure on $\boldsymbol{\delta}^1_{2n+1}$ by $h(d; h_1, \ldots, h_t, f)$. We represent $h(d; h_1, \ldots, h_t, f; \bar{h}_2, \ldots, \bar{h}_\ell)$, in general (where f does not appear if $I_a = (\bar{K}_2, \ldots, \bar{K}_a)$) w.r.t. $\bar{K}_{\ell+1}$ by the function $h(d; h_1, \ldots, h_t; f; \bar{h}_2, \ldots, \bar{h}_\ell, \bar{h}_{\ell+1})$ for $[\bar{h}_{\ell+1}] < \theta(\bar{K}_{\ell+1})$. We set $h(d; h_1, \ldots, h_t, f; \bar{h}_2, \ldots, \bar{h}_a) = H(d; h_1, \ldots, h_t, f, \bar{h}_2, \ldots, \bar{h}_a)$. By F^2_{2n+1}, this is well defined (where F^2_{2n+1} is established below).

We now define the ordering $<$ on the set of descriptions of the form $d^{(I_a)}$, for a fixed index I_a; defined and satisfying Condition C relative to fixed $K_1, \ldots, K_t$. If $d_1 = d_1^{(I_a)}$, $d_2 = d_2^{(I_a)}$, we set $d_1 < d_2$ if for almost all $h_1, \ldots, h_t, f, \bar{h}_2, \ldots, \bar{h}_a$: $H(d_1; h_1, \ldots, h_t, f, \bar{h}_2, \ldots, \bar{h}_a) < H(d_2; h_1, \ldots, h_t, f, \bar{h}_2, \ldots, \bar{h}_a)$. This is well defined by induction.

Before verifying F_{2n+1}, F^2_{2n+1}, we introduce a further condition:

Condition A: We let $d^{(I_a)}$ be defined and satisfy C w.r.t. $K_1, \ldots, K_t$. If $d = (k; d_2^{(I_a+1)})^s$ where $K_k = S^{\ell_k, m_k}_{2n+1}$ with $\ell_k > 1$, and where s may or may not appear, then there is a $\hat{d}^{(I_a)}$ satisfying C w.r.t. $K_1, \ldots, K_t$ such that for almost all $h_1, \ldots, h_t, f, \bar{h}_2, \ldots, \bar{h}_a$, $H(\hat{d}; h_1, \ldots, h_t, f, \bar{h}_2, \ldots, \bar{h}_k) = \sup_{\text{a.e.}[\bar{h}_{a+1}]} H(d_2; h_1, \ldots, h_t, f, \bar{h}_2, \ldots, \bar{h}_1, \bar{h}_{a+1})$, which makes sense by F^2_{2n+1}. We also require $k(\hat{d}) > k$, and all component tuples of d as well as $\hat{d}$ satisfy A.

We now verify (in our simultaneous induction) F_{2n+1} and F^2_{2n+1}. We note that $F^2_{2n+1} \to F_{2n+1}$, so we consider F^2_{2n+1}. We consider the following cases:

1) $d^{(I_a)}$ basic of type 1. Here $d = (d_2)^s$, where s may or may not appear. F^2_{2n+2} follows by

induction, F_{2s+1} for $s < n$, and the fact that d_2 or $(d_2)^s$ satisfies D w.r.t $\bar{K}_2, \ldots, \bar{K}_a$.

2) $d^{(I_a)}$ basic of type 0. Here $d = (r)$ for some $r \leq m$, where $I_a = (f_m; \bar{K}_2, \ldots, \bar{K}_a)$. This is immediate since $H(d; h_1, \ldots, h_t, f, \bar{h}_2, \ldots, \bar{h}_a)$ only depends on $f_m : m \to \delta^1_{2n+1}$.

3) $d^{(I_a)}$ basic of type -1.

i) $d = (k; \ )$, where $K_k = W^{m_k}_{2n+1}$. Since (as in the statement of F^2_{2n+1}) $h_k(1) = h_k(2)$ almost everywhere, $\sup_{\text{a.e.}} h_k(1) = \sup_{\text{a.e.}} h_k(2)$, and the result follows.

ii) $d = (k; \bar{d}_2)^s$, where s may or may not appear. Since d_2 or $(d_2)^s$ satisfies D w.r.t. $K_{b_1}, \ldots, K_{b_u}, \bar{K}_2, \ldots, \bar{K}_a$, it follows from induction, F_{2s+1} for $s < n$, and the definition of H that F^2_{2n+1} is satisfied.

4) $d^{(I_a)}$ non-basic where $d = (k; d_2^{(I_{a+1})})^s$, where s may or may not appear, and $\ell_k > 1$. By induction F^2_{2n+1} holds for d_2 relative to $h_1, \ldots, h_t, f, \bar{h}_2, \ldots, \bar{h}_{a+1}$. Hence, for almost all $h_1, \ldots, h_t, f, \bar{h}_2, \ldots, \bar{h}_a$, the function $g : \theta(\bar{K}_{a+1}) \to \delta^1_{2n+1}$ defined by $g([\bar{h}_{a+1}] = H(d_2; h_1, \ldots, h_t, f, \bar{h}_2, \ldots, \bar{h}_a, \bar{h}_{a+1})$ is well defined almost everywhere. Since $k(d_2) > k = k(d)$ it follows from the definition of H that we may assume that g has range almost everywhere in a c.u.b. set $C \subseteq \delta^1_{2n+1}$ defining a measure one set w.r.t. $\mu^{v_k}_{2n+1}$, where $v_k = v(K_k)$, where $h_k(1)$, $h_k(2)$ agree. If the symbol s does not appear, then from the definition of Condition C, we may assume g is almost everywhere strictly increasing of uniform cofinality ω. Since we may assume C is sufficiently closed, we may assume G is everywhere increasing of uniform cofinality ω, and has range in C. Hence $h_k(1)([g]) = h_k(2)([g])$, and F^2_{2n+1} follows. If the symbol s appears, $[g]$ is the sup of $[g']$ where g' is of the correct type. Since C may be taken sufficiently closed, it follows readily that $[g]$ is the sup of $[g']$ where g' is of the correct type having range in C. Hence $\sup_{\beta<[g]} h_k(1)(\beta) = \sup_{\beta<[g]} h_k(2)(\beta)$, and F^2_{2n+1} follows.

5) $d^{(I_a)}$ non-basic, where $d = (k; d^{(I_a)}, \ldots, d_r^{(I_a)})^s$, and $K_k = S^{1,m_k}_{2n+1}$, where $r \leq m$ and s may or may not appear. We again let $C \subseteq \delta^1_{2n+1}$ be a c.u.b. set defining a set of measure one w.r.t. the m-fold product of the ω-cofinal normal measure on δ^1_{2n+1} where $h_k(1)$ and $h_k(2)$ agree. Since $k(d_1), \ldots, k(d_r) > k$, it follows that for almost all $h_1, \ldots, h_t, f, \bar{h}_2, \ldots, \bar{h}_a$, $\beta_1 = h(d_1; h_1, \ldots, h_t, f, \bar{h}_2, \ldots, \bar{h}_a), \ldots, \beta_r = H(d_r; h_1, \ldots, h_t, f, \bar{h}_2, \ldots, \bar{h}_a)$ are in C. From the definition of Condition C, we also have $\beta_1 < \cdots < \beta_r < \beta_1$, and $\beta_1, \ldots, \beta_{r-1}$ have cofinality ω, as does β_r if s does not appear. F^2_{2n+1} now follows from the definition of H and the fact that C is sufficiently closed, so β_r is a sup of points in C of cofinality ω.

This completes the simultaneous induction definition $\mathcal{D}, h, H, <, C, D, A$, and establishing F_{2n+1}, F^2_{2n+1}.

We require a lemma concerning the ordering $<$:

LEMMA. *If $d_1^{(Ia)}$, $d_2^{(Ia)}$ are defined and satisfy conditions C, A relative to $K_1, \ldots, K_t$, then $d_1 < d_2$ iff one of the following is satisfied:*

1) d_1, d_2 are basic of type -1, so of the form a) $d_1 = (k_1;\ )$ or b) $d_1 = (k_1; (\bar{d}_2)_1)^s$, where s may or may not appear, and similarly for d_2 where $\bar{d}_2$ is an integer for $n = 0$. Then we require that one of the following is satisfied:

i) *$k_1 < k_2$.*

ii) *$k_1 = k_2$ and d_2 is of type a and d_1 of type b.*

iii) *$k_1 = k_2$, both are of type b and $(\bar{d}_2)_1 < (\bar{d}_2)_2$. (Where for $n = 0$, $<$ is the usual ordering.)*

iv) *$k_1 = k_2$, both are of type b, $(\bar{d}_2)_1 = (\bar{d}_2)_2$ and d_1 has the symbol s and d_2 does not.*

2) d_1, d_2 basic of type 0, so $d_1 = (r_1)$, $d_2 = (r_2)$, and $r_1 < r_2$.

3) d_1, d_2 basic of type 1, so $d_1 = ((\bar{d}_2)_1)^s$ or $(\ )$, and similarly for d_2, where s may or may not appear. Then we require $(d_2)_1 < (d_2)_2$ or $(d_2)_1 = (d_2)_2$ and d_1 involves the symbol s and d_2 does not, or $d_2 = (\ )$ and d_1 is of the first type.

4) At least one of d_1, d_2 non-basic. We then require that one of the following is satisfied:

I. *$k(d_1) > k(d_2)$, in which case $\hat{d}_2 \geq d_1$ if $K_{k(d_2)} = S_{2n+1}^{\ell_k, m_k}$ with $\ell_k > 1$, and if $\ell_k = 1$, then $d_1 \leq (d_2)_1$, where $d_2 = (k; (d_2)_1^{(Ia)}, (d_2)_2^{(Ia)}, \ldots, (d_2)_r^{(Ia)})^s$.*

II. *$k(d_1) < k(d_2)$, in which case $\hat{d}_1 < d_2$ if $K_{k(d_1)} = S_{2n+1}^{\ell_k, m_k}$ with $\ell_k > 1$, and if $\ell_k = 1$, then $(d_1)_1 < d_2$; or $K_{k(d_1)} = W_{2n+1}^{m_k}$.*

III. *$k(d_1) = k(d_2)$ in which case one of the following is satisfied:*

i) *$K_k = S_{2n+1}^{\ell_k, m_k}$ with $\ell_k > 1$, so $d_1 = (k; (\bar{d}_2)_1)^s$, $d_2 = (k; (\bar{d}_2)_2)^s$, where s may or may not appear in d_1, d_2. We require that $(d_2)_1 < (d_2)_2$ or $(d_2)_1 = (d_2)_s$ and d_1 involves s and d_2 does not.*

ii) *$K_k = S_{2n+1}^{\ell_k, m_k}$ with $\ell_k = 1$, so $d_1 = (k; (d_1)_1^{(Ia)}, \ldots, (d_{r_1})_1^{(Ia)})^s$, $d_2 = (k; (d_1)_2^{(Ia)}, \ldots, (d_{r_2})_2^{(Ia)})^s$, where s may or may not appear. We then require that:*

a) *$(d_1)_1 < (d_1)_2$*

or b) *$(d_1)_1 = (d_1)_2$ and there is a $p \leq \min\{r_1, r_2\}$ such that $(d_p)_1 \neq (d_p)_2$, and for the least* *such p,* $(d_p)_1 < (d_p)_2$.

or c) $(d_p)_1 = (d_p)_2$ *for* $1 \le p \le \min\{r_1, r_2\}$ *and* $r_1 < r_2$ *and* d_1 *involves s; or* $r_1 > r_2$ *and* d_2 *does not involve s; or* $r_1 = r_2$ *and* d_1 *involves s and* d_2 *does not.*

We also have:

LEMMA. it If $d_1 \neq d_2$, and d_1, d_2 satisfy C, A, then $d_1 < d_2$ or $d_2 < d_1$.

This lemma follows readily from the first lemma and the corresponding statement for the ordering $<'$ defined from the statement of Lemma 1, which follows from a consideration of cases.

To establish the first lemma, it is enough to show that $d_1 <' d_2$ implies $d_1 < d_2$. We assume this for d's in $\mathcal{D}_{2s+1}$ for $s < n$, and note that for almost all $h_1, \ldots, h_t$, we may assume that:

1) If $k_1 < k_2$ and $K_{k_1} = W_{2n+1}^{m_{k_1}}$, $K_{k_2} = W_{2n+1}^{m_{k_2}}$, then $\inf h_{k_2} > \sup h_{k_1}$.

2) If $K_b = S_{2n+1}^{\ell_k, m_k}$, and by $h_k(0)$ we denote the function defined by $h_k(0)(\beta) = \sup_f h_k([f])$, where f ranges over functions $f : \kappa(S_{2n+1}^{\ell_k, m_k}) \to \beta$ if $\ell_k > 1$; and if $\ell_k = 1$, $h_k(0)(\beta) = \sup_{\beta_2, \ldots, \beta_m < \beta} h_k(\beta_2, \ldots, \beta_m, \beta)$, then if $k_1 < k_2$ and $K_{k_1} = S_{2n+1}^{\ell_k, m_k}$, the range of h_{k_2} is closed under $h_{k_1}(0)$.

It then follows from the definition of H and induction that the lemma is satisfied.

We now prove some technical lemmas concerning the family of canonical measures, which we prove in greater detail than required for this paper, and which simplify the description of conditions C and D.

It is convenient to introduce an auxiliary family of canonical measures $\tilde{R}_{2n+1}$ and embeddings $\tilde{\Pi}$.

$\tilde{W}_1^m$: We set $\tilde{W}_1^m = W_1^m =$ the m-fold product of the ω-cofinal normal measure on ω_1, which we identify with an ordinal, as in W_1^m, by ordering by the largest component first, then the next largest, etc. We define $\tilde{\Pi}$ on these measures exactly as before on W_1^m, so $\Pi_k^m(\alpha_1, \ldots, \alpha_m) = (\alpha_{m-k+1}, \ldots, \alpha_m)$.

$\tilde{S}_1^{1,m}$: We define A to have measure one w.r.t. $\tilde{S}_1^{1,m}$ if there is a c.u.b. $C \subseteq \omega_1$ such that for all $h' :<^{m+1} \to C$ of the correct type (where $<^{m+1}$ is as in the definition of $S_1^{1,m+1}$), $[h] \in A$, where $h :<^m \to C$ is defined by $h(\alpha_2, \ldots, \alpha_m, \alpha_1) = \sup_{\alpha_{m+1} < \alpha_1} h'(\alpha_2, \ldots, \alpha_m, \alpha_{m+1}, \alpha_1)$. This is essentially the same as the definition of $S_1^{1,m}$ except the functions no longer have uniform cofinality ω.

We define $\tilde{\Pi}^{m_2}_{m_1} : \theta(\tilde{S}^{1,m_2}_1) \to \theta(\tilde{S}^{1,m_1}_1)$ (for $m_2 > m_1$) as follows: given α represented by $h :<^{m_2}\to \omega_1$, (induced by an $h' :<^{m_2+1}\to \omega_1$, as above), we let $\tilde{\Pi}^{m_2}_{m_1}(\alpha)$ be represented by $\bar{h} :<^{m_1}\to \omega$ defined by

$$\bar{h}(\alpha_2,\ldots,\alpha_{m_1},\alpha_1) = \sup_{\alpha_{m_1+1},\ldots,\alpha_{m_2}} h(\alpha_2,\ldots,\alpha_{m_1},\ldots,\alpha_{m_2},\alpha_1).$$

In general, we proceed as follows:

$\tilde{W}^m_{2n+1}$: We define A to have measure one w.r.t. $\tilde{W}^m_{2n+1}$ if there is a c.u.b. $C \subseteq \boldsymbol{\delta}^1_{2n+1}$ such that for all $f' : \theta(\tilde{S}^{\ell,m+1}_{2n-1}) \to C$ of the correct type (where, as in the definition of W^m_{2n+1}, $\ell = 2^n - 1$ is maximal), $[f] \in A$, where $f : \theta(\tilde{S}^{\ell,m}_{2n-1}) \to \boldsymbol{\delta}^1_{2n+1}$ is defined by $f(\alpha) = \sup_{\beta:\tilde{\Pi}^{m+1}_m(\beta)=\alpha} f'(\beta)$, where $\tilde{\Pi}^{m+1}_m$, of course, denotes the embedding from $\theta(\tilde{S}^{\ell,m+1}_{2n-1})$ to $\theta(\tilde{S}^{\ell,m}_{2n-1})$.

We define $\tilde{\Pi}^{m_2}_{m_1}$ from $\theta(W^{m_2}_{2n+1}) \to \theta(W^{m_1}_{2n+1})$, $m_2 > m_1$, as follows: given α represented by $f : \theta(\tilde{S}^{\ell,m_2}_{2n-1}) \to \boldsymbol{\delta}^1_{2n+1}$ induced by an f' as above, we let $\tilde{\Pi}^{m_2}_{m_1}(\alpha)$ be represented by $\bar{f} : \theta(\tilde{S}^{\ell,m_1}_{2n-1}) \to \boldsymbol{\delta}^1_{2n+1}$ defined by $\bar{f}(\alpha) = \sup_{\beta:\tilde{\Pi}^{m_2}_{m_1}(\beta)=\alpha} f(\beta)$, where $\tilde{\Pi}^{m_2}_{m_1}$ denotes the embedding from $\tilde{S}^{\ell,m_2}_{2n-1}$ to $\tilde{S}^{\ell,m_1}_{2n-1}$.

$\tilde{S}^{\ell,m}_{2n+1}$: If $\ell = 1$, we proceed as in the definition of $\tilde{S}^{1,m}_1$, using $\boldsymbol{\delta}^1_{2n+1}$ insteady of ω_1. We define $\tilde{\Pi}$ similarly here.

For $\ell > 1$, we let $\tilde{R}^{\ell',m}_{2s+1}$ denote the $\ell - 1^{\text{st}}$ measure in the enumeration of the measures $\tilde{W}^m_1, \tilde{S}^{1,m}_1, \tilde{W}^m_3, \tilde{S}^{1,m}_3, \tilde{S}^{2,m}_3, \ldots$, etc., similar to the definition of $S^{\ell,m}_{2n+1}$. We then define A to have measure one w.r.t. $\tilde{S}^{\ell,m}_{2n+1}$ if there is a c.u.b. $C \subseteq \boldsymbol{\delta}^1_{2n+1}$ such that for all $F : \boldsymbol{\delta}^1_{2n+1} \to C$ of the correct type, $[F]_{\mu_{2n+1}^{\tilde{R}^{\ell',m}_{2s+1}}} \in A$, where for $f : \theta(\tilde{R}^{\ell',m}_{2s+1}) \to \boldsymbol{\delta}^1_{2n+1}$ induced by an $f' : \theta(\tilde{R}^{\ell',m+1}_{2s+1}) \to \boldsymbol{\delta}^1_{2n+1}$ of the correct type, (via the corresponding embedding $\tilde{\Pi}^{m+1}_m$ for $\tilde{R}^{\ell',m+1}_{2s+1}$ and $\tilde{R}^{\ell',m}_{2s+1}$), $\bar{F}([f]) = \sup_{\beta:\tilde{\Pi}(\beta)=[f]} F([f])$, where $\tilde{\Pi}([f])$, for $f : \theta(\tilde{R}^{\ell',m+1}_{2s+1}) \to \boldsymbol{\delta}^1_{2n+1}$ is defined as in the case ℓ maximal.

We define $\tilde{\Pi}^{m_2}_{m_1}$ from $\theta(S^{\ell,m_2}_{2n+1})$ to $\theta(S^{\ell,m_1}_{2n+1})$ as follows: for α represented w.r.t. $\mu_{2n+1}^{\tilde{R}^{\ell',m_2}_{2s+1}}$ by $\bar{F}$, as above, we represent $\tilde{\Pi}^{m_2}_{m_1}(\alpha)$ w.r.t $\mu_{2n+1}^{\tilde{R}^{\ell',m_1}_{2s+1}}$ by G defined by: for g induced by an appropriate g', we set $G([g]) = \sup_f \bar{F}([f])$, where the sup ranges over $f : \theta(\tilde{R}^{\ell',m_2}_{2s+1}) \to \boldsymbol{\delta}^1_{2n+1}$ of the appropriate type (i.e. induced by f') such that $f(\theta) < g(\tilde{\Pi}^{m_2}_{m_1}(\theta))$ for all θ (where $\tilde{\Pi}^{m_2}_{m_1}$ is defined).

We extend the embeddings slightly to include embeddings Π^{m+1}_m from $R^{\ell,m+1}_{2n+1}$ to $\tilde{R}^{\ell,m}_{2n+1}$ as follows:

If $R^{\ell,m+1}_{2n+1} = W^m_{2n+1}$, and $f : \theta(S^{\ell,m+1}_{2n-1}) \to \boldsymbol{\delta}^1_{2n+1}$ of the correct type represents α, then we let $\Pi^{m+1}_m(\alpha)$ be represented by $\bar{f} : \theta(\tilde{S}^{\ell,m}_{2n-1}) \to \boldsymbol{\delta}^1_{2n+1}$ defined by $\bar{f}(\beta) = \sup_{\beta' : \Pi(\beta') = \beta} f(\beta')$, where Π denotes the embedding from $S^{\ell,m+1}_{2n-1}$ to $\tilde{S}^{\ell,m}_{2n-1}$, defined by induction. It follows readily that for almost all α (w.r.t. $\mu^{S^{\ell,m+1}_{2n-1}}_{2n+1}$) that $\Pi(\alpha)$ is of the appropriate type, that is, induced by a $f' : \theta(\tilde{S}^{\ell,m+1}_{2n-1}) \to \boldsymbol{\delta}^1_{2n+1}$ of the correct type.

If $R^{\ell,m+1}_{2n+1} = S^{\ell,m+1}_{2n+1}$, and $F : \boldsymbol{\delta}^1_{2n+1} \to \boldsymbol{\delta}^1_{2n+1}$ of the correct type represents α w.r.t. $\mu^{R^{\ell',m+1}_{2s+1}}_{2n+1}$, then we represent $\Pi^{m+1}_m(\alpha)$ by a function $\bar{F}$ defined as follows: given $\bar{f} : \theta(\tilde{R}^{\ell',m}_{2s+1}) \to \boldsymbol{\delta}^1_{2n+1}$ induced by an $f' : \theta(R^{\ell',m+1}_{2s+1}) \to \boldsymbol{\delta}^1_{2n+1}$ of the correct type, we set $\bar{F}([\bar{f}]) = \sup_{f : \tilde{\Pi}([f]) = [\bar{f}]} F([f])$.

We now state four lemmas which we prove simultaneous induction on n:

LEMMA 1. *Given any $H : \theta(W^m_{2n+1}) = \boldsymbol{\delta}^1_{2n+1} \to ON$ monotonically increasing almost everywhere (i.e. restricted to a measure one set), if $m > 1$ then there is a measure one set A w.r.t. W^m_{2n+1} s.t. $H \restriction A$ is increasing, or $H(\alpha)$ for $\alpha \in A$ depends only on $\Pi^m_{m-1}(\alpha)$. If $m = 1$, then we replace the last part with: $H(\alpha)$ depends only on $\sup_{\text{a.e.}} f$, for $[f] = \alpha$, for $n > 1$, and for $n = 1$, with: H is constant almost everywhere.*

LEMMA 2. *Given any $H : \theta(S^{\ell,m}_{2n+1}) \to ON$ monotonically increasing almost everywhere, there is a measure one set A w.r.t. $S^{\ell,m}_{2n+1}$ s.t. $H \restriction A$ is increasing, or:*

1) *If $\ell = 1$, $m > 1$, then $H(\alpha)$ depends only on $\Pi^m_{m-1}(\alpha)$ (for $\alpha \in A$).*

2) *If $\ell = 1$, $m = 1$, then $H \restriction A$ is constant.*

3) *If $\ell > 1$, $m > 1$, then $H \restriction A$ depends only on $\Pi^m_{m-1}(\alpha)$.*

4) *If $\ell > 1$, $m = 1$, $H(\alpha)$, for α represented by $F : \boldsymbol{\delta}^1_{2n+1} \to \boldsymbol{\delta}^1_{2n+1}$ of the correct type, depends only on $F(0)$.*

5) *If $m = 0$, and we let $S^{\ell,0}_{2n+1}$ be the measure induced from the strong partition relation on $\boldsymbol{\delta}^1_{2n+1}$, functions $F : \boldsymbol{\delta}^1_{2n+1} \to \boldsymbol{\delta}^1_{2n+1}$ of the correct type, and the cof $\theta(R^{\tilde{\ell},1}_{2s+1})$ cofinal normal measure on $\boldsymbol{\delta}^1_{2n+1}$, where $R^{\tilde{\ell},1}_{2s+1} = \nu(S^{\ell,1}_{2n+1})$ – then H as above is either increasing almost everywhere, or constant almost everywhere.* LEMMA 3. *If $H : \theta(W^m_{2n+1}) = \boldsymbol{\delta}^1_{2n+1} \to ON$ is pressing down almost everywhere, then there is a measure one set A such that for $m > 1$, $H(\alpha)$ for $\alpha \in A$ depends only on $\Pi^m_{m-1}(\alpha)$; and for $m = 1$, $H \restriction A$ is constant.*

LEMMA 4. *$H : \theta(S^{\ell,m}_{2n+1}) \to ON$ is pressing down almost everywhere, then there is a mea-*

sure one set A w.r.t. $S^{\ell,m}_{2n+1}$ *such that:*

1) *If* $\ell = 1,\ m > 1,\ H(\alpha)$ *for* $\alpha \in A$ *depends only on* $\Pi^m_{m-1}(\alpha)$.

2) *If* $\ell = 1,\ m = 1,\ H \restriction A$ *is constant.*

3) *If* $\ell > 1,\ m > 1,\ H(\alpha)$ *for* $\alpha \in A$ *depends only on* $\Pi^m_{m-1}(\alpha)$.

4) *If* $\ell > 1,\ m = 1, H(\alpha)$, *for* $\alpha \in A$ *depends only on* $F(0)$, *for* F *representing* α.

We require two additional lemmas for the proofs, which we prove first in a separate induction:

LEMMA 5. *Given any* $C \subseteq \boldsymbol{\delta}^1_{2n+1}$, *and* $f, g : \theta(S^{\ell,m}_{2n-1}) \to C$ *of the correct type (where* $\ell = 2^n - 1$ *is maximal) and* $[f] < [g]$, $\Pi^m_{m-1}([f]) = \Pi^m_{m-1}([g])$, *then there are* f_2, g_2 *satisfying:*

1) $[f_2] = [f],\ [g_2] = [g]$.

2) $f_2(\alpha) < g_2(\alpha) < f_2(\alpha + 1)$ *for all* α, *and* f_2, g_2 *are of the correct type.*

3) f_2, g_2 *have range in* C.

If $m = 1$, we require f, g satisfy $\sup_{\text{a.e.}} f = \sup_{\text{a.e.}} g$, and have the same conclusion.

LEMMA 6. *The same as Lemma 5 except we use* $F, G : \boldsymbol{\delta}^1_{2n+1} \to \boldsymbol{\delta}^1_{2n+1}$ *of the correct type with* $[F]_{\mu_{2n+1}^{R^{\ell,m}_{2s+1}}} < [G]_{\mu_{2n+1}^{R^{\ell,m}_{2s+1}}}$, *and* $\Pi^m_{m-1}([F]) = \Pi^m_{m-1}([G])$ *if* $m > 1$ *and if* $m = 0,\ F(0) = G(0)$. *We have for* $n = 0$, $[F]_{W^m_1} < [G]_{W^m_1}$ *in the hypothesis. We have the same conclusion as Lemma 5.*

We first prove Lemmas 5 and 6 by induction.

We first consider Lemma 5.

We let C be given, and f, g as in Lemma 5. We assume $m > 1$, the case $m = 1$ being similar.

We let C_1 be a c.u.b. subset of $\boldsymbol{\delta}^1_{2n-1}$ such that for α represented by $H : \boldsymbol{\delta}^1_{2n-1} \to C_1$ or $:<^m\!\to C_1$ of the correct type, depending on whether $\ell > 1$ or $\ell = 1$, $f(\alpha) < g(\alpha)$ and $\sup_{\beta : \Pi^m_{m-1}(\beta) = \Pi^m_{m-1}(\alpha)} f(\beta) = \sup_{\beta : \Pi^m_{m-1}(\beta) = \Pi^m_{m-1}(\alpha)} g(\beta)$. We first assume $\ell > 1$. We consider the following partition:

$\mathcal{P}$: We partition functions $H_1, H_2 : \boldsymbol{\delta}^1_{2n-1} \to \boldsymbol{\delta}^1_{2n-1}$ of the correct type with $H_1(\alpha) < H_2(\alpha) < H_1(\alpha + 1)$ for all α, according to whether or not $f([H_2]) > g([H_1])$, where $[H_1]$, $[H_2]$

mean, of course, with respect to the measure $\mu_{2n-1}^{v(S_{2n-1}^{\ell,m})}$.

It follows readily that for H_1, H_2 of the above type, $\Pi_{m-1}^m([H_1]) = \Pi_{m-1}^m([H_2])$. We claim that on the homogeneous side of the partition, the property stated in partition $\mathcal{P}$ holds. We suppose not, and let $C_2 \subseteq \boldsymbol{\delta}_{2n-1}^1$ be a c.u.b. set homogeneous for the other side. We let $C_3 = C_1 \cap C_2$, and $C_4 \subseteq C_3$ be contained in the closure points of C_3, that is, if $\alpha \in C_4$, and $\beta, \gamma < \alpha$, then the $\omega \cdot \gamma^{\text{th}}$ element of C_3 after β is $< \alpha$. (We use this notation throughout.) We fix $H_1 : \boldsymbol{\delta}_{2n-1}^1 \to C_4$ of the correct type. Since $C_4 \subseteq C_1$, it follows that there is an $H_2 : \boldsymbol{\delta}_{2n-1}^1 \to \boldsymbol{\delta}_{2n-1}^1$ such that $\Pi_{m-1}^m([H]) = \Pi_{m-1}^m([H_2])$ and $f([H_2]) > g([H_1])$. We may clearly assume that range $H_2 \subseteq C_3$, replacing H_2 by a larger function if necessary (e.g. the function $\tilde{H}_2(\alpha)$ = the next ω^{th} element of C_3 after $\max(H_2(\alpha), \sup_{\beta<\alpha} \tilde{H}_2(\beta)))$. By Lemma 6 and induction, it now follows that there are functions H_1', H_2' satisfying:

1) $[H_1'] = [H_1]$, $[H_2'] = [H_2]$, and hence $f([H_2']) > g([H_1'])$.

2) H_1', H_2' are of the correct type everywhere and ordered as in $\mathcal{P}$.

3) Range H_1', $H_2' \subseteq C_3$.

This, however, contradicts $C_3 \subseteq C_2$ and the definition of C_2.

Hence, on the homogeneous side of the partition, the property stated in $\mathcal{P}$ holds. We let $C_2 \subseteq C_1$ be homogeneous for $\mathcal{P}$. We let C_3 be contained in the closure points of C_2 (as above), and C_4 be contained in the closure points of C_3. We then define the functions f_2, g_2 as follows:

For α represented by $H : \boldsymbol{\delta}_{2n-1}^1 \to C_4$ of the correct type, we set $f_2(\alpha) = f(\alpha)$, $g_2(\alpha) = g(\alpha)$. For β not of this form, we let β' be the least ordinal $> \beta$ which is represented by an $H' : \boldsymbol{\delta}_{2n-1}^1 \to C_3$ of the correct type. For $\beta < \alpha$, α as above, it is easily seen that $\beta' < \alpha$ as well. We fix such an H' representing β', with $H' \subseteq C_3$ everywhere. We then define $H'' : \boldsymbol{\delta}_{2n-1}^1 \to C_2$ by: $H''(\gamma)$ = the $\omega \cdot H_\beta(\gamma)^{\text{th}}$ element of C_2 after $H'(\gamma)$, where H_β represents β. We let $[H''] = \beta''$. We then set $f_2(\beta) = f(\beta'')$, $g_2(\beta) = g(\beta'')$. This is well defined. We note that β'' is also of the correct type, since H' was, and range $H' \subseteq C_3$. For α as above, that is, represented by H of the correct type with range in C_4, it follows readily that if $\beta < \alpha$, then $\beta'' < \alpha$. From this, and the definitions of C_1, C_2, it now follows that f_2, g_2, have the desire properties.

The case $\ell = 1$ is similar, using functions $H : <^m \to \boldsymbol{\delta}_{2n-1}^1$ instead.

We now consider Lemma 6: We again consider $m > 1$, the case $m = 1$ being similar. We fix $F, G : \boldsymbol{\delta}_{2n+1}^1 \to \boldsymbol{\delta}_{2n+1}^1$ of the correct type as in Lemma 6. We consider the partition:

$\mathcal{P}$: We partition functions $f, g : \theta(R^{\ell,m}_{2s+1}) \to \boldsymbol{\delta}^1_{2n+1}$ of the correct type with $f(\alpha) < g(\alpha) < f(\alpha+1)$ everywhere, according to whether or not $F([g]) > G([f])$.

It follows as in the proof of Lemma 5, using induction and Lemma 5, that on the homogeneous side of the partition, the property stated in $\mathcal{P}$ holds.

We let $C_2 \subseteq C_1$ be homogeneous for $\mathcal{P}$, where $C_1 \subseteq \boldsymbol{\delta}^1_{2n+1}$ is c.u.b. such that for $f : \theta(R^{\ell,m}_{2s+1}) \to C_1$ of the correct type, $F([f]) < G([f])$, and $\sup_{\beta:\Pi^m_{m-1}(\beta)=\Pi^m_{m-1}([f])} F(\beta) = \sup_{\beta:\Pi^m_{m-1}(\beta)=\Pi^m_{m-1}([f])} G(\beta)$. We let C_3 be contained in the closure points of C_2, and C_4 be contained in the closure points of C_3. We then define $F_2(\alpha) = F(\alpha)$, $G_2(\alpha) = G(\alpha)$, for α represented by $f : \theta(R^{\ell,m}_{2s+1}) \to C_4$ of the correct type, and for other β, we defined β', β'' similarly to Lemma 5, and set $F_2(\beta) = F(\beta'')$, $G_2(\beta) = G(\beta'')$. It follows that F_2, G_2 have the desired properties.

We now consider Lemma 1. We let $H : \boldsymbol{\delta}^1_{2n+1} \to ON$ be monotonically increasing almost everywhere, say restricted to $[f]$ for $f : \theta(S^{\ell,m}_{2n-1}) \to C$ of the correct type, where we assume $n \geq 1$ here, the case $n = 0$ being similar. We consider the following partition:

$\mathcal{P}$: We partition function $f, g : \theta(S^{\ell,,m}_{2n-1}) \to \boldsymbol{\delta}^1_{2n+1}$ of the correct type with $f(\alpha) < g(\alpha) < f(\alpha+1)$ everywhere, according to whether or not $H([f]) < H([g])$ or not.

We let $C_1 \subseteq \boldsymbol{\delta}^1_{2n+1}$ be c.u.b. and homogeneous for $\mathcal{P}$, and let C_2 be contained in the closure points on $C \cap C_1$. We then claim that C_2 defines the measure one set as in the statement of Lemma 1. This follows from the definition of C_2 and Lemma 5. In the event we are on the homogeneous side in which $\mathcal{P}$ holds, we use the trivial fact that if f, g are of the correct type with range in C_2 and $\Pi^m_{m-1}([f]) < \Pi^m_{m-1}([g])$, then there is an f' of the correct type with range in $C \cap C_1$ such that $f < f' < g$ almost everywhere, and $\Pi^m_{m-1}([f]) = \Pi^m_{m-1}([g])$.

The proof of Lemma 2 is similar to that of Lemma 1, only we partition functions $F, G : \boldsymbol{\delta}^1_{2n+1} \to \boldsymbol{\delta}^1_{2n+1}$ instead, and we use Lemma 6.

We now consider Lemma 3.

We fix $H : \boldsymbol{\delta}^1_{2n+1} \to ON$ pressing down almost everywhere, say on a measure one set w.r.t. $\mu^{S^{\ell,m}_{2n-1}}_{2n-1}$ determined by $C \subseteq \boldsymbol{\delta}^1_{2n+1}$, where we assume $n \geq 1$, the case $n = 0$ being easier.

We consider the following partition:

$\mathcal{P}$: We partition $f, g : \theta(S^{\ell,m}_{2n-1}) \to \boldsymbol{\delta}^1_{2n+1}$ of the correct type with $f(\alpha) < g(\alpha) < f(\alpha+1)$ everywhere, according to whether or not $[f] > H([g])$.

It follows readily that on the homogeneous side of the partition, the property stated in $\mathcal{P}$ holds. We let C_1 be homogeneous for $\mathcal{P}$, and let C_2 be contained in the closure points of $C \cap C_1$.

It follows that if $f : \theta(S_{2n-1}^{\ell,m}) \to C_2$ is of the correct type, and $\bar{f}$ is defined by $\bar{f}(\alpha) =$ the next ω^{th} element of $C \cap C_1$ after $\sup_{\beta>\alpha} f(\beta)$, then $[\bar{f}] > H([f])$.

We now consider the following partition:

$\mathcal{P}_2$: We partition $f, g : \theta(S_{2n-1}^{\ell,m}) \to \boldsymbol{\delta}_{2n+1}^1$ of the correct type, with $f(\alpha) < g(\alpha) < f(\alpha+1)$ everywhere, according to whether or not $H([f]) < H([g])$.

If on the homogeneous side of the partition, the property stated in $\mathcal{P}_2$ fails, then the lemma follows readily from Lemma 5.

Otherwise, we let $C_3 \subseteq C_2$ be homogeneous for $\mathcal{P}_2$, and let C_4 be contained in the closure points of C_3. We fix $f : \theta(S_{2n-1}^{\ell,m}) \to C_4$ of the correct type, and let $\bar{f}$ be as above.

We next claim that if f_2 is any function $f_2 : \theta(S_{2n-1}^{\ell,m}) \to C_3$ of the correct type with $\Pi_{m-1}^m([f_2]) = \Pi_{m-1}^m([f])$, $[\bar{f}] > H([f_2])$. This will give a contradiction, since it will give an order-preserving map, H, from the set of such $[f_2]$ into $[\bar{f}]$, and the former is easily seen to have larger order type.

To prove the claim, we fix $f_2 : \theta(S_{2n-1}^{\ell,m}) \to \boldsymbol{\delta}_{2n+1}^1$ of the correct type, with $\Pi_{m-1}^m([f]) = \Pi_{m-1}^m([f_2])$. It is enough to establish that $\tilde{f} = \tilde{f}_2$ almost everywhere, where $\tilde{f}(\alpha) = \sup_{\beta<\alpha} \tilde{f}(\beta)$, and similarly for $\tilde{f}_2$. We suppose not, say $\tilde{f}_2 > \tilde{f}$ almost everywhere. We then define the function h on $\theta(S_{2n-1}^{\ell,m})$ by $h(\alpha) =$ the least $\beta < \alpha$ such that $f_2(\beta) > \tilde{f}(\alpha)$, defined almost everywhere. Since h is pressing down, it follows from induction and Lemma 4 that $h(\alpha)$ depends only on $\Pi_{m-1}^m(\alpha)$ if $m > 1$, and on $F(0)$, for F representing α, if $m = 1$. It follows that $\Pi_{m-1}^m([f]) < \Pi_{m-1}^m([f_2])$, contradicting our assumption. This establishes the claim, and completes the proof of Lemma 3.

The proof of Lemma 4 is similar, using induction and Lemma 3.

Finally, we remark that the above six lemmas are also true with the measure $\tilde{S}_{2n+1}^{\ell,m}$, $\tilde{W}_{2n+1}^m$, the proofs being essentially identical to those give above.

To state the next two lemmas, we require a technical definition:

DEFINITION. Given two functions $f, g : \theta(S_{2n-1}^{\ell,m}) \to ON$, we say that they are ordered of k-type, where $1 \leq k \leq m$, if they satisfy the following:

1) f, g are of the correct type.

2) If $\alpha, \beta \in \theta(S^{\ell,m}_{2n-1})$ are represented by F, G of the correct type, then $f(\alpha) < g(\beta)$ if $\Pi^m_k(\alpha \leq \Pi^m_k(\beta)$, and $f(\alpha) > g(\beta)$ if $\Pi^m_k(\alpha) > \Pi^m_k(\beta)$.

For other ordinals the ordering is more or less arbitrary; we use the following:

3) If $\alpha_1 < \alpha < \alpha_2$, where α_1, α_2 are of the correct type (i.e. represented by functions of the correct type) with $\Pi^m_k(\alpha_1) = \Pi^m_k(\alpha_2)$, and $\beta \geq$ the sup of all the α' of the correct type with $\Pi^m_k(\alpha') = \Pi^m_k(\alpha_1)$, then $f(\beta) > g(\alpha)$.

4) If α_2 is the least ordinal $> \alpha$ of the correct type, $\alpha < \inf \alpha'$ for all α' of the correct type with $\Pi^m_k(\alpha') = \Pi^m_k(\alpha_2)$, and $\alpha \geq \sup \alpha'$ for all α' of the correct type with $\Pi^m_k(\alpha') < \Pi^m_k(\alpha_2)$, and β satisfies the above (for the same α_2), then $f(\alpha) < g(\beta)$ if $\alpha \leq \beta$, and $f(\alpha) > g(\beta)$ if $\alpha > \beta$.

5) If α_2 is as in 4 above, corresponding to α, and $\beta > \inf \alpha'$ for α' of the correct type with $\Pi^m_k(\alpha') = \Pi^m_k(\alpha_2)$, then $f(\beta) > g(\alpha)$.

It is easily seen that this defines a unique way of ordering range $f \cup$ range g.

DEFINITION. Given $F, G : \theta(\mu^{R^{\ell,m}_{2s+1}}_{2n+1}) = \boldsymbol{\delta}^1_{2n+1} \to ON$, we say F, G are ordered of k type for $1 \leq k \leq m$ if they satisfy the same 5 conditions above, where α of the correct type means represented by an $f : \theta(R^{\ell,m}_{2s+1}) \to \boldsymbol{\delta}^1_{2n+1}$ of the correct type, and we use the corresponding Π^m_k defined for such α. Finally, for $F, G :<^m\to \boldsymbol{\delta}^1_{2n+1}$ of the correct type, F, G are ordered of k-type if for $\vec{\alpha} = (\alpha_2, \ldots, \alpha_m, \alpha_1)$, $\vec{\beta} = (\beta_2, \ldots, \beta_m, \beta_1)$ and $\alpha_r < \beta_r$ for the least index where they disagree, then if $r \leq k$, $F(\vec{\beta}) > G(\vec{\alpha})$, and if $r > k$, $G(\vec{\alpha}) > F(\vec{\beta})$.

We now state two lemmas which generalize Lemmas 5 and 6.

LEMMA 7. *Give any c.u.b.* $C \subseteq \boldsymbol{\delta}^1_{2n+1}$ *and* $f, g : \theta(S^{\ell,m}_{2n-1}) \to C$ *of the correct type with* $\Pi^m_k([f]) = \Pi^m_k([g])$ *for some* $1 \leq k \leq m-1$, *and* $\Pi^m_{k+1}([f]) < \Pi^m_{k+1}([g])$, *then there are* f_2, g_2 *satisfying:*

1) $[f_2] = [f]$, $[g_2] = [g]$

2) f_2, g_2 *are ordered of* $k+1$*-type.*

3) f_2, g_2 *have range in* C.

If $\Pi^m_1([f]) < \Pi^m_1([g])$ then f_2, g_2 are ordered of 1-type if $\sup f = \sup g$, and if $\sup f <$ sup g, $\inf g_2 > \sup f_2$.

LEMMA 8. *The same as Lemma 7 except we use* $F, G : \theta(\mu_{2n+1}^{R_{2s+1}^{\ell,m}}) = \boldsymbol{\delta}_{2n+1}^1 \to C$, *and the corresponding* Π_k^m*'s. If* $\Pi_1^m([F]) < \Pi_1^m([G])$ *here, we require* F_2, G_2 *to be ordered of 1-type if* $[F_0] = [G_0]$, *and if* $[F_0] < [G_0]$, *then we require that* $F_2(\alpha) < G_2(\beta)$ *if* $\sup_{\text{a.e.}} f_\alpha \leq \sup_{\text{a.e.}} f_\beta$, *and if* $\sup_{\text{a.e.}} f_\alpha > \sup_{\text{a.e.}} f_\beta$, $G_2(\beta) < F_2(\alpha)$ *for* f_α, f_β *representing* α, β.

Proof of Lemma 7. We fix such f, g of the correct type, and suppose $\Pi_k^m([f]) = \Pi_k^m([g])$ and $\Pi_{k+1}^m([f]) < \Pi_{k+1}^m([g])$, the case $\Pi_1^m([f]) < \Pi_1^m([g])$ being similar.

We consider the following partitions:

$\mathcal{P}_1$: We partition functions $H_1, H_2 : \boldsymbol{\delta}_{2n-1}^1 \to \boldsymbol{\delta}_{2n-1}^1$ ordered of $k+1$-type for $\ell > 1$, and $H_1, H_2 :<^m\to \boldsymbol{\delta}_{2n+1}^1$ ordered of $k+1$-type if $\ell = 1$, according to whether or not $f([H_2]) > g([H_1])$.

$\mathcal{P}_2$: We partition functions H_1, H_2 as above, ordered of $k+2$-type, according to whether or not $g([H_1]) > f([H_2])$. (If $k = m-1$, we don't consider $\mathcal{P}_2$.)

We then claim that on the homogeneous sides of these partitions, the properties stated in them hold. We consider first $\mathcal{P}_1$.

We suppose not, and fix a c.u.b. $C_1 \subseteq \boldsymbol{\delta}_{2n-1}^1$ homogeneous for the contrary side. We may assume C_1 is contained in a c.u.b. C such that for H having range in C of the correct type, $\sup_{\beta:\Pi_k^m(\beta)=\Pi_k^m([H])} f(\beta) = \sup_{\beta:\Pi_k^m(\beta)=\Pi_k^m([H])} g(\beta)$. We fix H_1 of the correct type with range in C_1. It follows that there is an H_2 of the correct type with range in C_1 such that $F([H_2]) > g([H_1])$ and with $\Pi_k^m([H_2]) = \Pi_k^m([H_1])$, and $\Pi_{k+1}^m([H_2]) > \Pi_{k+1}^m([H_1])$. It then follows from Lemma 8 and induction that there are H_1', H_2' ordered of $k+1$-type with range in C_1, and $[H_1'] = [H_1]$, $[H_2'] = [H_2]$. This contradicts the definitions of C_1.

We let C_1 be a c.u.b. subset of $\boldsymbol{\delta}_{2n-1}^1$ homogeneous for $\mathcal{P}_1$.

It similarly follows that there is a c.u.b. $C_2 \subseteq \boldsymbol{\delta}_{2n-1}^1$ homogeneous for $\mathcal{P}_2$.

We let $C_3 = C_1 \cap C_2$, C_4 be contained in the closure points of C_3, and A_3, A_4 be the measure one sets (w.r.t. $S_{2n-1}^{\ell,m}$) they define.

Restricted to A_3, it follows that f, g are ordered of $k+1$-type. For example, if α, β are represented by H_1, H_2 having range in C_3 of the correct type, and $\Pi_{k+1}^m(\beta) > \Pi_{k+1}^m(\alpha)$, then since we may assume without loss of generality that $\Pi_k^m(\beta) = \Pi_k^m(\alpha)$ (since f, g are increasing), we have from Lemma 8 and induction that $\alpha = [H_1']$, $\beta = [H_2']$ for some H_1', H_2' ordered of $k+1$-type with range in C_3; so since $C_3 \subseteq C_1$, $f(\beta) > g(\alpha)$. Similarly, if $\Pi_{k+1}^m(\alpha) = \Pi_{k+1}^m(\beta)$

and (without loss of generality), $\Pi^m_{k+2}(\alpha) \leq \Pi^m_{k+2}(\beta)$ for $m < k-1$, then from Lemma 8 and induction, and $C_3 \subseteq C_2$, $g(\alpha) > f(\beta)$.

We then define f_2, g_2 as follows: For $\alpha \in A_4$, we set $f_2(\alpha) = f(\alpha), g_2(\alpha) = g(\alpha)$. For $\alpha \notin A_4$, we define, by induction, $f_2(\alpha) =$ the next ω^{th} element in range of $f \cup g$ greater that $\sup\{f_2(\beta),\ g_2(\gamma)\}$, where β, γ range over the ordinals such that for any f', g' ordered of $k+1$-type, $f'(\beta) < f'(\alpha)$, $g'(\gamma) < f'(\alpha)$. We similarly define $g_2(\alpha)$, where β, γ now range over ordinals such that $f'(\beta) < g'(\alpha)$, $g'(\gamma) < g'(\alpha)$, f', g' as above. It then follows readily that f_2, g_2 are ordered of $k+1$-type. This is immediate once it is shown that f_2, g_2 are increasing, and this follows once it is shown that if $\alpha < \beta$, $\alpha \notin A_4$, $\beta \in A_4$, then $f_2(\alpha) < f_2(\beta)$, and similarly for g_2. This, in turn, follows from the definition of A_4 and $k+1$-type. We use here the facts that if $\gamma_1 < \gamma < \gamma_2$, γ_1, γ_2 of the correct type with $\Pi^m_{k+1}(\gamma_1) = \Pi^m_{k+1}(\gamma_2)$, then the definition of $k+1$-type requires no value $g(\delta)$ between $f(\gamma_1)$ and $f(\gamma_2)$, and no value $f(\delta)$ between $g(\gamma_1)$ and $g(\gamma_2)$. Also, if $f(\delta)$ or $g(\delta)$ is required to be less than inf $f(\gamma)$ for γ of the correct type with $\Pi^m_{k+1}(\gamma) = \Pi^m_{k+1}(\gamma_0)$ for some fixed γ_0, then δ is less than the least such γ.

This completes the proof of Lemma 7.

The proof of Lemma 8 is entirely similar, using induction and Lemma 7.

We require two additional technical lemmas.

LEMMA 9. *Given any $H : \theta(S^{\ell,m}_{2n-1}) \to ON$, there is a measure on set A restricted to which H is monotonically increasing.*

LEMMA 10. *Same as Lemma 9 using $H : \theta(W^m_{2n+1}) = \delta^1_{2n+1} \to ON$, where $n \geq 1$.*

Proof of Lemma 9. We fix $H : \theta(S^{\ell,m}_{2n-1}) \to ON$. For each $1 \leq k \leq m$, we consider the partition:

$\mathcal{P}_k$: We partition functions $H_1, H_2 : \delta^1_{2n-1} \to \delta^1_{2n-1}$ (or, if $\ell = 1$, $H_1, H_2 :<^m\to \delta^1_{2n-1}$), ordered of k-type according to whether $H([H_1]) \leq H([H_2])$. If $k = m$, then by order of m-type we mean $H_1(\alpha) < H_2(\alpha) < H_1(\alpha+1)$ for all α.

For $k = 0$, we also consider $\mathcal{P}_0$, where H_1, H_2 are ordered as in the last clause of Lemma 8 for $\ell > 1$; and for $\ell = 1$ and $\vec{\alpha} = (\alpha_2, \ldots, \alpha_m, \alpha_1)$, $\vec{\beta} = (\beta_2, \ldots, \beta_m, \beta_1)$, if $\alpha_1 < \beta_1$ then $G(\vec{\alpha}) < F(\vec{\beta})$ and if $\alpha_1 \geq \beta_1$, $G(\vec{\alpha}) > f(\vec{\beta})$.

It follows from an easy well-foundedness argument that on the homogeneous side of the partition, the property stated in $\mathcal{P}_k$ holds. We let C_k, for $0 \leq k \leq m$, be homogeneous for $\mathcal{P}_k$, and $C = \cap_k C_k$. If A is the measure one set determined by C, then we claim that $H \restriction A$ is monotonically increasing. For, let $\alpha < \beta$ be in A, say represented by H_1, H_2 of the correct type with range in C. We let $1 \leq k \leq m$ be maximal such that $\Pi_k^m(\alpha) = \Pi_k^m(\beta)$, if such a k exists. In this case, by Lemma 8, it follows that there are H_1', H_2' ordered of $k+1$-type with range in $C \subseteq C_k$ with $[H_1'] = [H_1]$, $[H_1'] = [H_2]$. Hence $H([H_1]) = H([H_1']) \leq H([H_2'])] = H([H_2])$. If $\Pi_1^m(\alpha) < \Pi_1^m(\beta)$, then H_1', H_2' are ordered as in the last part of Lemma 8 (depending on whether $[H_1(0)] < [H_2(0)]$ or not) for $\ell > 1$, and for $\ell = 1$ as in the paragraph above, and $H([H_1]) \leq H([H_2])$ follows.

The proof of Lemma 10 is similar.

We now state two lemmas which simplify conditions C and D.

LEMMA 11. *For any c.u.b.* $C \subseteq \boldsymbol{\delta}_{2n+1}^1$, *there is a c.u.b.* $C' \subseteq C$ *such that (for* $s \leq n$*) for any* $f : \theta(R_{2s-1}^{\ell,m}) \to C'$ *which is monotonically increasing,* $[f] = \sup[g]$, *where the sup ranges over* $g : \theta(R_{2s-1}^{\ell,m}) \to C$ *of the correct type.*

LEMMA 12. *For any c.u.b.* $C \subseteq \boldsymbol{\delta}_{2n+1}^1$, *there is a c.u.b.* $C' \subseteq C$ *such that for any* $F : \boldsymbol{\delta}_{2n+1}^1 \to \boldsymbol{\delta}_{2n+1}^1$ *monotonically increasing almost everywhere w.r.t.* $\mu_{2n+1}^{R_{2s+1}^{\ell,m}}$, $[F] = \sup[G]$, *where the sup ranges over* $G : \boldsymbol{\delta}_{2n+1}^1 \to C$ *of the correct type, provided* $[F]$ *is not minimal with respect to being non-constant almost everywhere.*

We first consider Lemma 11. We consider the case $R_{2s-1}^{\ell,m} \neq W_1^m$, this case following directly.

For a given C, we let $C' \subseteq C$ be contained in the closure points of C. We fix $f : \theta(R_{2s-1}^{\ell,m}) \to C'$ monotonically increasing almost everywhere, say on the measure one set determined by $C_1 \subseteq \boldsymbol{\delta}_{2s-1}^1$. Applying Lemmas 1 and 2, we conclude that either f is increasing almost everywhere, or $f(\alpha)$ depends only on $\Pi_{m-1}^m(\alpha)$ almost everywhere. In the latter case, we view f as a function on $\theta(\tilde{R}_{2s-1}^{\ell,m-1})$ (for $m > 1$), that is, there is a $f^{m-1} : \theta(\tilde{R}_{2s-1}^{\ell,m-1}) \to C'$ with $f(\alpha) = f^{m-1}(\Pi_{m-1}^m(\alpha))$ almost everywhere. Applying Lemmas 1 and 2 again to f^{m-1}, we get that either f^{m-1} is increasing almost everywhere, or there is an $f^{m-1} : \theta(\tilde{R}_{2s-1}^{\ell,m-2}) \to C'$ (for $m > 2$) such that $f(\alpha) = f^{m-2}(\Pi_{m-2}^m(\alpha))$ almost everywhere, where $\Pi_{m-2}^m(\alpha) \equiv \tilde{\Pi}_{m-2}^{m-1}(\Pi_{m-1}^m(\alpha))$, etc. Continuing in this manner, we conclude that either for some $0 \leq k \leq m$ there is an $f^k : \theta(\tilde{R}_{2s-1}^{\ell,k}) \to C'$ (or : $\theta(R_{2s-1}^{\ell,m}) \to C'$ if $k = m$) with $f(\alpha) = f^k(\Pi_k^m(\alpha))$ almost everywhere (where $\Pi_m^m =$ identity) and f^k is increasing almost everywhere, or else f is constant almost everywhere, where for $k = 0$ by $\tilde{S}_{2s-1}^{\ell,0}$ we mean the measure on $H : \boldsymbol{\delta}_{2s-1}^1 \to \boldsymbol{\delta}_{2s-1}^1$ induced by the measure $S_{2s-1}^{\ell,1}$ on $H' : \boldsymbol{\delta}_{2s-1}^1 \to \boldsymbol{\delta}_{2s-1}^1$ of the correct type and $H = H'(0)$; and by W_{2s-1}^0 we mean the measure (for $s > 1$) induced by $f : \kappa(W_{2s-1}^1) \to \boldsymbol{\delta}_{2s-1}^1$ of the correct type, the weak partition relation on δ^1_{2s-1}, and sup f. In the case f is constant almost everywhere the lemma easily follows, so we assume the former.

We fix such an $f^k : \theta(\tilde{R}^{\ell,k}_{2s-1}) \to C'$, where if $k = m$ we use $R^{\ell,m}_{2s-1}$, and let $C_2 \subseteq C_1$ be a c.u.b. subset of δ^1_{2s-1} defining a measure one set A_2 restricted to which f^k is increasing and $f(\alpha) = f^k(\Pi^m_k(\alpha))$ holds.

We consider the following two cases:

Case 1. There is a c.u.b. $C_3 \subseteq C_2$ defining a measure one set $A_3 \subseteq A_2$ restricted to which f^k is non-normal, that is, for $\alpha \in A_3$, $f^k(\alpha) > \sup_{\beta<\alpha, \beta\in A_3} f^k(\beta)$. We let $g : \theta(R^{\ell,m}_{2s-1}) \to \delta^1_{2n+1}$ be given with $[g] < [f]$, and we proceed to show that there is an $\bar{f}$ of the correct type with range in C with $[g] < [\bar{f}] < [f]$. From Lemmas 9 and 10, there is a $C_4 \subseteq C_3$ defining a measure one set $A_4 \subseteq A_3$ such that $g \restriction A_4$ is monotonic increasing. We then define $\bar{f}$ as follows: for $\alpha \in A_4$, we set $\bar{f}(\alpha) =$ the next ω^{th} point in C after $\max\{g(\alpha), \sup_{\beta<\alpha} \bar{f}(\beta)\}$. For $\alpha \notin A_4$, we define $\bar{f}(\alpha) =$ the next ω^{th} element in C after $\sup_{\beta<\alpha} \bar{f}(\beta)$. It follows readily that for $C_5 \subseteq$ the closure points of C_4, that $\bar{f} \restriction A_5$ is less that $f \restriction A_5$.

Case 2. There is a $C_3 \subseteq C_2$ defining a measure one set $A_3 \subseteq A_2$ restricted to which f^k is normal, that is, for $\alpha \in A_3$ $f^k(\alpha) = \sup_{\beta<\alpha, \beta\in A_3} f^k(\beta)$. Hence, for $[g] < [f]$, for almost all α w.r.t. $R^{\ell,m}_{2s-1}$, $g(\alpha) < f(\alpha) = f^k(\Pi^m_k(\alpha))$, and it follows that for almost all α, there is a $\beta \in A_3$, $\beta < \Pi^m_k(\alpha)$ such that $g(\alpha) < f^k(\beta)$. Applying Lemmas 3 and 4 repeatedly, we conclude that β depends only on $\Pi^m_{k-1}(\alpha)$, for $k > 1$, and if $k = 1$, then β is constant almost everywhere if $\ell = 1$, and for $\ell > 1$, depends only on $\Pi^m_0(\alpha)$ where $\Pi^m_0(\alpha)$ denotes $[F(0)]$ for F representing α if $R^{\ell,m}_{2s-1} = S^{\ell,m}_{2s-1}$, and denotes $[f]$ for f representing α if $R^{\ell,m}_{2s-1} = W^m_{2s-1}$. If $k = 0$, then β is constant almost everywhere. Hence, there is an $\bar{f}^{k-1} < f^{k-1}$ (where $f^{k-1}(\alpha) \equiv \sup_{\beta:\Pi^k_{k-1}(\beta)=\alpha} f^k(\beta)$) almost everywhere w.r.t. $\tilde{S}^{\ell,k-1}_{2s-1}$ (in the case $k > 1$, or $k \geq 1$ if $\ell > 1$) such that for almost all α w.r.t. $S^{\ell,m}_{2s-1}$, $g(\alpha) < \bar{f}^{k-1}(\Pi^m_{k-1}(\alpha))$. The result now follows easily from the choice of C^1. If $k = 1$, $\ell = 1$, or $k = 0$, $\ell > 1$, the result similarly follows easily.

Finally, to see that either Case 1 or Case 2 holds requires a routine partition argument (partitioning essentially $\omega + 1$ many functions $f_0, f_1, ..., f_\omega$ of the correct type with $f_\omega = \sup_n f_n$) which we omit.

This completes the proof of Lemma 11.

The proof of Lemma 12 is similar.

The last two lemmas allow us to simplify the description of conditions C and D. In condition C, if $d^{(I_a)}$ is non-basic of the form $d = (k; d_2^{(I_a;\bar{K}_{a+1})})^s$ where s appears, $\bar{K}_{a+1} = v(K_k)$ and $K_k = S_{2n+1}^{\ell_k,m_k}$, for almost all $h_1,\ldots,h_t,f,\bar{h}_2,\ldots,\bar{h}_a$, we consider the function $g:\theta(\bar{K}_{a+1}) \to \boldsymbol{\delta}_{2n+1}^1$ defined by $g([\bar{h}_{a+1}]) = H(d_2;h_1,\ldots,h_t;f,\bar{h}_2,\ldots,\bar{h}_{a+1})$, as in the definition of C. If $\ell > 2$, so $\bar{K}_{a+1}$ is not of the form W_1^m, then from lemmas 11 and 12 it follows that d satisfies C provided $[g]$ is not minimal amongst all $[g']$ with $\sup_{\text{a.e.}} g' = \sup_{\text{a.e.}} g$. Thus, for $d \in \mathcal{D}_{2n+1}$ with the symbol s appearing on all component tuples, condition C reduces to the following:

1) All basic component tuples of d satisfy C; which is just a condition on certain component tuples $\bar{d}_2 \in \mathcal{D}_{2n-1}$ of d.

2) The previous definition in the case d is non-basic with $K_k = S_{2n+1}^{\ell_k,m_k}$ and $\ell_k = 1$ or 2.

3) The above non-minimality condition on g, for $\ell_k > 2$.

As for condition D, we note the following: if $\ell > 1$, then $(d_{2n+1}^{\ell,m})^s$ satisfies D if $d_{2n+1}^{\ell,m}$ satisfies C (relative to $K_1,\ldots,K_t$), $I_a = (f;\ )$, and for almost all $h_1,\ldots,h_t$, $[h(d;h_1,\ldots,h_t)]$ is not minimal subject to being nonconstant almost everywhere w.r.t. $\mu_{2n+1}^{\bar{K}_1}$. This follows from Lemma 12 and the fact that for $I_a = (f(\bar{K}_1);\ )$, and $C \subseteq \boldsymbol{\delta}_{2n+1}^1$ a c.u.b. set, for almost all $h_1,\ldots,h_t$, $h(d;h_1,\ldots,h_t)$ is represented w.r.t. $\mu_{2n+1}^{\bar{K}_1}$ by a function F having range in C almost everywhere. This follows easily from the definition of h.

These facts will be of use later.

We now proceed to define the lowering operation $\mathcal{L}$ on $\mathcal{D}_{2n+1}$. We first require a preliminary definition.

DEFINITION. Given $d_1^{(I_a)}, d_2^{(I_{a+1})}$ satisfying C relative to $K_1,\ldots,K_t$ where $I_{a+1} = (I_a;\bar{K}_{a+1})$, we say condition $M(d_1^{(I_a)}, d_2^{(I_{a+1})})$ is satisfied if for almost all $h_1,\ldots,h_t,f,$ $\bar{h}_2,\ldots,\bar{h}_a$, $H(d_1;h_1,\ldots,h_t,f,\bar{h}_2,\ldots,\bar{h}_a) = \sup_{\text{a.e.}[\bar{h}_{a+1}]} H(d_2;h_1,\ldots,h_t,f,\bar{h}_2,\ldots,\bar{h}_{a+1})$. We say $M_2(d_1^{(I_a)}, d_2^{(I_{a+1})})$ is satisfied if $M(d_1,d_2)$ is satisfied and the function $g:\theta(\bar{K}_{a+1}) \to \boldsymbol{\delta}_{2n+1}^1$ defined by $g([\bar{h}_{a+1}]) = H(d_1;h_1,\ldots,h_t,f,\bar{h}_2,\ldots,\bar{h}_{a+1})$ is not minimal with respect to the set of $[g']$ for $g':\theta(\bar{K}_{a+1}) \to \boldsymbol{\delta}_{2n+1}^1$ with $\sup_{\text{a.e.}} g' = \sup_{\text{a.e.}} g$.

We will define the operation $\mathcal{L}$ on objects of the form (d) or $(d)^s$, where $d = d_{2n+1}^{\ell,m} \in \mathcal{D}_{2n+1}$, where d or $(d)^s$ satisfies D and d satisfies A relative to fixed $K_1,\ldots,K_t$. To do this, we first define a preliminary operation $\hat{\mathcal{L}}$ on d satisfying C and A. $\hat{\mathcal{L}}(d)$ will also satisfy C and A.

We introduce a notation. For $d^{(I_a)}$ satisfying C relative to $K_1,\ldots,K_t$, where $I_a = (f(\bar{K}_1);$ $\bar{K}_2,\ldots,\bar{K}_a)$ (or $f = f_k$ or does not appear) we let $\delta_d(\bar{K}_i)$ for $2 \le i \le a$ be the integer $\tau(\bar{K}_i) =$

j, where $1 \leq j \leq t$ such that $v(K_j) = \bar{K}_i$ used in the construction of d. We recall that (with a slight abuse of notation), the $\bar{K}_i$ are tagged with integers referring to such K_j – the function δ_d recovers this integer.

In defining $\hat{\mathcal{L}}$, we will actually define a more general operation $\hat{\mathcal{L}}^{\restriction k}(d)$, defined for d satisfying C, A relative to $K_1, \dots, K_t$ where $\delta_d(\bar{K}_a) < k \leq k(d)$, and d is not minimal w.r.t. the ordering $<$ restricted to the set of d with $k(d) \geq k$ satisfying C, A.

We consider the following cases:

1) $k = \infty$, so d basic of type 1 or 0.

a) d basic of type 1, so $d = (d_2)^s$, where s may or may not appear. If s does not appear, we set $\hat{\mathcal{L}}^{\restriction\infty}(d) = (\bar{d}_2)^s$. If s appears and $\bar{d}_2$ is not minimal w.r.t. the operation $\mathcal{L}$ on $\mathcal{D}_{2n-1}$, defined by induction, then we set $\hat{\mathcal{L}}^{\restriction\infty}(d) = \mathcal{L}(\bar{d}_2))^{(I_a)}$. If s appears and $\bar{d}_2$ is minimal w.r.t $\mathcal{L}$, then d is minimal w.r.t $\mathcal{L}^{\restriction\infty}$, and $\mathcal{L}^{\restriction\infty}(d)$ is not defined.

b) d basic of type 0, $d^{(I_a)} = d^{(f_m; \bar{K}_2, \dots, \bar{K}_a)}$, where $d = (r)$, $1 \leq r \leq m$. If $r > 1$, we set $\hat{\mathcal{L}}^{\restriction\infty}(d) = r - 1$, and if $r = 1$, d is minimal w.r.t. $\hat{\mathcal{L}}^{\restriction\infty}$ (that is, we do not define $\hat{\mathcal{L}}^{\restriction\infty}(d)$).

2) $k < \infty$, $k = k(d)$, and $K_k = W^{m_k}_{2n+1}$. So, d is basic of type -1 of the form $d = (k;\ )$ or $d = (k; \bar{d}_2)^s$, where s may or may not appear, and $I_a = (\bar{K}_2, \dots, \bar{K}_a)$. (for $n = 0$, $\bar{d}_2$ is an integer and s does not appear).

a) $d = (k;\ )$. We recall that $K_{b_1}, \dots, K_{b_u}$ enumerates the subsequence of $K_1, \dots, K_t$ of measures not of the form $S^{\ell,m}_{2n+1}$ or W^m_{2n+1} (i.e. those of the form $R^{\ell,m}_{2s+1}$ for $s < n$). We let $K_{b(k)}, \dots, K_{b_u}$ enumerate those in $K_{k+1}, \dots, K_t$. We assume, inductively on m, that relative to each fixed sequence of measures $K_1, \dots, K_t$ in R_{2m+1}, for which there is a $d = d^{(I_a)}$ or $(d)^s$, for $d \in \mathcal{D}_{2m+1}$, satisfying D, A that there is a canonically defined maximal (relative to $<$) description $\tilde{d}$ or $(\tilde{d})^s$ satisfying D, A; that is $\tilde{d} \geq d'$ if d' or $(d')^s$ satisfies D, A. This definition is, of course, relative to fixed $m, \bar{\ell}, \bar{m}$ such that $d \in \mathcal{D}^{\bar{\ell},\bar{m}}_{2m+1}$.

We let, in this case $v(K_k) = S^{\bar{\ell}_k, \bar{m}_k}_{2n-1}$ if $n \geq 1$. We let $\tilde{d} = \tilde{d}^{\bar{\ell}_k, \bar{m}_k}_{2n-1}$ be the maximal tuple relative to $K_{b(k)}, \dots, K_{b_u}, \bar{K}_1, \dots, \bar{K}_a$ if defined. If $\tilde{d}$ is not defined, we define d to be minimal with respect to $\hat{\mathcal{L}}^{\restriction k}$. If $\tilde{d}$ is defined, we set $\hat{\mathcal{L}}^{\restriction k}(d) = (k; \tilde{d})$ or $(k; \tilde{d})^s$, depending on whether $\tilde{d}$ satisfies D or not.

b) $d = (k; \bar{d}_2)^s$, where s may or may not appear.

If $n = 0$, so $d = (k; i)$, we set $\hat{\mathcal{L}}^{\restriction k}(d) = (k; i - 1)$ if $i > 1$, and if $i = 1$, d is minimal w.r.t $\hat{\mathcal{L}}^{\restriction k}$. We assume now $n > 0$.

If s does not appear, we set $\hat{\mathcal{L}}\restriction^k(d) = (k;\bar{d}_2)^s$.

If s appears and $\bar{d}_2$ is not minimal with respect to $\mathcal{L}$ relative to $K_{b(k)},\ldots,K_{b_u}$, $\bar{K}_2,\ldots,\bar{K}_a$, we set $\hat{\mathcal{L}}\restriction^k(d) = (k;\mathcal{L}'((\bar{d}_2)^s))$ or $(k;\mathcal{L}'((\bar{d}_2)^s))^s$ depending on whether $\mathcal{L}((\bar{d}_2)^s)$ does not or does involve s, where if $\mathcal{L}((\bar{d}_2)^s) = (d')$ or $(d')^s$, $\mathcal{L}'((\bar{d}_2)^s) \equiv d'$.

If s appears, and $\bar{d}_2$ is minimal with repsect to $\mathcal{L}$ relative to $K_{b(k)},\ldots,\bar{K}_a$, then d is minimal with respect to $\hat{\mathcal{L}}\restriction^k$.

3) $k < \infty$, $k = k(d)$, $K_k = S^{\ell_k,m_k}_{2n+1}$ with $\ell_k > 1$. Hence $d = (k; d_2^{(I_a;\bar{K}_{a+1})})^s$, where s may or may not appear. We let $\hat{d}$ be the tuple from condition A for d.

a) If s does not appear, then $\hat{\mathcal{L}}\restriction^k(d) = (k; d_2^{(I_a;\bar{K}_{a+1})})^s$.

b) $\ell > 2$, s appears and d_2 is not minimal w.r.t. $\hat{\mathcal{L}}\restriction^{k+1}$. We then set $\hat{\mathcal{L}}\restriction^k(d) = (k;\hat{\mathcal{L}}\restriction^{k+1}(d_2))$ if this tuple satisfies C and $M(\hat{d}_1,\hat{\mathcal{L}}\restriction^{k+1}(d_2))$ is satisfied. If this tuple does not satisfy C, we set $\hat{\mathcal{L}}\restriction^k(d) = (k;\hat{\mathcal{L}}\restriction^{k+1}(d_2))^s$ if $M_2(\hat{d},\hat{\mathcal{L}}\restriction^{k+1}(d_2))$ and C are satisfied, and otherwise $\hat{\mathcal{L}}\restriction^k(d) = \hat{d}$.

c) $\ell = 2$, s appears and d_2 not minimal w.r.t. $\hat{\mathcal{L}}\restriction^{k+1}$. We set $\hat{\mathcal{L}}\restriction^k(d) = (k;\hat{\mathcal{L}}^{k+1(q)}(d_2))^s$, where s appears if $(k;\hat{\mathcal{L}}^{k+1(q)}(d_2))$ does not satisfy C, and $\hat{\mathcal{L}}\restriction^{k+1(q)}(d_2)$ is the least iterate of $\hat{\mathcal{L}}\restriction^{k+1}$ such that $(k;\hat{\mathcal{L}}\restriction^{k+1(q)}(d_2))$ or $(k;\hat{\mathcal{L}}\restriction^{k+1(q)}(d_2))^s$ satisfies C. If no such q exists, we set $\hat{\mathcal{L}}\restriction^k(d) = \hat{d}$.

d) Otherwise we set $\hat{\mathcal{L}}\restriction^k(d) = \hat{d}$.

4) $k < \infty$, $k = k(d)$, $K_k = S^{1,m_k}_{2n+1}$. Hence, $d = (k; d_1^{(I_a)},\ldots,d_r^{(I_a)})^s$, where s may or may not appear, and $r \le m_k$.

a) s appears $r > 2$ and $\hat{\mathcal{L}}\restriction^{k+1}(d_r)$ is not defined or $\hat{\mathcal{L}}\restriction^{k+1}(d_r) \le d_{r-1}$. We set $\hat{\mathcal{L}}\restriction^k(d) = (k; d_1^{(I_a)},\ldots,d_{r-1}^{(I_a)})^s$.

b) s appears, $r = 2$, and $\hat{\mathcal{L}}\restriction^{k+1}(d_r)$ is not defined. We set $\hat{\mathcal{L}}\restriction^k(d) = d_1$.

c) s appear, $\hat{\mathcal{L}}\restriction^{k+1}(d_r)$ is defined, and if $r > 2$, $\hat{\mathcal{L}}\restriction^k(d_r) > d_{r-1}$. Then $\hat{\mathcal{L}}\restriction^k(d) = (k; d_1^{(I_a)},\ldots,d_{r-1}^{(I_a)},\ \hat{\mathcal{L}}\restriction^{k+1}(d_r^{(I_a)}))^s$, where s appears if without s the tuple does not satisfy C.

d) s does not appear and $r < m_k$.

i) $\hat{\mathcal{L}}\restriction^{k+1}(d_1)$ is defined, and if $r \ge 2$, $\hat{\mathcal{L}}\restriction^{k+1}(d_1) > d_r$. We set $\hat{\mathcal{L}}\restriction^k(d) = (k; d_1^{(I_a)},\ldots,d_r^{(I_a)}, \hat{\mathcal{L}}\restriction^{k+1}(d_1))^s$, where s appears if without s the tuple does not satisfy C.

ii) $\hat{\mathcal{L}}\restriction^{k+1}(d_1)$ defined, but i) fails. We set $\hat{\mathcal{L}}\restriction^k(d) = (k; d_1^{(I_a)},\ldots,d_r^{(I_a)})^s$.

iii) $\hat{\mathcal{L}}\restriction^{k+1}(d_1)$ not defined. It will follow in this case, from our main inductive hypothesis, that $r = 1$, since $\hat{\mathcal{L}}\restriction^{k+1}(d_1)$ not defined implies d_1 minimal w.r.t $<$ on the d' satisfying C, A and $k(d') > k$, which is not the case if $r > 1$. In this case, we set $\hat{\mathcal{L}}\restriction^{k}(d) = d_1$.

e) s does not appear, and $r = m_k$. We set $\hat{\mathcal{L}}\restriction^{k}(d) = (k; d_1^{(I_a)}, \ldots, d_r^{(I_a)})^s$ if $m_k > 1$ and $= d_1$ if $m_k = 1$.

5) $k < \infty,\ k < k(d),\ K_k = W_{2n+1}^{m_k}$.

If d is not minimal with respect to $\hat{\mathcal{L}}\restriction^{k+1}$, then we set $\hat{\mathcal{L}}\restriction^{k}(d) = \hat{\mathcal{L}}\restriction^{k+1}(d)$.

If d is minimal with respect to $\hat{\mathcal{L}}\restriction^{k+1}$, then we set $\hat{\mathcal{L}}\restriction^{k}(d) = (k;\ )$, a basic -1 description.

6) $k < \infty,\ k < k(d),\ K_k = S_{2n+1}^{1,m_k}$.

a) $\hat{\mathcal{L}}\restriction^{k+1}(d)$ defined and $(k; \hat{\mathcal{L}}\restriction^{k+1}(d))$ satisfies C. Then we set $\hat{\mathcal{L}}\restriction^{k}(d) = (k; \hat{\mathcal{L}}\restriction^{k+1}(d))$.

b) $\hat{\mathcal{L}}\restriction^{k+1}(d)$ defined, but a fails. We set $\hat{\mathcal{L}}\restriction^{k}(d) = \hat{\mathcal{L}}\restriction^{k+1}(d)$.

c) a) and b) fail. Then d is minimal with respect to $\hat{\mathcal{L}}\restriction^{k+1}(d)$.

7) $k < \infty,\ k < k(d),\ K_k = S_{2n+1}^{\ell,k,m_k}$ with $\ell_k > 1$.

a) $\hat{\mathcal{L}}\restriction^{k+1}(d)$ defined, and for almost all $h_1, \ldots, h_t, f, \bar{h}_2, \ldots, \bar{h}_a$, cof $H(\hat{\mathcal{L}}\restriction^{k+1}(d); h_1, \ldots, h_t, f, \bar{h}_2, \ldots, \bar{h}_a) =$ cof $\kappa(K_k)$. We set $\hat{\mathcal{L}}\restriction^{k}(d) = (k; \tilde{\hat{\mathcal{L}}}\restriction^{k+1}(d)^{(I_a;\bar{K}_{a+1})})^s$, where $\tilde{\hat{\mathcal{L}}}\restriction^{k+1}(d)$ denotes the tuple obtained by replacing in all component tuples of d, the index $(f(\bar{K}_1); \bar{K}_2, \ldots, \bar{K}_a, \underline{\qquad})$ by $(f(\bar{K}_1); \bar{K}_2, \ldots, \bar{K}_a, \bar{K}_{a+1}, \underline{\qquad})$.

b) $\hat{\mathcal{L}}\restriction^{k+1}(d)$ defined and a) fails. We set $\hat{\mathcal{L}}\restriction^{k}(d) = \hat{\mathcal{L}}\restriction^{k+1}(d)$.

c) $\hat{\mathcal{L}}\restriction^{k+1}(d)$ not defined. Then d is minimal with respect to $\hat{\mathcal{L}}\restriction^{k}$.

Finally for (d) or $(d)^s$ satisfying D, A, we set $\mathcal{L}(d) = (d)^s$, and $\mathcal{L}(d)^s = (\hat{\mathcal{L}}\restriction^{1(p)}(d))^s$, where s appears if the tuple without s does not satisfy D, and $\hat{\mathcal{L}}\restriction^{1(p)}$ denotes the pth iterate of $\hat{\mathcal{L}}\restriction^{1}$, and p is minimal such that $\hat{\mathcal{L}}\restriction^{1(p)}(d)$ or $(\hat{\mathcal{L}}\restriction^{1(p)}(d))^s$ satisfies D. If no such p exists, then $(d)^s$ is minimal with respect to $\mathcal{L}$.

LEMMA. *If d satisfied C, A relative to $K_1, \ldots, K_t$, and $\delta_d(\bar{K}_1) < k \le k(d)$, then $\hat{\mathcal{L}}\restriction^{k}(d)$ also satisfies C, A. Hence, if $d^{(I_a)}$ satisfies C, A where $I_a = (f(\bar{K}_1);\ )$, then $\hat{\mathcal{L}}(d) \equiv \hat{\mathcal{L}}\restriction^{1}(d)$ satisfies C, A and in this case, if d satisfies D, so does $\mathcal{L}(d)$.*

Proof. The first statement clearly implies the second. The proof of the first statement is routine upon consideration of the cases.

For example, we consider the case $d = (k; d_2^{(I_a;\bar{K}_{a+1})})^s$, where $K_k = S_{2n+1}^{\ell_k,m_k}$, and $v(K_k) = \bar{K}_{a+1}$. We consider $\hat{\mathcal{L}}^{\restriction k}(d)$, where $k = k(d)$, here. If $\hat{\mathcal{L}}^{\restriction k}(d) = (k; \hat{\mathcal{L}}^{\restriction k+1}(d_2))$, then C is satisfied by definition. By induction (reverse induction on $k(d)$), $\hat{\mathcal{L}}^{\restriction k+1}(d_2)$ satisfies A. Since $M(\hat{d}, \hat{\mathcal{L}}^{\restriction k+1}(d_2))$ is satisfied in this case, A is also satisfied by $\hat{\mathcal{L}}^{\restriction k}(d)$, since for $\hat{\mathcal{L}}^{\restriction k}(d)$ we may take $\hat{d}$. Similarly, if $\hat{\mathcal{L}}^{\restriction k}(d) = (k; \hat{\mathcal{L}}^{\restriction k+1}(d_2))^s$, C and A are satisfied. If $\hat{\mathcal{L}}^{\restriction k}(d) = \hat{d}$, C and A are satisfied by definition.

We consider the case with d as above and $k < k(d)$. If $\hat{\mathcal{L}}^{\restriction k}(d) = \hat{\mathcal{L}}^{\restriction k+1}(d)$, we are done by induction. Hence $\hat{\mathcal{L}}^{\restriction k}(d) = (k; \widetilde{\hat{\mathcal{L}}^{\restriction k+1}(d)}^{(I_a;\bar{K}_{a+1})})^s$. By induction, $\hat{\mathcal{L}}^{\restriction k+1}(d)$ satisfies C, A. It follows readily that $\widetilde{\hat{\mathcal{L}}^{\restriction k+1}(d)}$ also satisfies C, A (if $M(d_1, d_2)$ is satisfied for component tuples d_1, d_2 of $\hat{\mathcal{L}}^{\restriction k+1}(d)$, then $M(\tilde{d}_1, \tilde{d}_2)$ is also satisfied). Since for almost all $h_1, \ldots, h_t, f, \bar{h}_2, \ldots, \bar{h}_a$, the function $g([\bar{h}_{a+1}]) = H(\widetilde{\hat{\mathcal{L}}^{\restriction k+1}(d)}, h_1, \ldots, h_t, f, \bar{h}_2, \ldots, \bar{h}_a, \bar{h}_{a+1})$ is constant almost everywhere in this case, it follows readily that C is satisfied, and to satisfy A we may take $\hat{d} = \hat{\mathcal{L}}^{\restriction k+1}(d)$. The remaining cases follow similarly.

Finally, to complete the inductive definition of $\mathcal{L}$, we define the maximal tuple $\tilde{d}_{2n+1}^{\ell,m}$ relative to $K_1, \ldots, K_t$.

1) $\ell = -1$. In this case $\tilde{d}^{(I_a)} = \tilde{d}^{(\,)}$. We let $1 \leq w \leq t$ denote the largest integer such that K_w is of the form $K_w = W_{2n+1}^{m_w}$. We let $a_1, \ldots, a_p$ enumerate the integers from 1 to w with K_a of the form $S_{2n+1}^{\ell,m}$ and $\operatorname{cof}\kappa(S_{2n+1}^{\ell,m}) = \operatorname{cof}\kappa(W_{2n+1}^{m_w})$, (or of the form $S_1^{1,m}$ if $n = 0$) or equivalently, $v(S_{2n+1}^{\ell,m}) \sim v(W_{2n+1}^{m_w})$. We set $\tilde{d}_\infty^{(\,)} = (w;\ )$ (or $= (w; m)$ if $n = 0$). We set $\tilde{d}_p^{(\,)} = (a_p; \tilde{d}_\infty^{(\bar{K}_p)})^s$, where $\bar{K}_p = v(K_p)$. Similarly $\tilde{d}_{p-1}^{(\,)} = (a_{p-1}; \tilde{d}_p^{(\bar{K}_{p-1})})^s$ where $\tilde{d}_p^{(\bar{K}_{p-1})} = (a_p; \tilde{d}_\infty^{(\bar{K}_{p-1},\bar{K}_p)})^s$, etc. Continuing, we define $\tilde{d}_1^{(\,)}$. We define $\tilde{d} = (\tilde{d}_1^{(\,)})^s$. If w does not exist, $\tilde{d}$ is not defined.

2) $\ell = 1$. In this case $\tilde{d}^{(I_a)} = \tilde{d}^{(f_m)}$. We let $e_1, \ldots, e_p$ enumerate the measures in $K_1, \ldots, K_t$ of the form S_{2n+1}^{1,m_k}. We set $\tilde{d}_\infty^{(f_m;\ )} = (m)$. We let $\tilde{d}_i^{(f_m;\ )} = (e_i; \tilde{d}_{i+1}^{(f_m;\ )})$ for $1 \leq i \leq p$. We set $\tilde{d} = \tilde{d}_1$ or $(\tilde{d}_1)^s$, depending on whether $\tilde{d}_1$ does or does not satisfy D. If e_1 does not exist, $\tilde{d}$ is not defined.

3) $\ell > 1$, $\tilde{d}^{(I_a)} = \tilde{d}^{(f(\bar{K}_1);\ )}$. We let $e_1, \ldots, e_p$ enumerate the integers corresponding to measures K_k with $\operatorname{cof}\kappa(K_k) = \operatorname{cof}\kappa(S_{2n+1}^{\ell,m})$, or equivalently, $v(K_k) \sim v(S_{2n+1}^{\ell,m})$. Here K_k must be of the form $S_{2n+1}^{\ell_k,m_k}$ with $\ell_k > 1$. We let $\tilde{d}_\infty^{(f(\bar{K}_1);\ )} = (\,)^{(f(\bar{K}_1);\ )}$. We set $\tilde{d}_p^{(f(\bar{K}_1))} = (e_p, \tilde{d}_\infty^{(f(\bar{K}_1);\bar{K}_p)})^s$, where $\bar{K}_p = v(K_p)$. Similarly, $\tilde{d}_{p-1}^{(f(\bar{K}_1);\ )} = (e_{p-1}; \tilde{d}_p^{(f(\bar{K}_1);\bar{K}_{p-1})})^s$, where $\tilde{d}_p^{(f(\bar{K}_1);\bar{K}_{p-1})} = (e_p; \tilde{d}_\infty^{(f(\bar{K}_1);\bar{K}_{p-1},\bar{K}_p)})$ etc. We set $\tilde{d} = \tilde{d}_1$ or $(\tilde{d}_1)^s$ as above. If e_1 does not exist, $\tilde{d}$ is not defined.

We now introduce some additional notation required for the proof.

We fix $K_1, \ldots, K_t$ and an index I_a.

We let θ be an ordinal. We represent θ w.r.t. the measure K_1 by the function g. We write $g(h_1) = g([h_1])$ for $[h_1] \in \theta(K_1)$. Thus g is defined almost everywhere. For a fixed h_1, we represent $g(h_1)$ w.r.t. K_2 by the function $g(h_1, h_2)$. We emphasize what is defined only for a fixed choice of g, and fixed function representing $g(h_1)$. Continuing, we define $g(h_1, \ldots, h_t)$. We represent this w.r.t. $\mu_{2n+1}^{\bar{K}_1}$ or the m-fold product of the ω-cofinal normal measure on δ_{2n+1}^1 by the function $g(h_1, \ldots, h_t, f)$, where $f : \theta(\bar{K}_1) \to \delta_{2n+1}^1$ of the correct type, or $f : m \to \delta_{2n+1}^1$, if $I_a = (f(\bar{K}_1); \bar{K}_2, \ldots, \bar{K}_a)$ or $= (f_k; \bar{K}_2, \ldots, \bar{K}_a)$. Continuing, we represent with respect to the measures $\bar{K}_2, \ldots, \bar{K}_a$ to get $g(h_1, \ldots, h_t, f, \bar{h}_2, \ldots, \bar{h}_a)$. Again, $g(h_1, \ldots, h_t, f, \bar{h}_2, \ldots, \bar{h}_i, \bar{h}_{i+1})$ is only defined after having chosen a specific function representing $g(h_1, \ldots, \bar{h}_i)$.

Given d satisfying C, A relative to $K_1, \ldots, K_t$, we define a sequence of component tuples $d^0, d^1, \ldots, d^\ell$ for some $\ell \geq 0$ as follows:

We set $d^0 = d$. We assume d^i have been defined.

a) if $d^{i(I_{a_i})} = (k; d_2^{(I_{a_i+1})})$ where $K_k = S_{2n+1}^{\ell_k, m_k}$, with $\ell_k > 1$, we set $i = \ell$.

b) if $d^{i(I_{a_i})} = (k; d_1^{(I_{a_i})}, \ldots, d_r^{(I_{a_i})})$ where $K_k = S_{2n+1}^{1, m_k}$ and $r = m_k$, we set $i = \ell$.

c) if $d^i = (k; d_2^{(I_{a_i+1})})^s$, we set $d^{i+1} = d_2 = d_2^{(I_{a_i+1})}$.

d) if $d^i = (k; d_1, \ldots, d_r)^s$, we set $d^{i+1} = d^r$.

e) if $d^i = (k; d_1, \ldots, d_r)$ where $r < m_k$ $(K_k = S_{2n+1}^{1, m_k})$, we set $d^{i+1} = d_1 = d_1^{(I_{a_i})}$.

f) if d^i is basic, we set $i = \ell$.

We let $k_i = k(d^i)$, so $1 \leq k_i \leq t$, or $k_i = \infty$, and $k_0 < k_1 < \cdots < k_\ell$.

For a fixed ordinal θ or description $\tilde{d}^{(I_{a_i})}$satisfying C, A w.r.t. $K_1, \ldots, K_t$, where the indices I_{a_i} are as above, we define the ordinal $h(d; (d^i \to \theta))$ or $h(d; (d^i \to \tilde{d}))$ as follows: we represent with respect to $K_1, \ldots, K_t$, $\mu_{2n+1}^{\bar{K}_1}$ or the m-fold product of the ω-cofinal normal measure on δ_{2n+1}^1, and $\bar{K}_2, \ldots, \bar{K}_a$ by $H(d; h_1, \ldots, h_t, f, \bar{h}_2, \ldots, \bar{h}_a; (d^i \to \theta))$ or $H(d; h_1, \ldots, h_t, f, \bar{h}_2, \ldots, \bar{h}_a;\ (d^i \to \tilde{d}^i))$, where these are defined exactly as $H(d; h_1, \ldots, h_t, f, \bar{h}_2, \ldots, \bar{h}_a)$ except that in defining $H(d^{i-1(I_{a_{i-1}})}; h_1, \ldots, h_t, f, \bar{h}_2, \ldots, \bar{h}_{a_{i-1}})$, $H(d^i; h_1, \ldots, h_t, f, \bar{h}_2, \ldots, \bar{h}_{a_i})$ is replaced by $g(h_1, \ldots, h_t, f, \bar{h}_2, \ldots, \bar{h}_{a_i})$ or $H(\tilde{d}^i; h_1, \ldots, h_t, f, \bar{h}_2, \ldots, \bar{h}_{a_i})$.

If C is a c.u.b. subset of δ_{2n+1}^1, we let $N_c(\alpha)$ = the next ω^{th} element of C after α. If g :

$\boldsymbol{\delta}^1_{2n+1} \to \boldsymbol{\delta}^1_{2n+1}$, $N_g(\alpha) \equiv$ the ωth element in the range of g after α. We abbreviate $N_{h_i}(\alpha)$ by $N_i(\alpha)$.

For $g; \boldsymbol{\delta}^1_{2n+1} \to \boldsymbol{\delta}^1_{2n+1}$, we let $h(d; (d^i \to g \circ \tilde{d}^i))$ be defined as above, replacing $H(d^i; h_1, \ldots, h_t, f, \bar{h}_2, \ldots, \bar{h}_{a_i})$ by $g(H(\tilde{d}^i; h_1, \ldots, h_t, f, \bar{h}_2, \ldots, \bar{h}_{a_i}))$.

If $d = d^{\ell,m}_{2n+1}$ or $(d)^s$ is given and satisfies D, A w.r.t. $K_1, \ldots, K_t$, and $g : \boldsymbol{\delta}^1_{2n+3} \to \boldsymbol{\delta}^1_{2n+3}$ we define the ordinal $(g; d; K_1, \ldots, K_t)$. We represent this with respect to $\mu^{\mathcal{V}}_{2n+3}$, where $\mathcal{V} = S^{\ell,m}_{2n+1}$ if $\ell \geq 1$, and $\mathcal{V} = W^m_{2n+1}$ if $\ell = -1$, by the function $(g; f; d; K_1, \ldots, K_t)$, for $f : \theta(\mathcal{V}) \to \boldsymbol{\delta}^1_{2n+3}$ of the correct type. We represent this, in turn, w.r.t. K_1 by the function $(g; f; d; h_1, K_2, \ldots, K_t)$, for $[h_1] \in \theta(K_1)$. We represent, in turn, with respect to $K_2, \ldots, K_t$ by $(g; f; d; h_1, h_2, K_3, \ldots, K_t), \ldots, (g; f; d; h_1, \ldots, h_t)$. Finally, we set $(g; f; d; h_1, \ldots, h_t) = g(f(h(d; h_1, \ldots, h_t)))$. For $(d)^s$, the procedure is similar, except $(g; f; (d)^s; h_1, \ldots, h_t) = g(\sup_{\beta < h(d; h_1, \ldots, h_t)} f(\beta))$.

We now state the main inductive hypothesis, H_{2n+1}.

$H_{2n+1}(a)$: We let $d^{(I_a)}$ or $(d)^s$, where $d = d^{\ell,m}_{2n+1} \in \mathcal{D}_{2n+1}$, be defined and satisfy D, A w.r.t. $K_1, \ldots, K_t$. We let $\mathcal{V} = S^{\ell,m}_{2n+1}$ or W^m_{2n+1} depending on whether $\ell \geq 1$ or $\ell = -1$. We assume $I_a = (f(\bar{K}_1); \ )$ or $(f_m; \ )$ or $(\ )$, and assume $\mathcal{L}(d)$ is defined. We let $F : \boldsymbol{\delta}^1_{2n+3} \to \delta^1_{2n+3}$ be given and satisfy $F < (id; d; K_1, \ldots, K_t)$ almost everywhere w.r.t. $\mu^{\mathcal{V}}_{2n+3}$. Then there is a $g : \boldsymbol{\delta}^1_{2n+3} \to \boldsymbol{\delta}^1_{2n+3}$ such that $F < (g; \mathcal{L}(d); K_1, \ldots, K_t)$ almost everywhere w.r.t. $\mu^{\mathcal{V}}_{2n+3}$. Similarly for $(d)^s$.

(b): As above, except we assume now that $\mathcal{L}(d)$ is not defined. We then have that for some $\alpha < \boldsymbol{\delta}^1_{2n+3}$, $F(\beta) = \alpha$ for almost all β w.r.t. $\mu^{\mathcal{V}}_{2n+3}$.

(c): If for almost all $f : \theta(\mathcal{V}) \to \boldsymbol{\delta}^1_{2n+3}$ of the correct type, and $h_2, \ldots, h_t$, $F(f, h_1, \ldots, h_t) < \sup_{\text{a.e.}} f$, and the maximal tuple $\tilde{d}^{\ell,m}_{2n+1}$ is defined, then for almost all f $F([f]) < (g; f; \tilde{d}; K_1, \ldots, K_t)$, for some $g : \boldsymbol{\delta}^1_{2n+3} \to \boldsymbol{\delta}^1_{2n+3}$.

(d): as above, except $\tilde{d}$ is not defined. We then have that F is constant almost everywhere.

The rest of this chapter is devoted to a proof of H_{2n+1}. We proceed by induction on n, so we assume H_{2m+1} for $m < n$. We require the following lemma:

Cofinality Lemma. We let $d^{(I_a)}$, where $d = d^{\ell,m}_{2n+1} \in \mathcal{D}_{2n+1}$ be given and satisfy C, A relative to $K_1, \ldots, K_t$. We let $\theta \in ON$ be represented by $g(h_1, \ldots, h_t, f, \bar{h}_2, \ldots, \bar{h}_a)$ as defined previously. We assume that for almost all $h_1, \ldots, h_t, f, \bar{h}_2, \ldots, \bar{h}_a$ that $g(h_1, \ldots, h_t, f, \bar{h}_2, \ldots, \bar{h}_a) < H(d; h_1, \ldots, h_t, f; \bar{h}_2, \ldots, \bar{h}_a)$. We then have that there is an ordinal θ_ℓ such that if $g_\ell(h_1, \ldots, h_t,$

$f, \bar{h}_2, \ldots, \bar{h}_{a_\ell})$ represents θ_ℓ (where ℓ is as in the definition of $d^0, d^1, \ldots, d^\ell$), then $g_\ell(h_1, \ldots, h_t, f, \bar{h}_2, \ldots, \bar{h}_{a_\ell}) < H(d^\ell; h_1, \ldots, h_t, f, \bar{h}_2, \ldots, \bar{h}_{a_\ell})$ almost everywhere, and for almost all $h_1, \ldots, h_t, f, \bar{h}_2, \ldots, \bar{h}_a, g(h_1, \ldots, h_t, f, \bar{h}_2, \ldots, \bar{h}_a) < H(d; (d^\ell \to \theta_\ell); h_1, \ldots, h_t, f, \bar{h}_2, \ldots, \bar{h}_a)$.

Proof. By induction on $0 \le i \le \ell$, we establish that there is a θ^i such that for almost all $h_1, \ldots, h_t, f, \bar{h}_2, \ldots, \bar{h}_{a_i}$, $\theta^i(h_1, \ldots, h_t, f, \bar{h}_2, \ldots, \bar{h}_{a_i}) < H(d; (d^i); h_1, \ldots, h_t, f, \bar{h}_2, \ldots, \bar{h}_{a_i})$, and $g(h_1, \ldots, h_t, f, \bar{h}_2, \ldots, \bar{h}_a) < H(d; (d^i \to \theta^i); h_1, \ldots, h_t f, \bar{h}_2, \ldots, \bar{h}_a)$. For $i = 0$ this is true by assumption. Assuming true for i, it follows upon consideration of the cases (one of which we consider below) that there is a θ^{i+1} such that $\theta^{i+1}(h_1, \ldots, h_t, f, \bar{h}_2, \ldots, \bar{h}_{a_{i+1}}) < H(d^{i+1}; h_1, \ldots, h_t, f, \bar{h}_2, \ldots, \bar{h}_{a_{i+1}})$ and $\theta^i(h_1, \ldots, h_t, f, \bar{h}_2, \ldots, \bar{h}_{a_i}) < H(d^i; (d^{i+1} \to \theta^{i+1}); h_1, \ldots, h_t, f, \bar{h}_2, \ldots, \bar{h}_{a_i})$ hold almost everywhere. Hence $g(h_1, \ldots, h_t, f, \bar{h}_2, \ldots, \bar{h}_a) < H(d; (d^{i+1} \to \theta^{i+1}); h_1, \ldots, h_t, f, \bar{h}_2, \ldots, \bar{h}_a)$ follows.

As an example, we consider the case $d^i = (k; d^{i+1})^s$, where $K_k = S^{\ell_k, m_k}_{2n+1}$ with $\ell_k > 1$. So, for almost all $h_1, \ldots, h_t, f, \bar{h}_2, \ldots, \bar{h}_{a_i}$, $\theta^i(h_1, \ldots, h_t, f, \bar{h}_2, \ldots, \bar{h}_{a_i}) < H(d^i; h_1, \ldots, h_t, f, \bar{h}_2, \ldots, \bar{h}_{a_i}) = \sup_{\beta < [g]} h_k(\beta)$, where $g([h_{a_{i+1}}]) = H(d^{i+1}; h_1, \ldots, h_t, f, \bar{h}_2, \ldots, \bar{h}_{a_{i+1}})$ almost everywhere. Hence, for almost all $h_1, \ldots, \bar{h}_{a_i}$, there is a $g' < g$ almost everywhere with $\theta^i(h_1, \ldots, \bar{h}_{a_i}) < \eta_k([g'])$. Equivalently, for almost all $h_1, \ldots, \bar{h}_{a_{i+1}}$, there is a $\theta^{i+1}(h_1, \ldots, \bar{h}_{a_{i+1}}) < H(d^{i+1}; h_1, \ldots, \bar{h}_{a_{i+1}})$ such that $\theta^i(h_1, \ldots, \bar{h}_{a_{i+1}}) < H(d^i; (d^{i+1} \to \theta^{i+1}); h_1, \ldots, h_t, f, \bar{h}_2, \ldots, \bar{h}_{a_i})$. The remaining cases are similar.

The main part of the proof of the main inductive hypothesis consists of establishing the main inductive lemma, which we now state:

Main Inductive Lemma. For $1 \le i \le t$ or $i = \infty$ we consider the statement $\bar{H}(i)$: we let $d^{(I_a)}$ for $d = d^{\ell,m}_{2n+1}$ be given, where d is defined and satisfies C, A relative to $K_1, \ldots, K_t$, and d is not minimal with respect to $\hat{\mathcal{L}}^{\restriction i}$ where $\delta_d(\bar{K}_a) < i \le k(d)$, and let θ be an ordinal. We assume that for almost all $h_1, \ldots, h_t, f, \bar{h}_2, \ldots, \bar{h}_a$ that $\theta(h_1, \ldots, h_t, f, \bar{h}_2, \ldots, \bar{h}_a) < H(d^{(I_a)}; h_1, \ldots, h_t, f, \bar{h}_2, \ldots, \bar{h}_a)$. Then for almost all $h_1, \ldots, h_{i-1}$, there is a c.u.b. $C_i \subseteq \boldsymbol{\delta}^1_{2n+1}$ such that for almost all $h_i, \ldots, h_t$ we have that $\theta(h_1, \ldots, h_t) < h(\mathcal{L}^{\restriction i}(d); (\mathcal{L}^{\restriction i}(d) \to N_{C_i}(\mathcal{L}^{\restriction i}(d))); h_1, \ldots, h_t)$.

If $d^{(I_a)}$ is minimal with respect to $\hat{\mathcal{L}}^{\restriction i}$ (i.e. $\hat{\mathcal{L}}^{\restriction i(d)}$ is not defined), then we require that for almost all $h_1, \ldots, h_{i-1}$, there is an $\alpha_i < \boldsymbol{\delta}^1_{2n+1}$ such that for almost all $h_i, \ldots, h_t$, $\theta(h_1, \ldots, h_t) < \alpha_i$.

We prove the main inductive lemma by reverse induction on i. We consider the necessary cases.

I) $i = \infty$. Hence $d = d^{\ell,m}_{2n+1}$ is basic of type 0 or 1.

a) d basic of type 1, so $d^{(I_a)} = (d_2)^{s(f(\bar{K}_a);\bar{K}_2,\ldots,\bar{K}_a)}$, where s may or may not appear, $d_2 \in \mathcal{D}^{\bar{\ell},\bar{m}}_{2s+1}$ and $\bar{K}_1 = R^{\bar{\ell},\bar{m}}_{2s+1}$.

i) $d(= d^{\ell})$ not minimal with respect to $\hat{\mathcal{L}}^{\restriction\infty}$. We let θ be given such that for almost all $h_1,\ldots,h_t,f,\bar{h}_2,\ldots,\bar{h}_a$, $\theta(h_1,\ldots,h_t,f,\bar{h}_2,\ldots,\bar{h}_a) < H(d;h_1,\ldots,h_t,f,\bar{h}_2,\ldots,\bar{h}_a)$. We have in this case that $\hat{\mathcal{L}}^{\restriction\infty}(d) = \mathcal{L}((d_2))$ or $\mathcal{L}((d_2)^s)$ depending on whether s does not or does appear. By induction and $H_{2n-1}(a)$, it follows that for almost all $h_1,\ldots,h_t$, there is a c.u.b. $C \subseteq \boldsymbol{\delta}^1_{2n+1}$ such that for almost all $f,\bar{h}_2,\ldots,\bar{h}_a$, $\theta(h_1,\ldots,h_t,f,\bar{h}_2,\ldots,\bar{h}_a) < (N_C;f;\mathcal{L}((d_2)^s);\bar{h}_2,\ldots,\bar{h}_a) = H(\hat{\mathcal{L}}^{\restriction\infty}(d);\ (\hat{\mathcal{L}}^{\restriction\infty}(d) \to N_C \circ \hat{\mathcal{L}}^{\restriction\infty}(d));\ h_1,\ldots,h_t,\ f,\bar{h}_2,\ldots,\bar{h}_a)$, and similarly if s does not appear.

ii) d is minimal with respect to $\hat{\mathcal{L}}^{\restriction\infty}$. Hence, the symbol s appears, and $(d_2)^s$ is minimal with respect to $\mathcal{L}$. For almost all $h_1,\ldots,h_t,f,\bar{h}_2,\ldots,\bar{h}_a$, $\theta(h_1,\ldots,h_t,\ f,\bar{h}_2,\ldots,\bar{h}_a) < H(d;h_1,\ldots,h_t,f,\bar{h}_2,\ldots,\bar{h}_a) = (f;(d_2)^s;\bar{h}_2,\ldots,\bar{h}_a)$. Hence it follows by induction and $H_{2n-1}(b)$ that for almost all $h_1,\ldots,h_t$, there is an $\alpha < \boldsymbol{\delta}^1_{2n+1}$ such that for almost all $f_1,\bar{h}_2,\ldots,\bar{h}_a$, $\theta(h_1,\ldots,h_t,f,\bar{h}_2,\ldots,\bar{h}_a) < \alpha_t$.

b) $d(= d^{\ell})$ basic of type 0. Hence $d^{(I_a)} = (k)^{(f_m;\bar{K}_2,\ldots,\bar{K}_a)}$, where $k \leq m$.

i) d not minimal w.r.t. $\hat{\mathcal{L}}^{\restriction\infty}$, so $k > 1$. Then for almost all $h_1,\ldots,h_t$, $f : m \to \boldsymbol{\delta}^1_{2n+1}$, $\bar{h}_2,\ldots,\bar{h}_a$, $\theta(h_1,\ldots,h_t,f,\bar{h}_2,\ldots,\bar{h}_a) < f(k)$. It follows readily that for almost all $h_1,\ldots,h_t$ there is a c.u.b. $C \subseteq \boldsymbol{\delta}^1_{2n+1}$ such that for almost all $f_m,\bar{h}_2,\ldots,\bar{h}_a$, $\theta(h_1,\ldots,h_t,f,\bar{h}_2,\ldots,\bar{h}_a) < N_c(f_m(k-1)) = H(\hat{\mathcal{L}}^{\restriction\infty}(d);(\hat{\mathcal{L}}^{\restriction\infty}(d) \to N_C \circ \hat{\mathcal{L}}^{\restriction\infty}(d));\ h_1,\ldots,h_t,f_m,\bar{h}_2,\ldots,\bar{h}_a)$. We use here a simple partition (of $f_m : m \to \boldsymbol{\delta}^1_{2n+1}$ with an extra value inserted between $f(k-1)$ and $f(k)$) and the countable additivity of the measures $\bar{h}_2,\ldots,\bar{h}_a$.

ii) d minimal w.r.t. $\hat{\mathcal{L}}^{\restriction\infty}(d)$. Similar to i).

II) $i < \infty$ and $i < k(d)$.

a) $K_i = S^{\ell_i,m_i}_{2n+1}$ with $\ell_i > 1$, and d not minimal w.r.t. $\hat{\mathcal{L}}^{\restriction i}$. Hence, d is non-minimal w.r.t. $\hat{\mathcal{L}}^{\restriction i+1}$. We let θ be given such that for almost all $h_1,\ldots,h_t,f,\bar{h}_2,\ldots,\bar{h}_a$, $\theta(h_1,\ldots,h_t,f,\bar{h}_2,\ldots,\bar{h}_a) < H(d^{(I_a)};\ h_1,\ldots,h_t,f,\bar{h}_2,\ldots,\bar{h}_a)$. By induction, for almost all $h_1,\ldots,h_i$, there is a C_{i+1} such that for almost all $h_{i+1},\ldots,h_t$, $\theta(h_1,\ldots,h_t) < h(\hat{\mathcal{L}}^{\restriction i+1}(d);\ \hat{\mathcal{L}}^{\restriction i+1}(d) \to N_{C_{i+1}}(\hat{\mathcal{L}}^{\restriction i+1}(d));\ h_1,\ldots,h_t)$. We let κ be such that for almost all $h_1,\ldots,h_t,f,\bar{h}_2,\ldots,\bar{h}_a$, $\kappa = \operatorname{cof} H(\hat{\mathcal{L}}^{\restriction i+1}(d);\ h_1,\ldots,h_t,\ f,\bar{h}_2,\ldots,\bar{h}_a)$.

i) $\kappa = \operatorname{cof} \kappa(K_i)$.

For fixed $h_1, \ldots, h_{i-1}$ such that $\theta(h_1, \ldots, h_{i-1}, K_i, \ldots, K_t) < h(d; h_1, \ldots, h_{i-1}, K_i, \ldots, K_t)$, we consider the partition $\mathcal{P}_i(h_1, \ldots, h_{i-1})$: we partition functions $h_i : \boldsymbol{\delta}^1_{2n+1} \to \boldsymbol{\delta}^1_{2n+1}$ of the correct type with the extra value $g(\alpha)$ inserted between $\sup_{f: f < \alpha \text{almost everywhere}} h_i([f]) = h_i(0)(\alpha)$ and the next element in the range of h_i after $h_i(0)(\alpha)$, where g has uniform cofinality ω, according to whether or not for almost all $h_{i+t}, \ldots, h_t$, $\theta(h_1, \ldots, h_t) < h(\hat{\mathcal{L}}\restriction^{i+1}(d); \hat{\mathcal{L}}\restriction^{i+1}(d) \to g \circ \hat{\mathcal{L}}\restriction^{i+1}(d);\ h_1, \ldots, h_t)$. It follows from a sliding argument (as in Lemma 8 of this chapter) that on the homogeneous side of the partition, the property stated in $\mathcal{P}(h_1, \ldots, h_{i-1})$ holds. We let C_i be homogeneous for this partition. We therefore have that for almost all $h_1, \ldots, h_{i-1}$ there is a C_i such that for almost all h_i (say with range in the closure points of C_i), $h_{i+1}, \ldots, h_t$, $\theta(h_1, \ldots, h_t) < h(\hat{\mathcal{L}}\restriction^{i+1}(d); \hat{\mathcal{L}}\restriction^{i+1}(d) \to (N_{C_i} \circ h_i(0)) \circ \hat{\mathcal{L}}\restriction^{i+1}(d);\ h_1, \ldots, h_t) = h(\hat{\mathcal{L}}\restriction^{i}(d); \hat{\mathcal{L}}\restriction^{i}(d) \to N_{C_i} \circ \hat{\mathcal{L}}\restriction^{i}(d); h_1, \ldots, h_t)$. The last equality follows form $\hat{\mathcal{L}}\restriction^{i}(d) = (i; \hat{\mathcal{L}}\restriction^{i+1}(d))^s$ and the definition of H in this case.

ii) $\kappa \neq \text{cof}\,(K_k)$.

For fixed $h_1, \ldots, h_{i-1}$, we consider $\mathcal{P}_i(h_1, \ldots, h_{i-1})$ as above, and have that on the homogeneous side of the partition, $\mathcal{P}_i(h_1, \ldots, h_{i-1})$ holds. However, for almost all $h_1, \ldots, h_t, f$, $\bar{h}_2, \ldots, \bar{h}_a$, $\sup_{f': f' < H(\hat{\mathcal{L}}\restriction^{i+1}(d); h_1, \ldots, h_t, f, \bar{h}_2, \ldots, \bar{h}_a) \text{a.e.}} h_i([f']) = H(\hat{\mathcal{L}}\restriction^{i+1}(d); h_1, \ldots, h_t, f, \bar{h}_2, \ldots, \bar{h}_a)$. Hence, $\theta(h_1, \ldots, h_t) < h(\hat{\mathcal{L}}\restriction^{i+1}(d); \hat{\mathcal{L}}\restriction^{i+1}(d) \to N_{C_i} \circ \hat{\mathcal{L}}\restriction^{i+1}(d); h_1, \ldots, h_t) = h(\hat{\mathcal{L}}\restriction^{i}(d); \hat{\mathcal{L}}\restriction^{i}(d) \to N_{C_i} \circ \hat{\mathcal{L}}\restriction^{i}(d); h_1, \ldots, h_t)$ for almost all $h_1, \ldots, h_t$ as $\hat{\mathcal{L}}\restriction^{i}(d) = \hat{\mathcal{L}}\restriction^{i+1}(d)$ in this case.

b) $K_i = S^{\ell_i, m_i}_{2n+1}$ with $\ell_i > 1$ and d minimal with respect to $\hat{\mathcal{L}}\restriction^{i}$. Hence, d is minimal with respect to $\hat{\mathcal{L}}\restriction^{i+1}$. Hence, by induction, for almost all $h_1, \ldots, h_i$, there is an $\alpha_{i+1} < \boldsymbol{\delta}^1_{2n+1}$ such that for almost all $h_{i+1}, \ldots, h_t$, $\theta(h_1, \ldots, h_t) < \alpha_{i+1}$. Hence, by $\boldsymbol{\delta}^1_{2n+1}$ additivity of K_i, it follows that for almost all $h_1, \ldots, h_{i-1}$, there is an $\alpha_i < \boldsymbol{\delta}_{2n+1}$ s.t. for almost all $h_i, \ldots, h_t$, $\theta(h_1, \ldots, h_t) < \alpha_i$.

c) $K_i = S^{l, m_i}_{2n+1}$ and d not minimal w.r.t. $\hat{\mathcal{L}}\restriction^{i}$. The proof is similar to a, where for fixed $h_1, \ldots, h_{i-1}$ as in a, we consider the partition $\mathcal{P}_i(h_1, \ldots, h_{i-1})$ where we partition $h_i : <^{m_i} \to \boldsymbol{\delta}^1_{2n+1}$ of the correct type with the extra value $g(\alpha)$ inserted between $h_i(0)(\alpha) = \sup_{\beta_2 < \cdots < \beta_{m_i} < \alpha_i} h_i(\beta_2, \ldots, \beta_{m_i}, \alpha)$ and the next element in the range of h_i after $h_i(0)(\alpha)$ according to whether or not for almost all $h_{i+1}, \ldots, h_t$, $\theta(h_1, \ldots, h_t) < h(\hat{\mathcal{L}}\restriction^{i+1}(d);\ \hat{\mathcal{L}}\restriction^{i+1}(d) \to g \circ \hat{\mathcal{L}}\restriction^{i+1}(d); h_1, \ldots, h_t)$. It follows that on the homogeneous side of the partition, $\mathcal{P}_i(h_1, \ldots, h_{i-1})$ holds for almost all $h_1, \ldots, h_{i-1}$. Taking cases on whether or not cof $H(\hat{\mathcal{L}}\restriction^{i+1}(d); h_1, \ldots, h_t, f, \bar{h}_2, \ldots, \bar{h}_a) = \omega$ and $\hat{\mathcal{L}}\restriction^{i}(d) = (i; \hat{\mathcal{L}}\restriction^{i+1}(d)^{(I_a)})$ or $\hat{\mathcal{L}}\restriction^{i}(d) = \hat{\mathcal{L}}\restriction^{i+1}(d)$ correspondingly, the result follows.

d) $K_i = S^{\ell_i, m_i}_{2n+1}$ with $\ell_i > 1$ and d minimal w.r.t. $\hat{\mathcal{L}}\restriction^{i}$. Similar to b) above.

e) $K_i = W^{m_i}_{2n+1}$ and d not minimal w.r.t. $\hat{\mathcal{L}}^{\restriction i+1}$. Once again, by induction for almost all $h_1, \ldots, h_i$ three is a C_{i+1} s.t. for almost all $h_{i+1}, \ldots, h_t, \theta(h_1, \ldots, h_t) < h(\hat{\mathcal{L}}^{\restriction i+1}(d); \hat{\mathcal{L}}^{\restriction i+1}(d) \rightarrow N_{C_{i+1}}(\hat{\mathcal{L}}^{\restriction i+1}(d)); \; h_1, \ldots, h_t)$. By an easy partition and sliding argument (partitioning h_i followed by $g : \boldsymbol{\delta}^1_{2n+1} \rightarrow \boldsymbol{\delta}^1_{2n+1}$, where $\inf g > \sup h_i$) for almost all $h_i, \ldots, h_{i-1}$ there is a C_i such that for almost all $h_i, \ldots, h_t, \theta(h_1, \ldots, h_t) < h(\hat{\mathcal{L}}^{\restriction i+1}(d); \hat{\mathcal{L}}^{\restriction i+1}(d) \rightarrow N_{C_i}(\hat{\mathcal{L}}^{\restriction i+1}(d)); \; h_1, \ldots, h_t)$, and the result follows since $\hat{\mathcal{L}}^{\restriction i}(d) = \hat{\mathcal{L}}^{\restriction i+1}(d)$ in this case.

f) $K_i = W^{m_i}_{2n+1}$ and d minimal w.r.t. $\hat{\mathcal{L}}^{\restriction i+1}$. By induction, for almost all $h_1, \ldots, h_i$, there is an $\alpha_{i+1} < \boldsymbol{\delta}^1_{2n+1}$ s.t. for almost all $h_1, \ldots, h_t, \; \theta(h_1, \ldots, h_t) < \alpha_{i+1}$. For almost all $h_1, \ldots, h_{i-1}$ we consider $\mathcal{P}_i(h_1, \ldots, h_{i-1})$ where we partition $h_i : v(K_i) \rightarrow \boldsymbol{\delta}^1_{2n+1}$ of the correct type (or $h_i : m \rightarrow \boldsymbol{\delta}^1_1$ if $n = 0$) with the extra value α after $\sup h_i$ according to whether or not for almost all $h_{i+1}, \ldots, h_t, \; \theta(h_1, \ldots, h_t) < \alpha$. On the homogeneous side, $\mathcal{P}_i$ holds, and, for almost all $h_1, \ldots, h_{i-1}$, we let C_i be homogeneous for $\mathcal{P}_i$. We then have for almost all $h_i, \ldots, h_t$ that $\theta(h_1, \ldots, h_t) < N_{C_i}(\sup_{\text{a.e.}} h_i)$, and the result follows since $\hat{\mathcal{L}}^{\restriction i}(d) = (i; \;)$ in this case, so $H(\hat{\mathcal{L}}^{\restriction i}(d); h_1, \ldots, h_t) = \sup_{\text{a.e.}} h_i$. The case $n = 0$ is similar.

g) $K_i \in \bigcup_{m<n} R_{2m+1}$. The result follows easily from induction.

III) $i < \infty, \; i = k(d) = k$.

a) $K_k = S^{\ell_k, m_k}_{2n+1}$ where $\ell_k > 1$.

1) $d^{(I_a)} = (k; d_2^{(I_a; \bar{K}_{a+1})})^{s(I_a)}$, where s appears. Here $\bar{K}_{a+1} = v(K_k)$. We let θ be such that for almost all $h_1, \ldots, h_t, \; \theta(h_1, \ldots, h_t) < h(d; h_1, \ldots, h_t)$. By the cofinality lemma, there is a θ_2 such that for almost all $h_1, \ldots, h_t, f, \bar{h}_2, \ldots, \bar{h}_{a+1}, \theta_2(h_1, \ldots, h_t, f, \bar{h}_2, \ldots, \bar{h}_{a+1}) < H(d_2; h_1, \ldots, h_t, f, \bar{h}_2, \ldots, \bar{h}_{a+1})$ and $\theta(h_1, \ldots, h_t) < h(d; (d_2 \rightarrow \theta_2); \; h_1, \ldots, h_t)$. We note that d is not minimal w.r.t. to $\hat{\mathcal{L}}^{\restriction i}$ in this case. We assume that d_2 is not minimal with respect to $\hat{\mathcal{L}}^{\restriction i+1}$. We will show in case iv) below that this is the case. By induction, it then follows that for almost all $h_1, \ldots, h_k$ there is a C_{k+1} s.t. for almost all $h_{k+1}, \ldots, h_t, \; \theta_2(h_1, \ldots, h_t) < h(\hat{\mathcal{L}}^{\restriction k+1}(d_2); \; \hat{\mathcal{L}}^{\restriction k+1}(d_2) \rightarrow N_{C_{k+1}} \circ \hat{\mathcal{L}}^{\restriction k+1}(d_2); \; h_1, \ldots, h_t)$, hence $\theta(h_1, \ldots, h_t) < h(d; d_2 \rightarrow N_{C_{k+1}} \circ \hat{\mathcal{L}}^{\restriction k+1}(d_2); \; h_1, \ldots, h_t)$. We let $\hat{d}$ be the tuple corresponding to d as in A.

i) $M(\hat{d}; \hat{\mathcal{L}}^{\restriction k+1}(d_2))$ is satisfied and $(k; \hat{\mathcal{L}}^{\restriction k+1}(d_2))$ satisfies C; that is for almost all $h_1, \ldots, h_t, \; f, \bar{h}_2, \ldots, \bar{h}_a, \; g$ defined by $g([\bar{h}_{a+1}]) = H(\hat{\mathcal{L}}^{\restriction k+1}(d_2); h_1, \ldots, h_t, \; f, \bar{h}_2, \ldots, \bar{h}_{a+1})$ is strictly increasing of uniform cofinality ω. In this case, $\hat{\mathcal{L}}^{\restriction k}(d) = (k; \hat{\mathcal{L}}^{\restriction k+1}(d))$. For almost all $h_1, \ldots, h_{k-1}$, we consider the partition $\mathcal{P}_k(h_1, \ldots, h_{k-1})$ where we partition functions $h_k : \boldsymbol{\delta}^1_{2n+1} \rightarrow \boldsymbol{\delta}^1_{2n+1}$ of the correct type, with the extra value $g(\alpha)$ inserted between $h_k(\alpha)$ and $h_k(\alpha+1)$, and g has uniform cofinality ω, according to whether or not for almost all $h_{k+1}, \ldots, h_t, \; \theta(h_1, \ldots, h_t) <$

$h(\hat{\mathcal{L}}^{\restriction k}(d);\ \hat{\mathcal{L}}^{\restriction k}(d) \to N_g \circ \hat{\mathcal{L}}^{\restriction k}(d);\ h_1, \ldots, h_t)$.

We claim that on the homogeneous side of the partition, the property stated in $\mathcal{P}_k$ $(h_1, \ldots, h_{k-1})$ holds. We suppose not and let $C_k \subseteq \boldsymbol{\delta}^1_{2n+1}$ be c.u.b. and homogeneous for the contrary side, and let C^2_k be such that for $h_k : \boldsymbol{\delta}^1_{2n+1} \to C^2_k$ of the correct type, there is a C_{k+1} such that for almost all $h_{k+1}, \ldots, h_t$, $\theta(h_1, \ldots, h_t) < h(d; d_2 \to N_{C_{k+1}} \circ \hat{\mathcal{L}}^{\restriction k+1}(d_2);\ h_1, \ldots, h_t)$. We let C^3_k be contained in the closure points of $C_k \cap C^2_k$. We fix $h_k : \boldsymbol{\delta}^1_{2n+1} \to C^3_k$ of the correct type, and fix C_{k+1} as above. We then get h^2_k and g satisfying:

1) $h^2_k = h_k$ almost everywhere w.r.t. $\mu^{v(K_k)}_{2n+1}$.

2) h^2_k, g are of the correct type and ordered as in $\mathcal{P}_k$.

3) h^2_k, g have range in C^3_k (in fact a subset of the range of h_k).

4) $g([f]) > h_k([N_{C_{k+1}} \circ f])$ for all $f : \kappa(K_k) \to \boldsymbol{\delta}^1_{2n+1}$.

The construction of h^2_k, g follows as in the previous technical lemmas and will be omitted. We then elect $h_{k+1}, \ldots, h_t$ such that $\theta(h_1, \ldots, h_t) < h(d; d_2 \to N_{c_{k+1}} \circ \hat{\mathcal{L}}^{\restriction k+1}(d_2);\ h_1, \ldots, h_t)$, and hence $< h(\hat{\mathcal{L}}^{\restriction k}(d); \hat{\mathcal{L}}^{\restriction k}(d) \to N_g \circ \hat{\mathcal{L}}^{\restriction k}(d); h_1, \ldots, h_t)$ by 4) above; and also $\theta(h_1, \ldots, h^2_k, \ldots, h_t) > h(\hat{\mathcal{L}}^{\restriction k}(d); \hat{\mathcal{L}}^{\restriction k}(d) \to N_g \circ \hat{\mathcal{L}}^{\restriction k}(d); h_1, \ldots, h_t)$, by (2), (3). Also, since $\theta(h_1, \ldots, h_k) = \theta(h_1, \ldots, h^2_k)$ by (1), we may assume that $\theta(h_1, \ldots, h_k, \ldots, h_t) = \theta(h_1, \ldots, h^2_k, \ldots, h_t)$. This contradiction establishes that $\mathcal{P}_k(h_1, \ldots, h_{k-1})$ holds.

We let C_k be homogeneous for $\mathcal{P}_k$, for fixed $h_1, \ldots, h_{k-1}$, and let C^2_k be contained in the closure points of C_k. It then follows that for almost all h_k, namely $h_k : \boldsymbol{\delta}^1_{2n+1} \to C^2_k$ of the correct type, that for almost all $h_{k+1}, \ldots, h_t$, $\theta(h_1, \ldots, h_t) < h(\hat{\mathcal{L}}^{\restriction k}(d); \hat{\mathcal{L}}^{\restriction k}(d) \to N_{C_k} \circ \hat{\mathcal{L}}^{\restriction k}(d); h_1, \ldots, h_t)$.

ii) $M_2(\hat{d}, \hat{\mathcal{L}}^{\restriction k+1}(d))$ holds, case i) fails, and $(k; \hat{\mathcal{L}}^{\restriction k+1}(d_2))^s$ satisfies C. In this case $\hat{\mathcal{L}}^{\restriction k}(d) = (k; \hat{\mathcal{L}}^{\restriction k+1}(d_2))^s$. For almost all $h_1, \ldots, h_{k-1}$, we consider the partition $\mathcal{P}_k(h_1, \ldots, h_{k-1})$ where we partition $h_k : \boldsymbol{\delta}^1_{2n+1} \to \boldsymbol{\delta}^1_{2n+1}$ of the correct type with the extra value $g(\alpha)$ inserted between $\sup_{\beta<\alpha} h_k(\beta)$ and $h_k(\alpha)$ according to whether or not for almost all $h_{k+1}, \ldots, h_t$, $\theta(h_1, \ldots, h_t) < h(\hat{\mathcal{L}}^{\restriction k}(d);\ \hat{\mathcal{L}}^{\restriction k}(d) \to N_g \circ \hat{\mathcal{L}}^{\restriction k}(d);\ h_1, \ldots, h_t)$. If $\mathcal{P}_k$ fails, we fix C_k homogeneous for the contrary side. We define C^2_k, C^3_k as above, fix $h_k : \boldsymbol{\delta}^1_{2n+1} \to C^3_k$ of the correct type, and fix C_{k+1} as above. We then get h^2_k, g satisfying:

1) $h^2_k = h_k$ almost everywhere w.r.t. $\mu^{v(K_k)}_{2n+1}$.

2) h^2_k, g are ordered as in $\mathcal{P}_k$ for this case.

3) h_k^2, g have range in C_k^3.

4) $g([f]) > h_k([N_{C_{k+1}} \circ f])$ for all $f : \kappa(K_k) \to \boldsymbol{\delta}^1_{2n+1}$ not of the correct type. We then proceed as in case i) above.

iii) Cases i) and ii) fail. In the case, $\ell > 2$, $\hat{\mathcal{L}}^{\restriction k}(d) = \hat{d}$. We first note that $M(\hat{d}, \hat{\mathcal{L}}^{\restriction k+1}(d_2))$ is satisfied. For if not, then for almost all $h_1, \dots, h_t f, \bar{h}_2, \dots, \bar{h}_a$, we consider the functions $g, \tilde{g}$ defined almost everywhere, where $g([h_{a+1}]) = H(d_2; h_1, \dots, h_t, f, \bar{h}_2, \dots, \bar{h}_{a+1})$ and $\tilde{g} : \theta(\bar{K}_{a+1}) \to \boldsymbol{\delta}^1_{2n+1}$ is minimal subject to $\sup_{\text{a.e.}} \tilde{g} = \sup_{\text{a.e.}} g$. Hence, $[\tilde{g}] < [g]$ for almost all $h_1, \dots, \bar{h}_a$, since $d = (k; d_2)^s$ satisfies C. Hence, by induction, for almost all $h_1, \dots, h_k$ there is a C_{k+1} s.t. for almost all $h_{k+1}, \dots, h_t, f, \bar{h}_2, \dots, \bar{h}_a$, $[\tilde{g}] < [g_2]$ where $g_2([\bar{h}_{a+1}]) = N_{C_{k+1}}(H(\hat{\mathcal{L}}^{\restriction k+1}(d_2); h_1, \dots, h_t, f, \bar{h}_2, \dots, \bar{h}_{a+1}))$, a contradiction since $\sup_{\text{a.e.}} \tilde{g} > \sup_{\text{a.e.}} g_2$ for g_2 with range in a set closed under $N_{C_{k+1}}$, which happens almost everywhere. Hence $M(\hat{d}, \hat{\mathcal{L}}^{\restriction k+1}(d_2))$ holds.

We first consider the case $\ell > 2$. We claim in this case that $M_2(\hat{d}, \hat{\mathcal{L}}^{\restriction k+1}(d_2))$ is not satisfied. We suppose not. For almost all $h_1, \dots, h_t f, \bar{h}_2, \dots, \bar{h}_a$, we consider g where $g(\bar{h}_{a+1}) = H(\hat{\mathcal{L}}^{\restriction k+1}(d_2);\ h_1, \dots, h_t, f, \bar{h}_2, \dots, \bar{h}_{a+1})$. From our technical lemmas, $[g] = \sup[g']$ where g' ranges over functions from $\theta(\bar{K}_{a+1})$ into $\boldsymbol{\delta}^1_{2n+1}$ of the correct type. Since $M_2(\hat{d}, \hat{\mathcal{L}}^{\restriction k+1}(d_2))$ is satisfied, $[g] = \sup[g'']$, where g'' ranges over functions from $\theta(\bar{K}_{a+1})$ into $\boldsymbol{\delta}^1_{2n+1}$ with $\sup_{\text{a.e.}} g'' = \sup_{\text{a.e.}} g$. Hence $(k; \hat{\mathcal{L}}^{\restriction k+1}(d_2))^s$ satisfies C, contrary to the assumption of this case. Hence $M_2(\hat{d}, \hat{\mathcal{L}}^{\restriction k+1}(d_2))$ fails. By induction, for almost all $h_1, \dots, h_k$ there is a C_{k+1} such that for almost all $h_{k+1}, \dots, h_t$, $\theta(h_1, \dots, h_t) < h(d; d_2 \to N_{C_{k+1}} \circ \hat{\mathcal{L}}^{\restriction k+1}(d_2); h_1, \dots, h_t)$. For such fixed $h_1, \dots, h_{k-1}$, we consider the partition $\mathcal{P}_k(h_1, \dots, h_{k-1})$ where we partition $h_k : \boldsymbol{\delta}^1_{2n+1} \to \boldsymbol{\delta}^1_{2n+1}$ of the correct type with the extra value $g(\alpha)$ inserted between $\sup_{\text{a.e.}} h_k(0)(\beta)$ $(= \sup_{f : \sup_{\text{a.e.}} f < \alpha} h_k([f]))$ and the next element in the range of h_k after $\sup_{\beta < \alpha} h_k(0)(\beta)$, where g is of uniform cofinality ω, according to whether or not for almost all $h_{k+1}, \dots, h_t$, $\theta(h_1, \dots, h_t) < h(\hat{d}; \hat{d} \to N_g \circ \hat{d}; h_1, \dots, h_t)$. If $\mathcal{P}_k$ fails, we fix C_k homogeneous for the contrary side, let C_k^2 be such that for $h_k : \boldsymbol{\delta}^1_{2n+1} \to C_k^2$ of the correct type there is a C_{k+1} such that the above inequality is satisfied, and let C_k^3 be contained in the closure points of $C_k \cap C_k^2$. We fix $h_k : \boldsymbol{\delta}^1_{2n+1} \to C_k^3$ of the correct type and C_{k+1}, and get h_k^2, g satisfying:

1) $h_k^2 = h_k$ a.e.w.r.t. $\mu^{v(K_k)}_{2n+1}$.

2) h_k^2, g are of the correct type and ordered as in $\mathcal{P}_k$.

3) h_k^2, g have range in C_k^3.

4) $g(\alpha) > h_k([N_{C_{k+1}} \circ f])$, where f represents α w.r.t. $v(K_k)$.

The construction of h_k^2, g is similar to that in the proofs of the technical lemmas. We then have that for almost all $h_{k+1},\ldots,h_t,f,\bar{h}_2,\ldots,\bar{h}_a$, $\theta(h_k,\ldots,h_t,\ f,\bar{h}_2,\ldots,\bar{h}_a) < h_k([N_{C_{k+1}} \circ \bar{g}])$, $\bar{g} : \theta(\bar{K}_{a+1}) \rightarrow \boldsymbol{\delta}^1_{2n+1}$ as above, and $[\bar{g}] = \sup_{\text{a.e.}} \bar{g}$, hence $\theta(h_1,\ldots,h_t,\ f,\bar{h}_2,\ldots,\bar{h}_a) < N_g(\sup_{\text{a.e.}}\bar{g}) = N_g(h(\hat{d};h_1,\ldots,h_t,f,\bar{h}_2,\ldots,\bar{h}_a))$. However, $\theta(h_1,\ldots,\ h_k^2,\ldots,h_t,f,\bar{h}_2,\ldots,\bar{h}_a) = \theta(h_1,\ldots,h_k,\ldots,h_t,f,\bar{h}_2,\ldots,\bar{h}_a)$. This contradiction establishes $\mathcal{P}_k$. The result then follows readily as in previous cases.

We now consider the case $\ell = 2$. Here $v(K_k) =$ the m-fold product of the normal measure on $\omega_1 = v$ say. We recall $\theta(v)$ is identified with an ordinal ($= \omega_1$) by ordering by the largest ordinal first, then the next largest, etc. For almost all $h_1,\ldots,h_t,f,\bar{h}_2,\ldots,\bar{h}_a$, we again consider the function $g : \theta(v) \rightarrow \boldsymbol{\delta}^1_{2n+1}$ defined by $g([\bar{h}_{a+1}]) = H(\hat{\mathcal{L}}^{\restriction k+1}(d_2);\ h_1,\ldots,h_t,f,\bar{h}_2,\ldots,\bar{h}_{a+1})$. Here $\bar{h}_{a+1}$ is an m-tuple of ordinals $< \omega_1$. We first assume $M_2(\hat{d},\hat{\mathcal{L}}^{\restriction k+1}(d_2))$ is satisfied. For fixed $h_1,\ldots,h_{k-1}$, we then consider the partition $\mathcal{P}_k$: we partition $h_k : \boldsymbol{\delta}^1_{2n+1} \rightarrow \boldsymbol{\delta}^1_{2n+1}$ of the correct type according to whether or not for almost all $h_{k+1},\ldots,h_t$, $\theta(h_1,\ldots,h_t) < h((k;\hat{\mathcal{L}}^{\restriction k+1}(d_2));\ h_1,\ldots,h_t)$. We note that $(k;\hat{\mathcal{L}}^{\restriction k+1}(d_2))$ does not satisfy C, but this still makes sense. We then claim that on the homogeneous side of the partition the property stated in $\mathcal{P}_k$ holds. We suppose not, and fix C_k, C_k^2, C_k^3, h_k, C_{k+1} as in the previous cases. We then get an h_k^2 satisfying:

1) $h_k^2 = h_k$ a.e.w.r.t. μ^v_{2n+1}.

2) h_k is of the correct type.

3) h_k has range in C_k^3.

4) for α represented by $g : \omega_1^m \rightarrow \boldsymbol{\delta}^1_{2n+1}$ of the correct type, $h_k^2(\alpha) > h_k[N_{C_{k+1}} \circ g_\alpha]$, where $g_\alpha : \omega_1^m \rightarrow \boldsymbol{\delta}^1_{2n+1}$ represents α w.r.t. W_1^m. We then proceed to a contradiction as in the previous cases, establishing that for almost all $h_1,\ldots,h_t$, $\theta(h_1,\ldots,h_t) < h((k;\hat{\mathcal{L}}^{\restriction k+1}(d_2));h_1,\ldots,h_t)$. We then consider, for fixed $h_1,\ldots,h_{t-1}$, the partition $\mathcal{P}_k$ where we partition $h_k : \boldsymbol{\delta}^1_{2n+1} \rightarrow \boldsymbol{\delta}^1_{2n+1}$ of the correct type according to whether or not for almost all $h_{k+1},\ldots,h_t$, $\theta(h_1,\ldots,h_t) < h((k;\hat{\mathcal{L}}^{\restriction k+1}(d_2))^s;h_1,\ldots,h_t)$. If $\mathcal{P}_k$ fails, we again get C_k, C_k^2, C_k^3, and h_k. We then get h_k^2 satisfying 1)-3) as above and 4) for α represented by $g_\alpha : \omega_1^m \rightarrow \boldsymbol{\delta}^1_{2n+1}$ not monotonically increasing a.e.w.r.t. W_1^m, $\sup_{\beta<\alpha} h_k^2 > h_k(\alpha)$. The construction is similar to that of previous cases, using here the fact that if $g : \omega_1^m \rightarrow \boldsymbol{\delta}^1_{2n+1}$ is not monotonically increasing almost everywhere, then $[g]$ is not the sup of $[g']$ for g' of the correct type. We then proceed to a contradiction establishing that for almost all $h_1,\ldots,h_t$, $\theta(h_1,\ldots,h_t) < h((k;\hat{\mathcal{L}}^{\restriction k+1}(d_2))^s;\ h_1,\ldots,h_t)$.

If $(k;\hat{\mathcal{L}}^{\restriction k+1(q)}(d_2))$ or $(k;\hat{\mathcal{L}}^{\restriction k+1(q)}(d_2))^s$ satisfies C for some q, we then repeat the above argument to establish $\bar{H}_k$.

Hence, we may assume without loss of generality that $M_2(\hat{d}, \hat{\mathcal{L}}^{\restriction k+1}(d_2))$ is not satisfied. Hence $\hat{\mathcal{L}}^{\restriction k+1}(d) = \hat{d}$. The result now follows as in the corresponding case for $\ell > 2$.

iv) d_2 minimal with respect to $\hat{\mathcal{L}}^{\restriction k+1}$. We show that this case does not occur. We show that $M_2(\hat{d}; d_2)$ is not satisfied. If it were, then define θ_2 such that for almost all $h_1, \ldots, h_t, f$, $\bar{h}_2, \ldots, \bar{h}_a$, $\theta_2(h_1, \ldots, h_t, f, \bar{h}_2, \ldots, \bar{h}_a) = [\tilde{g}]$ where $\tilde{g} : \theta(\bar{K}_{a+1}) \to \boldsymbol{\delta}^1_{2n+1}$ is minimal s.t. $\sup_{\text{a.e.}} g = H(\hat{d}; h_1, \ldots, h_t, f, \bar{h}_2, \ldots, \bar{h}_a)$. We then have that for almost all $h_1, \ldots, \bar{h}_{a+1}$, $\theta_2(h_1, \ldots, \bar{h}_{a+1}) < H(d_2; h_1, \ldots, \bar{h}_{a+1})$, however, for almost all $h_1, \ldots, h_k$ there is no $\alpha_k < \boldsymbol{\delta}^1_{2n+1}$ such that for almost all $h_{k+1}, \ldots, h_t, f, \bar{h}_2, \ldots, \bar{h}_a)$, $H(\hat{d}; h_1, \ldots, \bar{h}_a) < \alpha_k$ (which follows easily from $k(\hat{d}) > k$). This contradicts the minimality of d_2 and induction.

2) $d = (k; d_2^{(I_a; \bar{K}_{a+1})})$, where s does not appear. Here $\hat{\mathcal{L}}^{\restriction k}(d) = (k; d_2^{(I_a; \bar{K}_{a+1})})^s$. Since d satisfies C, it follows readily that $M_2(\hat{d}, d_2)$ is satisfied. It also follows as above that d_2 is not minimal with respect to $\hat{\mathcal{L}}^{\restriction k+1}$. For fixed $h_1, \ldots, h_{k-1}$, we consider the partition $\mathcal{P}_k(h_1, \ldots, h_{k-1})$ where we partition $h_k : \boldsymbol{\delta}^1_{2n+1} \to \boldsymbol{\delta}^1_{2n+1}$ of the correct type with the extra value $g(\alpha)$ inserted between $\sup_{\beta<\alpha} h_k(\beta)$ and $h_k(\alpha)$, where g has uniform cofinality ω, according to whether or not for not for almost all $h_{k+1}, \ldots, h_t$, $\theta(h_1, \ldots, h_t) < h(\hat{\mathcal{L}}^{\restriction k}(d); \hat{\mathcal{L}}^{\restriction k}(d) \to N_g \circ \hat{\mathcal{L}}^{\restriction k}(d);\ h_1, \ldots, h_t)$. We claim that on the homogeneous side of the partition $\mathcal{P}_k$ holds. If not, we let C_k be homogeneous for the contrary side, and let C_k^2 be such that for $h_k : \boldsymbol{\delta}^1_{2n+1} \to C_k^2$ of the correct type, for almost all $h_{k+1}, \ldots, h_t$, $\theta(h_1, \ldots, h_t) < H(d; h_1, \ldots, h_t)$. We let C_k^3 be contained in the closure points of $C_k \cap C_k^2$. We fix $h_k : \boldsymbol{\delta}^1_{2n+1} \to C_k^3$ of the correct type. We let $\tilde{h}_k : \omega \cdot \boldsymbol{\delta}^1_{2n+1} \to C_k \cap C_k^2$ exhibit the uniform cofinality ω of h_k. Then for almost all $h_{k+1}, \ldots, h_t, f, \bar{h}_2, \ldots, \bar{h}_a$, there is an $n < \omega$ such that $\theta(h_1, \ldots, h_t, f, \bar{h}_2, \ldots, \bar{h}_a) < \tilde{h}_k(n, [\bar{g}])$, where $\bar{g}$ is defined by $\bar{g}([\bar{h}_{a+1}]) = H(d_2; h_1, \ldots, h_t, f, \bar{h}_2, \ldots, \bar{h}_{a+1})$. By countable additivity of the measures, it follows that there is a $g : \boldsymbol{\delta}^1_{2n+1} \to C_k \cap C_k^2$ of the correct type with h_k, g ordered as in $\mathcal{P}_k$ and such that $g(\alpha) > \tilde{h}_k(n, \alpha)$ for all α, where n is fixed so that the above inequality holds almost everywhere. This contradiction establishes $\mathcal{P}_k$. The result then follows as in the previous cases.

b) $K_k = S_{2n+1}^{\ell_k, m_k}$ where $\ell_k = 1$.

1) $d^{(I_a)} = (k; d_1^{(I_a)}, \ldots, d_r^{(I_a)})^{s(I_a)}$, where $2 \le r \le m_k$, and s appears. By the cofinality lemma it follows that there is a θ_2 such that for almost all $h_1, \ldots, h_t, f, \bar{h}_2, \ldots, \bar{h}_a$, $\theta(h_1, \ldots, h_t, f, \bar{h}_2, \ldots, \bar{h}_a) < H(d_r^{(I_a)}; h_1, \ldots, h_t, f, \bar{h}_2, \ldots, \bar{h}_a)$ and such that $\theta(h_1, \ldots, h_t, f, \bar{h}_2, \ldots, \bar{h}_a) < H(d; (d_r \to \theta_2);\ h_1, \ldots, h_t, f, \bar{h}_2, \ldots, \bar{h}_a)$.

i) $d_r^{(I_a)}$ non-minimal with respect to $\hat{\mathcal{L}}^{\restriction k+1}$, $\hat{\mathcal{L}}^{\restriction k+1}(d_r) > d_{r-1}$ if $r > 2$, and cof $H(\hat{\mathcal{L}}^{\restriction k_1}(d_r)$;

$h_1,\ldots,h_t,f,\bar h_2,\ldots,\bar h_a) = \omega$ almost everywhere. In this case, $\hat{\mathcal{L}}{\restriction}^k(d) = (k; d_1^{(Ia)},\ldots,d_{r-1}^{(Ia)}, \hat{\mathcal{L}}{\restriction}^{k+1}(d_r^{(Ia)}))$. By induction, for almost all $h_1,\ldots,h_k$ there is a C_{k+1} such that for almost all $h_{k+1},\ldots,h_t$, $\theta(h_1,\ldots,h_t) < h(d;(d_r \to N_{C_{k+1}} \circ \hat{\mathcal{L}}{\restriction}^{k+1}(d_r)); h_1,\ldots,h_t)$. For such fixed $h_1,\ldots,h_{k-1}$, we consider the partition $\mathcal{P}_k(h_1,\ldots,h_{k-1})$: we partition $h_k :<^m\to \boldsymbol{\delta}^1_{2n+1}$ of the correct type with the extra value $g(\alpha_2,\ldots,\alpha_r,\alpha_1)$ inserted between $h_k(r)(\alpha_2,\ldots,\alpha_r,\alpha_1) \equiv \sup_{\beta_{r+1}<\cdots<\beta_{m_k}<\alpha_1} h_k(\alpha_2,\ldots,\alpha_r,\beta_{r+1},\ldots,\beta_{m_k},\alpha_1)$ and the next element in the range of h_k after $h_k(r)(\alpha_2,\ldots,\alpha_r,\alpha_1)$, where g has uniform cofinality ω, according to whether or not for almost all $h_k,\ldots,h_t$ $\theta(h_1,\ldots,h_t) < H(\hat{\mathcal{L}}{\restriction}^k(d); \hat{\mathcal{L}}{\restriction}^k(d) \to N_g \circ \hat{\mathcal{L}}{\restriction}^k(d); h_1,\ldots,h_t)$. If $\mathcal{P}_k$ fails, then we let C_k be homogeneous for the contrary side. We let C_k^2 be such that for $h_k :<^m\to C_k^2$ of the correct type there is a C_{k+1} such that for almost all $h_{k+1},\ldots,h_t$ the above inequality (with $N_{C_{k+1}}$) is satisfied. We let C_k^3 be contained in the closure points of $C_k \cap C_k^2$, and fix $h_k :<^m\to C_k^3$ of the correct type, and C_{k+1} as above. We then get h_k^2, g such that:

1) $h_k^2 = h_k$ a.e.w.r.t. the m_k-fold product of the ω-cofinal normal measure on $\boldsymbol{\delta}^1_{2n+1}$.

2) h_k^2, g are of the correct types and ordered as in $\mathcal{P}_k(h_1,\ldots,h_{k-1})$.

3) h_k^2, g have range in C_k^3.

4) $g(\alpha_2,\ldots,\alpha_r,\alpha_1) > h_k(r)(\alpha_2,\ldots,\alpha_{r-1}, N_{C_{k+1}}(\alpha_r),\alpha_1)$, for almost all $\alpha_2,\ldots,\alpha_r,\alpha_1$.

The existence of h_k^2, g follows from an easy sliding argument which we omit. We then elect $h_{k+1},\ldots,h_t$ such that $\theta(h_1,\ldots,h_k,\ldots,h_t) = \theta(h_1,\ldots,h_k^2,\ldots,h_t)$, $\theta(h_1,\ldots,h_t) < h(d; d_r \to N_{C_{k+1}} \circ \hat{\mathcal{L}}{\restriction}^{k+1}(d_r); h_1,\ldots,h_t)$, and $\theta(h_1,\ldots,h_k^2,\ldots,h_t) > h(\hat{\mathcal{L}}{\restriction}^k(d); \hat{\mathcal{L}}{\restriction}^k(d) \to N_g \circ \hat{\mathcal{L}}{\restriction}^k(d); h_1,\ldots,h_k^2,\ldots,h_t)$. This contradiction establishes $\mathcal{P}_k$, and the result, $\bar H(k)$, then follows as in the previous cases.

ii) $d_r^{(Ia)}$ non-minimal w.r.t. $\hat{\mathcal{L}}{\restriction}^{k+1}, \hat{\mathcal{L}}{\restriction}^{k+1}(d_r) > d_{r-1}$ if $r > 2$, and cof $H(\hat{\mathcal{L}}{\restriction}^{k+1}(d_r); h_1,\ldots,h_t,f,\bar h_2,\ldots,\bar h_a) \neq \omega$ almost everywhere. In this case, $\hat{\mathcal{L}}{\restriction}^k(d) = (k; d_1^{(Ia)},\ldots,d_{r-1}^{(Ia)}, \hat{\mathcal{L}}{\restriction}^{k+1}(d_r^{(Ia)}))^s$. We proceed as in the previous case, and for fixed $h_1,\ldots,h_{k-1}$ consider $\mathcal{P}_k(h_1,\ldots,h_{k-1})$: we partition $h_k :<^m\to \boldsymbol{\delta}^1_{2n+1}$ of the correct type with the extra value $g(\alpha_2,\ldots,\alpha_r,\alpha_1)$ inserted between $\sup_{\beta<\alpha_r} h_k(r)(\beta)$ and the next element in the range of h_k after this, where g has uniform cofinality ω, according to whether or not for almost all $h_{k+1},\ldots,h_t$, $\theta(h_1,\ldots,h_t) < H(\hat{\mathcal{L}}{\restriction}^k(d);\ \hat{\mathcal{L}}{\restriction}^k(d) \to N_g \circ \hat{\mathcal{L}}{\restriction}^k(d);\ h_1,\ldots,h_t)$. If $\mathcal{P}_k$ fails, we proceed as in the previous case, fixing C_k, C_k^2, C_k^3, h_k, C_{k+1} respectively, and get h_k^2, g satisfying 1-4 as in that case. We then proceed to get a contradiction as in that case, which establishes $\mathcal{P}_k(h_1,\ldots,h_{k-1})$, from which $\bar H_{2n+1}(k)$ readily follows.

iii) $d_r^{(Ia)}$ non-minimal with respect to $\hat{\mathcal{L}}{\restriction}^{k+1}$, and $\hat{\mathcal{L}}{\restriction}^{k+1}(d_r) \leq d_{r-1}$ (here $r > 2$), and d_r is minimal with respect to $\hat{\mathcal{L}}\restriction^{k+1}$ and $r > 2$. In this case, $\hat{\mathcal{L}}\restriction^{k}(d) = (k; d_1^{(I_a)}, \ldots, d_{r-1}^{(I_a)})^s$. The second case can not arise since d satisfies C, so $d_{r-1} < d_r$ and (since $k(d_{r-1}) > k$), for almost all $h_1, \ldots, h_k$ there is no $\alpha_{k+1} < \boldsymbol{\delta}^1_{2n+1}$ such that for almost all $h_{k+1}, \ldots, h_t, \bar{h}_2, \ldots, \bar{h}_a$, $H(d_{r-1}; h_1, \ldots, h_t, f, \bar{h}_2, \ldots, \bar{h}_a) < \alpha_{k+1}$.

For almost all $h_1, \ldots, h_{k-1}$, we consider $\mathcal{P}_k(h_1, \ldots, h_{k-1})$ where we partition $h_k :<^m\rightarrow \boldsymbol{\delta}^1_{2n+1}$ of the correct type with the extra value $g(\alpha_2, \ldots, \alpha_{r-1}, \alpha_1)$ inserted between $\sup_{\beta<\alpha_{r-1}} h_r(r-1)(\alpha_2, \ldots, \alpha_{r-2}, \beta, \alpha_1)$ and the next element in the range of h_k after this, where g has uniform cofinality ω, according to whether or not $\theta(h_1, \ldots, h_t) < h(\hat{\mathcal{L}}\restriction^{k}(d); \hat{\mathcal{L}}\restriction(d) \rightarrow N_g \circ \hat{\mathcal{L}}\restriction^{k}(d); h_1, \ldots, h_t)$. If $\mathcal{P}_k$ fails, we fix C_k, C_k^2, C_k^3, h_k, C_{k+1} as in the previous case, and get h_k^2, g satisfying 1-3 as in that case, and 4) $g(\alpha_2, \ldots, \alpha_{r-1}, \alpha_1) > h_k(\alpha_2, \ldots, \alpha_{r-2}, \alpha_{r-1}, N_{C_{k+1}}(\alpha_{r-1}, \alpha_1)$ almost everywhere. We then elect $h_{k+1}, \ldots, h_t, f, \bar{h}_2, \ldots \bar{h}_a$ such that $\theta(h_1, \ldots, h_k, \ldots, \bar{h}_a) = \theta(h_1, \ldots, h_k^2, \ldots, \bar{h}_a)$, $\theta(h_1, \ldots, h_k, \ldots, \bar{h}_a) < H(d; d_r \rightarrow N_{C_{k+1}} \circ \hat{\mathcal{L}}\restriction^{k+1}(d_1); h_1, \ldots, h_k, \ldots, \bar{h}_a) = \sup_{\beta_{r+1}<\cdots<\beta_{m_k}<H(d_1;h_1,\ldots,\bar{h}_a)} h_k(H(d_2; h_1, \ldots, \bar{h}_a), \ldots, H(d_{r-1}, h_1, \ldots, \bar{h}_a), N_{C_{k+1}}(H(\hat{\mathcal{L}}\restriction^{k+1}(d_r); h_1, \ldots, \bar{h}_a)), \beta_{r+1}, \ldots, \beta_{m_k}; H(d_1; h_1, \ldots, \bar{h}_a)) \leq g(H(d_2, \ldots), \ldots, H(d_{r-1}, \ldots), H(d_1, \ldots)) = N_g(\sup_{(\beta_{r-1}<H(d_{r-1};h_1,\ldots,h_k 2,\ldots,\bar{h}_a)} h_k(r-1)\ (H(d_2, \ldots), \ldots, H(d_{r-2} \cdots), \beta_{r-1}, H(d_1, \ldots))) = H(\hat{\mathcal{L}}\restriction^{k}(d), \hat{\mathcal{L}}\restriction^{k}(d) \rightarrow N_g \circ \hat{\mathcal{L}}\restriction^{k}(d); h_1, \ldots, h_k^2, \ldots, h_t, f, \bar{h}_2, \ldots, \bar{h}_a)$, and also such that $\theta(h_1, \ldots, h_k^2, \ldots, h_t) > h(\hat{\mathcal{L}}\restriction^{k}(d); \hat{\mathcal{L}}\restriction^{k}(d) \rightarrow N_g \circ \hat{\mathcal{L}}\restriction^{k}(d); h_1, \ldots, h_k^2, \ldots, h_t)$. This contradiction establishes $\mathcal{P}_k$, and the result follows as in the previous cases.

iv) $d_r^{(I_a)}$ is minimal w.r.t. $\hat{\mathcal{L}}\restriction^{k+1}$, and $r = 2$. We have $\hat{\mathcal{L}}\restriction^{k}(d) = d_1$. By induction, for almost all $h_1, \ldots, h_k$ there is an $\alpha_{k+1} < \boldsymbol{\delta}^1_{2n+1}$ s.t. for almost all $h_{k+1}, \ldots, h_t$ $\theta(h_1, \ldots, h_t) < h(d; d_r \rightarrow \alpha_{k+1}, \ldots, h_1, \ldots, h_t)$. For almost all $h_1, \ldots, h_{k-1}$, we consider $\mathcal{P}_k(h_1, \ldots, h_{k-1})$ where we partition $h_k :<^m\rightarrow \boldsymbol{\delta}^1_{2n+1}$ of the correct type with $g(\alpha)$ inserted between $\sup_{\beta<\alpha} h_k(1)(\beta)$ and the next element in the range of h_k after this, according to whether or not for almost all $h_{k+1}, \ldots, h_t$, $\theta(h_1, \ldots, h_t) < (\hat{\mathcal{L}}\restriction^{k}(d); \hat{\mathcal{L}}\restriction^{k}(d) \rightarrow N_g \circ \hat{\mathcal{L}}\restriction^{k}(d); h_1, \ldots, h_t)$. If $\mathcal{P}_k$ fails, we elect C_k, C_k^2, C_k^3, h_k, α_{k+1} similarly to the previous cases. We then get h_k^2, g satisfying the usual 1-3 and 4) $g(\alpha) > \sup_{\beta_3<\cdots<\beta_{m_k}<\alpha} h_k(\alpha_{k+1}, \beta_3, \ldots, \beta_{m_k}, \alpha)$ for all $\alpha (> \alpha_{k+1})$. We then elect $h_{k+1}, \ldots, h_t, f, \bar{h}_2, \ldots, \bar{h}_a$ such that $\theta(h_1, \ldots, h_k^2, \ldots, h_t, f, \bar{h}_2, \ldots, \bar{h}_a) = \theta(h_1, \ldots, h_t, f, \bar{h}_2, \ldots, \bar{h}_a) < h(d; d_r \rightarrow \alpha_{k+1}, h_1, \ldots, h_t, f, \bar{h}_2, \ldots, \bar{h}_a) \leq N_g \sup_{\beta_1<H(d_1;h_1,\ldots,\bar{h}_a), \beta_2<\cdots<\beta_{m_k}<\beta_1} h_k^2(\beta_2, \ldots, \beta_{m_k}, \beta_1)) = N_g(H(d_1; h_1, \ldots, h_k^2, \ldots, h_t, f, \bar{h}_2, \ldots, \bar{h}_a) = H(\hat{\mathcal{L}}\restriction^{k}(d); \hat{\mathcal{L}}\restriction^{k}(d) \rightarrow N_g \circ \hat{\mathcal{L}}\restriction^{k}(d); h_1, \ldots, h_k^2, \ldots, h_t, f, \bar{h}_2, \ldots, \bar{h}_a)$. This contradiction establishes $\mathcal{P}_k$ from which the result follows.

2) $d^{(I_a)} = (k; d_1^{(I_a)}, \ldots, d_r^{(I_a)})^{(I_a)}$, where s does not appear, and $r = m_k$. We consider

the case $m_k > 1$, the case $m_k = 1$ being similar. In this case $\hat{\mathcal{L}}^{\restriction k}(d) = (k;\ d_1,\ldots,d_r)^s$. As in the previous case, we consider $\mathcal{P}_k(h_1,\ldots,h_{k-1})$ for almost all $h_1,\ldots,h_{k-1}$, where here the extra value $g(\alpha_2,\ldots,\alpha_{m_k},\alpha_1)$ is inserted between $\sup_{\beta<m_k} h_k(\alpha_2,\ldots,\beta,\alpha_1)$ and the next element in the range of h_k after this. If $\mathcal{P}_k$ fails, we fix C_k homogeneous for the contrary side, let C_k^2 be such that for $h_k :<^m\to C_k^2$ of the correct type, for almost $h_{k+1},\ldots,h_t$, $\theta(h_1,\ldots,h_t) < h(d;h_1,\ldots,h_t)$, and let C_k^3 be contained in the closure points of $C_k \cap C_k^2$. We again fix $h_k : \delta^1_{2n+1} \to C_k^3$ of the correct type, and let $\tilde{h}_k(\alpha_2,\ldots,\alpha_{m_k},n,\alpha_1) : \omega\times <^m\to \delta^1_{2n+1}$ exhibit the uniform cofinality ω of h_k. By countable additivity there is an $n<\omega$ s.t. for almost $h_{k+1},\ldots,h_t$, $f,\bar{h}_2,\ldots,\bar{h}_a$, $\theta(h_1,\ldots,\bar{h}_a) < \tilde{h}_k(H(d_1,\ldots),\ldots,H(d_r,\ldots),n,H(d_1,\ldots,))$. We then get h_k^2, g satisfying the usual 1-3 and 4.) $g(\alpha_2,\ldots,\alpha_{m_k},\alpha_1) > \tilde{h}_k(\alpha_2,\ldots,\alpha_{m_k},n,\alpha_1)$ for all $\alpha_1,\ldots,\alpha_{m_k}$. The result then follows as in the previous cases.

3) $d^{(I_a)} = (k;d_1^{(I_a)},\ldots,d_r^{(I_a)})^{(I_a)}$, where s does not appear, and $r < m_k$.

i) d_1 is not minimal w.r.t $\hat{\mathcal{L}}^{\restriction k+1}$, $\hat{\mathcal{L}}^{\restriction k+1}(d_1) > d_r$ if $r \geq 2$, and cof $H(\hat{\mathcal{L}}^{\restriction k+1}(d_1);h_1,\ldots,h_t, f,\bar{h}_2,\ldots,\bar{h}_a) = \omega$ almost everywhere. In this case $\hat{\mathcal{L}}^{\restriction k}(d) = (k;d_1,\ldots,d_r,\hat{\mathcal{L}}^{\restriction k+1}(d_1))$. By the cofinality lemma, there is a θ_2 such that for almost all $h_1,\ldots,h_t,f,\bar{h}_2,\ldots,\bar{h}_a$, $\theta_2(h_1,\ldots,h_t,f, \bar{h}_2,\ldots,\bar{h}_a) < H(d_1;h_1,\ldots,h_t, f,\bar{h}_2,\ldots,\bar{h}_a)$ and $\theta(h_1,\ldots,\bar{h}_a) < H(\hat{\mathcal{L}}^{\restriction k}(d);\hat{\mathcal{L}}^{\restriction k+1}(d_1) \to \theta_2;h_1,\ldots,\bar{h}_a)$. Hence, by induction, for almost all $h_1,\ldots,h_k$ there is a C_{k+1} such that for almost all $h_{k+1},\ldots,h_t$, $\theta(h_1,\ldots,h_t) < H(\hat{\mathcal{L}}^{\restriction k}(d);\ \hat{\mathcal{L}}^{\restriction k+1}(d_1) \to N_{C_{k+1}} \circ \hat{\mathcal{L}}^{\restriction k}(d_1);\ h_1,\ldots,h_t)$. For fixed $h_1,\ldots,h_{k-1}$, we consider $\mathcal{P}_k(h_1,\ldots,h_{k-1})$ where we partition h_k of the correct type with $g(\alpha_1,\ldots,\alpha_{r+1},\alpha_1)$ inserted between $h_k(r+1)(\alpha_2,\ldots,\alpha_{r+1},\alpha_1)$ and the next element in the range of h_k according to whether or not for almost all $h_{k+1},\ldots,h_t,\theta(h_1,\ldots,h_t) < H(\hat{\mathcal{L}}^{\restriction k}(d); \hat{\mathcal{L}}^{\restriction k}(d) \to N_g \circ \hat{\mathcal{L}}^{\restriction k}(d);h_1,\ldots,h_t)$. If $\mathcal{P}_k$ fails, we fix C_k, C_k^2, C_k^3, h_k, C_{k+1}, as in previous cases, and get h_k^2, g satisfying the usual 1-3 and 4) $g(\alpha_2,\ldots,\alpha_{r+1},\alpha_1) > h_k(r+1)(\alpha_2,\ldots,\alpha_r, N_{C_{k+1}}(\alpha_{r+1}),\alpha_1)$. We then proceed as in the previous cases to establish $\bar{H}_{2n+1}(k)$.

ii) d_1 not minimal w.r.t. $\hat{\mathcal{L}}^{\restriction k+1}$, $\hat{\mathcal{L}}^{\restriction k+1}(d_1) > d_r$ if $r \geq 2$, and cof $H(\hat{\mathcal{L}}^{\restriction k+1}(d_1); h_1,\ldots,h_t,f,\bar{h}_2,\ldots,\bar{h}_a) \neq \omega$ almost everywhere. In this case $\hat{\mathcal{L}}^{\restriction k}(d) = (k;d_1,\ldots,d_r\ \ \hat{\mathcal{L}}^{\restriction k+1}(d_1))^s$. We proceed as in i) except that in $\mathcal{P}_k(h_1,\ldots,h_{k-1})$ we insert $g(\alpha_2,\ldots,\ \alpha_{r+1},\alpha_1)$ between $\sup_{\beta<\alpha_{r+1}} h_k(r+1)(\alpha_2,\ldots,\alpha_r,\beta,\alpha_1)$ and the next element in the range of h_k after this. We proceed as in i), getting h_k^2, g which satisfy 1-4 of that case, and proceed to establish $\bar{H}_{2n+1}(k)$ as in that case.

iii) d_1 not minimal with respect to $\hat{\mathcal{L}}^{\restriction k+1}$, and $\hat{\mathcal{L}}^{\restriction k+1}(d_1) \leq d_r$ (where $r \geq 2$). In this case $\hat{\mathcal{L}}^{\restriction k}(d) = (k;d_1,\ldots,d_r)^s$. We proceed as in the above cases, where in $\mathcal{P}_k(h_1,\ldots,h_{k-1})$, $g(\alpha_2,\ldots,\alpha_r,\alpha_1)$ is inserted between $\sup_{\beta<\alpha_r} h_k(r)(\alpha_2,\ldots,\alpha_{r-1},\beta,\alpha_1)$ and the next element in the range of h_k, and we get h_k^2, g satisfying 1-3 as these and 4) $g(\alpha_2, \ldots, \alpha_r, \alpha_1) > h_k(r+1)(\alpha_2, \ldots, \alpha_r, N_{C_{k+1}}(\alpha_r), \alpha_1)$ for almost all $\alpha_2, \ldots, \alpha_r, \alpha_1$. We then proceed as in the previous cases.

iv) d_1 minimal w.r.t. $\hat{\mathcal{L}}\restriction^{k+1}$. It follows readily that $r = 1$. In this case $\hat{\mathcal{L}}\restriction^k(d) = d_1$. By the cofinality lemma and induction, we have that for almost all $h_1, \ldots, h_k$ there is an $\alpha_{k+1} < \boldsymbol{\delta}^1_{2n+1}$ such that for almost all $h_{k+1}, \ldots, h_t$, $\theta(h_1, \ldots, h_t) < h(\tilde{h}; \tilde{\tilde{d}} \to \alpha_{k+1}; h_1, \ldots, h_t)$, where we define $\tilde{d} = (k; d_1, \tilde{\tilde{d}})$. We assume here that $m_k \geq 2$, the case $m_k = 1$ following similarly. As in the previous cases, we consider $\mathcal{P}_k(h_1, \ldots, h_{k-1})$ where the value $g(\alpha)$ is inserted between $\sup_{\beta<\alpha} h_k(1)(\beta)$ and the next element in the range of h_k. If $\mathcal{P}_k$ fails, we fix C_k, C_k^2, C_k^3, h_k, α_{k+1}, and then get h_k^2, g satisfying the usual 1-3 and 4) $g(\alpha) > h_k(2)(\alpha_{k+1}, \alpha) = \sup_{\beta_3<\cdots<\beta_{m_k}<\alpha} h_k(\alpha_{k+1}, \beta_3, \ldots, \beta_{m_k}, \alpha)$. We then proceed as in the previous cases to establish $\bar{H}(k)$.

c) $K_k = W^{m_k}_{2n+1}$. In the following we assume $n > 0$, the case $n = 0$ following easily.

1) $d^{(I_a)} = (k;\ )^{(I_a)}$. We consider the sequence of measures $K_{b(k)}, \ldots, K_{b_u}$, $\bar{K}_2, \ldots, \bar{K}_a$, where we recall $K_{b(k)}, \ldots, K_{b_u}$ enumerates the subsequence of $K_{k+1}, \ldots, K_t$ of measures in $\bigcup_{m<n} R_{2m+1}$. We let $S^{\bar{\ell}_k, \bar{m}_k}_{2n-1} = v(K_k)$ (where $\bar{\ell}_k = 2^n - 1$ is maximal here).

i) the maximal tuple $\tilde{d}^{\bar{\ell}_k, \bar{m}_k}_{2n-1}$ relative to $K_{b(k)}, \ldots, K_{b_u}, \bar{K}_2, \ldots, \bar{K}_a$ is defined. Hence, for almost all $h_1, \ldots, h_{k-1}$, for almost all $h_k, \ldots, h_t, f, \bar{h}_2, \ldots, \bar{h}_a$ there is a $\theta_2(h_k, \ldots, h_t, f, \bar{h}_2, \ldots, \bar{h}_a) < \sup_{\text{a.e.}} h_k$ such that $\theta(h_k, \ldots, h_t, f, \bar{h}_2, \ldots, \bar{h}_a) < \theta_2$. It follows readily (by $\boldsymbol{\delta}^1_{2n+1}$-additivity of W^m_{2n+1} and $S^{\ell,m}_{2n+1}$) that for almost all $h_1, \ldots, h_{k-1}$ there is a θ_2 such that for almost all $h_k, h_{b(k)}, \ldots, h_{b_u}, \bar{h}_2, \ldots, \bar{h}_a$, $\theta_2(h_k, \ldots, \bar{h}_a) < \sup_{\text{a.e.}} h_k$ and for almost all $h_k, \ldots, h_t, f$, $\bar{h}_2, \ldots, \bar{h}_a$, $\theta(h_1, \ldots, h_t, f, \bar{h}_2, \ldots, \bar{h}_a) < \theta_2(h_1, \ldots, h_{k-1})(h_k, h_{b(k)}, \ldots, h_{b_u}, \ldots, \bar{h}_a)$. By $H_{2n-1}(c)$ it follows that for almost all $h_1, \ldots, h_{k-1}$ there is a c.u.b. $C_k \subseteq \boldsymbol{\delta}^1_{2n+1}$ such that for almost all $h_k, h_{b(k)}, \ldots, h_{b_u}, \bar{h}_2, \ldots, \bar{h}_a$, $\theta_2(h_1, \ldots, h_{k-1})(h_k, h_{b(k)}, \ldots, \bar{h}_a) < (N_{C_k};\ h_k;\ \tilde{d}^{\bar{\ell}_k, \bar{m}_k}_{2n-1};\ h_{b(k)}, \ldots, \bar{h}_a)$ $= H(\hat{\mathcal{L}}\restriction^k(d);\ \hat{\mathcal{L}}\restriction^k(d) \to N_{C_k} \circ \hat{\mathcal{L}}\restriction^k(d);\ h_1, \ldots, h_t,\ f, \bar{h}_2, \ldots, \bar{h}_a)$. This establishes $\bar{H}(k)$ in this case.

ii) the maximal tuple $d^{\bar{\ell}_k, \bar{m}_k}_{2n-1}$ relative to $K_{b(k)}, \ldots, K_{b_u}, \bar{K}_2, \ldots, \bar{K}_a$ is not defined. In this case d is minimal w.r.t. $\hat{\mathcal{L}}\restriction^k$. As above, for almost all $h_1, \ldots, h_{k-1}$, there is $\theta_2 = \theta_2(h_1, \ldots, h_{k-1})$ such that for almost all $h_k, h_{b(k)}, \ldots, \bar{h}_a$, $\theta_2(h_1, \ldots, h_{k-1})\ (h_k, h_{b(k)}, \ldots, \bar{h}_a) < \sup_{\text{a.e.}} h_k$ and for almost all $h_k, \ldots, h_t, f, \bar{h}_2, \ldots, \bar{h}_a$, $\theta(h_1, \ldots, h_t, f,\ \bar{h}_2, \ldots, \bar{h}_a) < \theta_2(h_1, \ldots, h_{k-1})(h_k, h_{b(k)}, \ldots, \bar{h}_a)$. By $H_{2n-1}(d)$ it follows that for almost all $h_1, \ldots, h_{k-1}$ there is an $\alpha_{k+1} < \boldsymbol{\delta}^1_{2n+1}$ such that for almost all $h_k, \ldots, h_t, f, \bar{h}_2, \ldots, \bar{h}_a$, $\theta(h_1, \ldots, h_t, f, \bar{h}_2, \ldots, \bar{h}_a) < \theta_2(h_1, \ldots, h_{k-1})$ $(h_k, h_{b(k)}, \ldots, \bar{h}_a) < \alpha_{k+1}$, which establishes $\bar{H}(k)$ in this case.

2) $d = (k; \bar{d}_2)^s$, where s appears. Here $\bar{d}_2$ is defined relative to $K_{b(k)}, \ldots, K_{b_u}, \bar{K}_2, \ldots, \bar{K}_a$ and $(\bar{d}_2)^s$ satisfies D, A.

i) $(\bar{d}_2)^s$ is non-minimal w.r.t. $\mathcal{L}$ relative to $K_{b(k)}, \ldots, K_{b_u}, \bar{K}_2, \ldots, \bar{K}_a$. In this case $\hat{\mathcal{L}}^{\restriction k}(d) = (k; \mathcal{L}'(\bar{d}_2)^s)^s$, where s appears if it appears in $\mathcal{L}((d_2)^s)$ (where if $\mathcal{L}((d_2)^s) = (d')^s$, $\mathcal{L}'((d_2)^s) = d'$). Proceeding as above, we have that for almost all $h_1, \ldots, h_{k-1}$ there is a $\theta_2 = \theta_2(h_1, \ldots, h_{k-1})$ such that for almost all $h_k, h_{b(k)}, \ldots, h_{b_u}, \bar{h}_2, \ldots, \bar{h}_a$, $\theta_2(h_1, \ldots, h_{k-1})(h_k, \ldots, \bar{h}_a) < a\ (\text{id}; h_k; (\bar{d}_2)^s; h_{b(k)}, \ldots, h_{b_u}, \bar{h}_2, \ldots, \bar{h}_a)$ and for almost all $h_k, \ldots, h_t, f, \bar{h}_2, \ldots, \bar{h}_a$, $\theta(h_1, \ldots, \bar{h}_a) < \theta_2(h_1, \ldots, h_{k-1})(h_k, h_{b(k)}, \ldots, \bar{h}_a)$. Hence by $H_{2n-1}(a)$ it follows that for almost all $h_1, \ldots, h_{k-1}$, there is a $C_k \subseteq \boldsymbol{\delta}^1_{2n+1}$ such that for almost all $h_k, h_{b(k)}, \ldots, h_{b_u}$, $\bar{h}_2, \ldots, \bar{h}_a$, $\theta_2(h_1, \ldots, h_{k-1})\ (h_k, h_{b(k)}, \ldots, \bar{h}_a) < (N_{C_k};\ h_k;\ \mathcal{L}((\bar{d}_2)^s); h_{b(k)}, \ldots, \bar{h}_a) = H(\hat{\mathcal{L}}^{\restriction k}(d); \hat{\mathcal{L}}^{\restriction k}(d) \to N_{C_k} \circ \hat{\mathcal{L}}^{\restriction k}(d); h_1, \ldots, h_t, f, \bar{h}_2, \ldots, \bar{h}_a)$, which establishes $\bar{H}(k)$ in this case.

ii) $(\bar{d}_2)^s$ is minimal w.r.t. $\mathcal{L}$ relative to $K_{b(k)}, \ldots, K_{b_u}, \bar{K}_2, \ldots, \bar{K}_a$. In this case d is minimal w.r.t. $\hat{\mathcal{L}}^{\restriction k}$. We proceed as in the previous case, using $H_{2n-1}(b)$ and a simple partition and sliding argument (partitioning h_k with the extra value α_{k+1} inserted before inf h_k).

3) $d = (k; \bar{d}_2)$, where s does not appear. Here $\hat{\mathcal{L}}^{\restriction k}(d) = (k; \bar{d}_2)^s$. We proceed as in case 2) i) above.

This establishes $\bar{H}_{2n+1}(k)$ in all cases.

We now consider H_{2n+1}.

We fix the measure $S^{\ell,m}_{2n+1} = \mathcal{V}$, say, on $\kappa = \theta(S^{\ell,m}_{2n+1})$, and consider $\mu^{\mathcal{V}}_{2n+3}$. We assume $F : \boldsymbol{\delta}^1_{2n+3} \to \boldsymbol{\delta}^1_{2n+3}$ is given and for almost all $[f]$ w.r.t. $\mu^{<\mathcal{V}}_{2n+3}$, $F([f]) < (\text{id}; (d)^s;\ f;\ K_1, \ldots, K_t)$, where s may or may not appear, and d is defined and satisfies D, A relative to $K_1, \ldots, K_t$. Here $d = d^{\ell,m}_{2n+1}$, and $I_a = (f(\bar{K}_1);\ )$.

We first consider $H_{2n+1}(a)$:

1) s does not appear. In this case $\mathcal{L}((d)) = (d)^s$. We consider the partition $\mathcal{P}$: We partition functions $f : \kappa \to \boldsymbol{\delta}^1_{2n+3}$ of the correct type with the extra value $g(\alpha)$ inserted between $\sup_{\beta<\alpha} f(\beta)$ and $f(\alpha)$, with g of uniform cofinality ω, according to whether or not $F([f]) < (N_g; \mathcal{L}((d));\ f;\ K_1, \ldots, K_t)$. It follows readily by countable additivity of the measures $K_1, \ldots, K_t$ that on the homogeneous side of the partition, $\mathcal{P}$ holds. We let C be a c.u.b. subset of $\boldsymbol{\delta}^1_{2n+3}$ homogeneous for $\mathcal{P}$. It then follows that for almost all F, $F([f]) < (N_C;\ \mathcal{L}((d));\ f;$

$K_1, \ldots, K_t)$.

2) s appears. We first establish that there is a c.u.b. $C \subseteq \boldsymbol{\delta}^1_{2n+3}$ such that for almost all f, $F([f]) < (N_C;\ \hat{\mathcal{L}}(d);\ f;\ K_1, \ldots, K_t)$. We have that for almost all f, for almost all $h_1, \ldots, h_t$ there is a $\theta(h_1, \ldots, h_t) < h(d; h_1, \ldots, h_t)$ such that $F(f;\ h_1, \ldots, h_t) < f(\theta(h_1, \ldots, h_t))$. Hence, by $\bar{H}_{2n+1}(1)$, for almost all f there is a c.u.b. $C \subseteq \boldsymbol{\delta}^1_{2n+1}$ such that for almost all $h_1, \ldots, h_t$, $\theta(h_1, \ldots, h_t) < h(\hat{\mathcal{L}}(d);\ \hat{\mathcal{L}}(d) \to N_C \circ \hat{\mathcal{L}}(d);\ h_1, \ldots, h_t)$. We consider the partition $\mathcal{P}$: We partition functions $f : \kappa \to \boldsymbol{\delta}^1_{2n+3}$ of the correct type with the extra value $g(\alpha)$ inserted between $f(\alpha)$ and $f(\alpha + 1)$, with g of uniform cofinality ω, according to whether or not $F([f]) < (N_g;\ \hat{\mathcal{L}}(d);\ f; K_1, \ldots, K_t)$. We claim that on the homogeneous side of the partition the property stated in $\mathcal{P}$ holds. If not, we fix a c.u.b. C^1 homogeneous for the contrary side, and let $C^2 \subseteq \boldsymbol{\delta}^1_{2n+3}$ be such that for $f : \kappa \to C^2$ of the correct type there is a $C \subseteq \boldsymbol{\delta}^1_{2n+1}$ as above. We let C^3 be contained in the closure points of $C^1 \cap C^2$. We fix $f : \kappa \to C^3$ of the correct type, and fix a c.u.b. $C \subseteq \boldsymbol{\delta}^1_{2n+1}$ as above. We then get f^2, g satisfying:

1) $f^2 = f$ almost everywhere w.r.t. $\mathcal{V}$.

2) f^2, g are of the correct type and ordered as in $\mathcal{P}$.

3) f^2, g have range in C^3 (in fact have range a subset of the range of f).

4) If $h : \boldsymbol{\delta}^1_{2n+1} \to \boldsymbol{\delta}^1_{2n+1}$ (or $h :<^m\to \omega_1$ if $n = 0$) represents $[h]$ w.r.t. $\mu_{2n+1}^{\mathcal{V}(S^{\ell,m}_{2n+1})}$, then $g([h]) > f([N_c \circ h])$. The construction of f^2, g is similar to that given previously and will be omitted. We then fix $h_1, \ldots, h_t$ such that $F(f; h_1, \ldots, h_t) = F(f^2; h_1, \ldots, h_t)$ and $\theta(h_1, \ldots, h_t) < H(\hat{\mathcal{L}}(d); \hat{\mathcal{L}}(d) \to N_C \circ \hat{\mathcal{L}}(d);\ h_1, \ldots, h_t)$ and $F(f^2;\ h_1, \ldots, h_t) > (N_g; \hat{\mathcal{L}}(d);\ f^2;\ h_1, \ldots, h_t) = N_g(f^2(h(\hat{\mathcal{L}}(d);\ h_1, \ldots, h_t))) > f(h(\hat{\mathcal{L}}(d);\ \hat{\mathcal{L}}(d) \to N_C \circ \hat{\mathcal{L}}(d); h_1, \ldots, h_t))$. This contradiction establishes $\mathcal{P}$. If $C \subseteq \boldsymbol{\delta}^1_{2n+3}$ is homogeneous for $\mathcal{P}$, it follows readily that for almost all $f, F([f]) < (N_C;\ \hat{\mathcal{L}}(d);\ f;\ K_1, \ldots, K_t)$.

We next claim that if d' satisfies C, A relative to $K_1, \ldots, K_t$, but (d') does not satisfy D, and if there is a c.u.b. $C \subseteq \boldsymbol{\delta}^1_{2n+3}$ such that for almost all f, $F([f]) < (N_C;\ d';\ f;\ K_1, \ldots, K_t)$, then for almost all $f, F([f]) <$ (if; $d'; f;\ K_1, \ldots, K_t)$. We consider the partition $\mathcal{P}$ where we partition $f : \kappa \to \boldsymbol{\delta}^1_{2n+3}$ of the correct type according to whether or not $F([f]) <$ (id; d'; f; $K_1, \ldots, K_t)$. If $\mathcal{P}$ fails, we fix a c.u.b. $C^1 \subseteq \boldsymbol{\delta}^1_{2n+3}$, where $C^1 \subseteq C$ and C^1 is homogeneous for the contrary side, let $C^2 \subseteq \boldsymbol{\delta}^1_{2n+3}$ be such that for $f : \kappa \to C^2$ of the correct type $F([f]) < (N_C;\ d';\ f;\ K_1, \ldots, K_t)$ holds, and let C^3 be contained in the closure points of $C^1 \cap C^2$. We fix $f : \kappa \to C^3$ of the correct type. We fix A of measure one w.r.t. $S^{\ell,m}_{2n+1}$ such that for almost all $h_1, \ldots, h_t$, $h(d';\ h_1, \ldots, h_t) \notin A$. We let C' be a c.u.b. subset of $\boldsymbol{\delta}^1_{2n+1}$ such that for h :

$\delta^1_{2n+1} \to C'$ of the correct type (or $h :<^m\to C'$ if $n = 0$), $[h] \in A$. We let C'' be contained in the closure points of C'. We then get f^2 satisfying:

1) for h of the correct type having range in C'', $f^2([h]) = f([h])$.

2) f^2 is of the correct type and has range in C^3 (in fact a subset of the range of f).

3) for $[h]$ not represented by h as in 1, $f^2([h]) > N_f(f([h]))$, hence $f^2([h]) > N_C(f([h]))$.

The existence of f^2 follows from an easy sliding argument. We then fix $h_1, \ldots, h_t$ such that $F(f;\ h_1, \ldots, h_t) = F(f^2; h_1, \ldots, h_t)$, $F(f;\ H_1, \ldots, h_t) < (N_C;\ d';\ f; h_1, \ldots, h_t) = N_C(f(h(d';\ h_1, \ldots, h_t)))$, $F(f^2;\ h_1, \ldots, h_t) > (\text{id};\ d';\ f^2;\ h_1, \ldots, h_t) = f^2(h(d':\ h_1, \ldots, h_t))$, and $h(d';\ h_1, \ldots, h_t) \notin A$. Hence $F(f^2;\ h_1, \ldots, h_t) > N_C(f(h(d';\ h_1, \ldots, h_t)))$. This contradiction establishes $\mathcal{P}$. Hence, for almost all f, $F([f]) < (\text{id};\ \hat{\mathcal{L}}(d);\ f;\ K_1, \ldots, K_t)$.

If $(\hat{\mathcal{L}}(d))$ satisfies D, then our first considerations apply, and we are done. If not, then we are in a position to repeat the argument of 1) to get a $C \subseteq \delta^1_{2n+3}$ s.t. for almost all f, $F([f]) < (N_C;\ (\hat{\mathcal{L}}(d))^s;\ f;\ K_1, \ldots, K_t)$. If $(\hat{\mathcal{L}}(d))^s$ satisfies D, we are done. If not, an argument similar to the above establishes that for almost all f, $F([f]) < (\text{id};\ (\hat{\mathcal{L}}(d))^s;\ f;\ K_1, \ldots, K_t)$. The proof proceeds by considering the partition $\mathcal{P}$ as above, and constructing f^2 satisfying 1,2 as above and 3) for $[h]$ not represented by a function of the correct type, $f^2([h]) > f$ (the next ordinal represented by a function of the correct type after $[h]$). We are now in a position to repeat the previous argument. Repeating this argument, we eventually get a c.u.b. $C \subseteq \delta^1_{2n+1}$ such that for almost all f, $F([f]) < (N_C;\ \mathcal{L}(d);\ f;\ K_1, \ldots, K_t)$, which establishes $H_{2n+1}(a)$.

We now consider $H_{2n+1}(b)$.

We assume that $(d)^s = (d^{\ell,m}_{2n+1})^s$ is minimal w.r.t. $\mathcal{L}$ relative to $K_1, \ldots, K_t$.

1) d is minimal w.r.t. $\hat{\mathcal{L}}$ relative to $K_1, \ldots, K_t$. We have that for almost all f, $h_1, \ldots, h_t$, there is a $\theta(h_1, \ldots, h_t) < h(d;\ h_1, \ldots, h_t)$ such that $F(f;\ h_1, \ldots, h_t) < f(\theta(h_1, \ldots, h_t))$. By $\bar{H}_{2n+1}(1)$, for almost all f there is an $\alpha < \kappa$ such that for almost all $h_1, \ldots, h_t$, $\theta(h_1, \ldots, h_t) < \alpha$. We consider the partition $\mathcal{P}$ where we partition $f : \kappa \to \delta^1_{2n+3}$ of the correct type with the extra value γ (of cofinality ω) inserted before inf f according to whether or not $F([f]) < \gamma$. If $\mathcal{P}$ fails, we let C^1 be homogeneous for the contrary side, let C^2 be such that for $F : \kappa \to C^2$ of the correct type $F(f;\ h_1, \ldots, h_t) < f(\alpha)$ almost everywhere (where $\alpha(f)$ is as above), and let $C^3 = C^1 \cap C^2$. We fix $f : \kappa \to C^3$ of the correct type, and get γ, f^2 satisfying:

1) $f^2 = f$ a.e.w.r.t. μ^{γ}_{2n+1}.

2) $\gamma < \inf f^2$, cof $\gamma = \omega$, f^2 is of the correct type.

3) γ, f^2 have range in C^3

4) $\gamma > f(\alpha)$.

Then for amost all $h_1, \ldots, h_t$, $F(f;\ h_1, \ldots, h_t) = F(f^2;\ h_1, \ldots, h_t)$, $F(f;\ h_1, \ldots, h_t) < f(\alpha)$, and $F(f^2;\ h_1, \ldots, h_t) > \gamma > f(\alpha)$. This contradiction established $\mathcal{P}$. It follows readily that $H_{2n+1}(b)$ is satisfied.

2) d is not minimal w.r.t. $\hat{\mathcal{L}}$ relative to $K_1, \ldots, K_t$. It follows as in the proof of $H_{2n+1}(a)$ that for almost all $f, h_1, \ldots, h_t$, $f(f;\ h_1, \ldots, h_t) < (\text{id};\ (\hat{\mathcal{L}}(d))^s;\ f;\ h_1, \ldots, h_t)$, and in fact $F(f;\ h_1, \ldots, h_t) < (\text{id};\ (\hat{\mathcal{L}}^p(d))^s;\ f; h_1, \ldots, h_t)$, where $\hat{\mathcal{L}}^p(d)$ denotes the pth iterate of $\hat{\mathcal{L}}$. We elect p minimal such that $\hat{\mathcal{L}}^{p+1}(d)$ is not defined, and then proceed as in 1. This establishes $H_{2n+1}(b)$.

We consider $H_{2n+1}(c)$.

We assume that F is given s.t. for almost all $f : \kappa \to \boldsymbol{\delta}^1_{2n+3}$ of the correct type, for almost all $h_1, \ldots, h_t$, $F([f]) < \sup f$, and the maximal tuple $\tilde{d}^{\ell,m}_{2n+1}$ is defined. For almost all $f, h_1, \ldots, h_t$, there is an $\alpha < \kappa$ such that $F(f;\ h_1, \ldots, h_t) < f(\alpha)$. It follows readily that for almost all f, $h_1, \ldots, h_t$ there is a c.u.b. $C_\infty \subseteq \boldsymbol{\delta}^1_{2n+1}$ such that $\alpha(f;\ h_1, \ldots, h_t) < h(\tilde{d}^\infty;\ \tilde{d}^\infty \to N_C \circ \tilde{d}^\infty;\ h_1, \ldots, h_t)$, where $\tilde{d}^\infty$ is the basic type 1 description ( ) with index $\mathcal{I}_a = (f(\bar{K}_1);\ )$ for $n > 0$. For $n = 0$, $\tilde{d}^\infty = (m)$, a basic type 0 description. This follows form an easy partition argument (partitioning $f : \kappa \to \boldsymbol{\delta}^1_{2n+3}$ with the extra value γ after sup f). We recall here that $h(\tilde{d}^\infty;\ h_1, \ldots, h_t)$ is represented by the function $H(\tilde{d}^\infty;\ h_1, \ldots, h_t, f) \underset{\text{a.e.}}{=} \sup f$. We then proceed as in the proof of $\bar{H}_{2n+1}(k)$ to establish that for $1 \le k \le t$, for almost all f, $h_1, \ldots, h_{k-1}$, there is a c.u.b. $C_k \subseteq \boldsymbol{\delta}^1_{2n+1}$ such that for almost all $h_{k+1}, \ldots, h_t$, $\alpha(f;\ h_1, \ldots, h_k) < h(\tilde{d}^k;\ \tilde{d}^k \to N_{C_k} \circ \tilde{d}^k;\ h_1, \ldots, h_t)$, $\tilde{d}^k$ as in the definition of the maximal tuple $\tilde{d}(= \tilde{d}^1)$. In particular, for almost all f there is a $C_1 \subseteq \boldsymbol{\delta}^1_{2n+1}$ such that for almost all $h_1, \ldots, h_t$, $\alpha(h_1, \ldots, h_t) < h(\tilde{d};\ \tilde{d} \to N_C \circ \tilde{d};\ h_1, \ldots, h_t)$. We then proceed as in $H_{2n+1}(a)$ to establish that there is a c.u.b. $C \subseteq \boldsymbol{\delta}^1_{2n+3}$ such that for almost all f, $h_1, \ldots, h_t$, $F(f;\ h_1, \ldots, h_t) < (N_C;\ (\tilde{d}^{\ell,m}_{2n+1})^s;\ f;\ h_1, \ldots, h_t)$ where s appears if $(\tilde{d}^{\ell,m}_{2n+1})$ does not satisfy D. This establishes $H_{2n+1}(c)$.

We now consider $H_{2n+1}(d)$.

We again assume that for almost all $f, h_1, \ldots, h_t$, $F(f;\ h_1, \ldots, h_t) \underset{\text{a.e.}}{<} \sup f$, but that $\tilde{d}^{\ell,m}_{2n+1}$ is not defined. In this case, e_1 (as in the definition of $\tilde{d}^{\ell,m}_{2n+1}$) does not exist. As above, we have that for almost all $f, h_1, \ldots, h_t$, there is a c.u.b. $C_\infty \subseteq \boldsymbol{\delta}^1_{2n+1}$ such that $\alpha(f; h_1, \ldots, h_t) < h(\tilde{d}^\infty \to N_{C_\infty} \circ \tilde{d}^\infty;\ h_1, \ldots, h_t)$, and $F(f; h_1, \ldots, h_t) < f(\alpha)$. Here $\tilde{d}^\infty = (\ )$, a basic type 1 description (for $n = 0$, $\tilde{d}^\infty = (m)$ a basic type 0 description). We then proceed as in $\bar{H}_{2n+1}(k)$, and since e^1 does not exist, we have that for almost all f there is a c.u.b. $C_1, \subseteq \boldsymbol{\delta}^1_{2n+1}$ such that for almost all $h_1, \ldots, h_t$, $\alpha(f; h_1, \ldots, h_t) < h(\tilde{d}^\infty;\ \tilde{d}^\infty \to N_{C_1} \circ \tilde{d}^\infty; h_1, \ldots, h_t)$. We then proceed as in $H_{2n+1}(b)$ to establish $H_{2n+1}(d)$.

This establishes H_{2n+1} in all cases.

VI. The Main Theorem. We define an ordering $<_r$ on tuples $((d)^s; K_1, \ldots, K_t)$, where s may or may not appear, and $(d)^s$ satisfies D, A relative to $K_1, \ldots, K_t$ (or (d) if s does not appear). We let $<_r$ be the transitive relation generated by the relations:

(1) $((d_1)^s; K_1, \ldots, K_t) <_r ((d_2)^s, K_1, \ldots, K_t)$ where $(d_1)^s < (d_2)^s$, where s may not appear in d_1 or d_2. Here $(d_1)^s < (d_2)^S$ if $d_1 < d_2$ or $d_1 = d_2$ and d_1 involves s, d_2 does not (i.e. $(d_1)^s < (d_1)$).

(2) $(\mathcal{L}((d)^s; K_1, \ldots, K_t); K_1, \ldots, K_t, K_{t+1}) <_r (d; K_1, \ldots, K_t)$, for all $K_{t+1} \in \bigcup_{m \leq n} R_{2m+1}$.

It is easy to see that $<_r$ is well founded. To be specific, if $((d_1)^2; K_1, \ldots, K_{t_1}) >_r ((d_2)^s; K_1, \ldots, K_{t_2}) >_r \ldots$, where $t_1 < t_2 < \cdots$, and s may not appear in any tuple, then for any p, for almost $f, \mathcal{H}_1, \ldots, \mathcal{H}_{t_p}$, where $f : \theta(S^{l,m}_{2n+1}) \to \boldsymbol{\delta}^1_{2n+3}$ and $d_i \in \mathcal{D}^{l,m}_{2n+1}$ for all i, we have that $(id; (d_1)^s; f; \mathcal{H}_1, \ldots, \mathcal{H}_{t_1}) > (id; (d_2)^s; f; \mathcal{H}_1, \ldots, \mathcal{H}_{t_2}) > \cdots$. Letting p become arbitrarily large, we define an infinite decreasing sequence of ordinals. Hence $<_r$ is well founded.

Definition: We let $f((d)^s; K_1, \ldots, K_t)$ be the rank of $((d)^s; K_1, \ldots, K_t)$ w.r.t. $<_r$ computed according to

$$f((d)^s; K_1, \ldots, K_t) = \Big(\sup_{((\bar{d})^s; \bar{K}_1, \ldots, \bar{K}_{\bar{t}}) <_r ((d)^s; K_1, \ldots, K_t)} f((\bar{d})^s; \bar{K}_1, \ldots, \bar{K}_{\bar{t}}) \Big) + 1.$$

We introduce the following notation: we let $E(0, \alpha) = \alpha$ and $E(n+1; \alpha) = \omega^{E(n;\alpha)}$. Also, we let $E(n) = E(n; 1)$.

We now state the main theorem of this section:

Theorem: *Let $(d)^s$, where s may or may not appear, satisfy D, A relative to $K_1, \ldots, K_t$, where $d = d^{l,m}_{2n+1}$, and let $\mathcal{V} = S^{l,m}_{2n+1}$. We let $F : \boldsymbol{\delta}^1_{2n+3} \to \boldsymbol{\delta}^1_{2n+3}$ be defined w.r.t. $\mu^{\overset{\mathcal{V}}{<}}_{2n+3}$ by $F([f]) = (id; (d)^s; f; K_1, \ldots, K_t)$ for almost all $f : \kappa(S^{l,m}_{2n+1}) \to \boldsymbol{\delta}^1_{2n+3}$. Then in the ultrapower of $\boldsymbol{\delta}^1_{2n+3}$ by the measure $\mu^{\overset{\mathcal{V}}{<}}_{2n+3}$ on $\boldsymbol{\delta}^1_{2n+3}$, F represents an ordinal $\leq \aleph_{E(2n+1)+f((d)^s; K_1, \ldots, K_t))}$.*

Proof. (1) $f((d)^s; K_1, \ldots, K_t) = 1$, hence $(d)^s$ is minimal w.r.t. $\mathcal{L}$ relative to $K_1, \ldots, K_t$. Hence, by $H_{2n+1}(b)$, $(id; (d)^s; K_1, \ldots, K_t)$ represents an ordinal $\leq \boldsymbol{\delta}^1_{2n+3} = \aleph_{E(2n+1)+1}$, by the inductive hypothesis I_{2n+1}.

(2) $f((d)^s; K_1, \ldots, K_t) > 1$. We may assume that $(d)^s$ is non-minimal w.r.t. $\mathcal{L}$. We then

have that if $F([f]) < (id; (d)^s; f; K_1, \ldots, K_t)$ for almost all f, there is a $G : \boldsymbol{\delta}^1_{2n+3} \to \boldsymbol{\delta}^1_{2n+3}$ such that for almost all f, $F([f]) < (G; \mathcal{L}((d)^s); f; K_1, \ldots, K_t)$.

By K_{2n+3}, there is a tree T on $\boldsymbol{\delta}^1_{2n+3}$ and a c.u.b. $C \subseteq \boldsymbol{\delta}^1_{2n+3}$ such that for $\alpha \in C$, $G(\alpha) < |T \restriction \sup_J J(\alpha)|$, the sup ranging over embeddings J from measures in $\bigcup_{m \leq n} R_{2m+1}$. We fix such a T and C. Hence, for almost all f,

$$F([f]) < (|T \restriction \sup_J J|; \mathcal{L}((d)^s); K_1, \ldots, K_t) < (\sup_J J; \mathcal{L}((d)^s); K_1, \ldots, K_t)^+.$$

To prove this, we define an ordering $\triangleleft$ on $(\sup_J J; \mathcal{L}((d)^s); K_1, \ldots, K_t)$ as follows: We set $G_1 \triangleleft G_2$ if for almost all $f, \mathcal{H}_1, \ldots, \mathcal{H}_t$,

$$\begin{aligned} &|T \restriction (\sup J; \mathcal{L}((d)^s); f; \mathcal{H}_1, \ldots, \mathcal{H}_t)(G_1(f; \mathcal{H}_1, \ldots, \mathcal{H}_t))| \\ &\quad < |T \restriction (\sup J; \mathcal{L}((d)^s); f; \mathcal{H}_1, \ldots, \mathcal{H}_t)(G_2(f; \mathcal{H}_1, \ldots, \mathcal{H}_t))|. \end{aligned}$$

If now $G < (|T \restriction \sup J|; \mathcal{L}((d)^s); K_1, \ldots, K_t)$, there is a $\bar{G} < (\sup J; \mathcal{L}((d)^s); K_1, \ldots, K_t)$ such that for almost all $f, \mathcal{H}_1, \ldots, \mathcal{H}_t$,

$$G(f; \mathcal{H}_1, \ldots, \mathcal{H}_t) = |T| (\sup J, \mathcal{L}((d)^s); f; \mathcal{H}_1, \ldots, \mathcal{H}_t)(\bar{G}(f; \mathcal{H}_1, \ldots, \mathcal{H}_t))|.$$

The map $G \to \bar{G}$ is order-preserving from $((|T \restriction \sup J|; \mathcal{L}((d)^s); K_1, \ldots, K_t); <)$ to $((\sup_J J; \mathcal{L}((d^s)); K_1, \ldots, K_t); \triangleleft)$. Hence $(id; (d)^s; K_1, \ldots, K_t) \leq (\sup J; \mathcal{L}((d)^s); K_1, \ldots, K_t)^+$.

By countable additivity, $(\sup J; \mathcal{L}((d)^s); f; K_1, \ldots, K_t)) = \sup_J (J; \mathcal{L}((d)^s); K_1, \ldots, K_t)$, where again J ranges over the embeddings corresponding to measures in $\bigcup_{m \leq n} R_{2m+1}$. For K_{t+1} in $\bigcup_{m \leq n} R_{2m+1}$, it follows readily that $(J_{K_{t+1}}; \mathcal{L}((d)^s); K_1, \ldots, K_t) = (id; \mathcal{L}((d)^s); K_1, \ldots, K_t, K_{t+1})$.

Hence, $(id; (d)^s; K_1, \ldots, K_t) \leq [\sup_{K_{t+1}} (id; \mathcal{L}((d)^s); K_1, \ldots, K_t, K_{t+1})]^+$, which by induction, is

$$\leq \left[\aleph_{E(2n+1)+\sup_{K_{t+1}} f(\mathcal{L}((d)^s; K_1, \ldots, K_{t+1})}\right]^+) + \aleph_{\tau(2n+1)+f((d)^s; K_1, \ldots, K_t)},$$

which completes the proof of the theorem.

VII. A Rank Computation. The main goal of this section is to compute the bound

$$\sup_{d, K_1, \ldots, K_t} f(d; K_1, \ldots, K_t) \leq E(2n+3).$$

We recall that for all $d^{(I_a)} = d^{(f(\bar{K}_1);)}$ satisfying C, A relative to $K_1, \ldots, K_t$, and the component tuples $d_2^{(f(\bar{K}_1); \bar{K}_2, \ldots, \bar{K}_a)}$ of d, the map $\delta_{d_2} : a \to \{1, \ldots, t\}$ is defined, where $\delta_{d_2}(i)$ is the integer j associated to $\bar{K}_i$ such that $v(K_j) = \bar{K}_i$. It is easy to see that δ_d satisfies the following:

(1) δ_d is strictly increasing.

(2) $v(K_{\delta_d(i)}) = \bar{K}_i$.

(3) for all component tuples $d'^{(I'_a)}$ of $d^{(I_a)}$ (where I'_a extends I_a), $\delta_{d'}$ extends δ_d.

(4) $\delta_d(\bar{K}_a) < k(d)$.

Definition: We define $<^k_p$ on $d_1^{(I_a)}$, $d_2^{(I_a)}$ satisfying C, A relative to $K_1, \ldots, K_t$ and $\delta_{d_1}(a) < k \leq k(d_1)$, (where $k \geq 1$ if $\delta_{d_1}(a)$ does not exist) $\delta_{d_2}(a) < k \leq k(d_2)$ to be the ordering generated by the relations:

(1) $(d_1; K_1, \ldots, K_t) <^k_p (d_2; K_1, \ldots, K_t)$ for $d_1 < d_2$.

(2) $(\hat{\mathcal{L}}^{\restriction k}(d_1; K_1, \ldots, K_t); K_1, \ldots, K_{t-p}, K_{t+1}, ..., K_t) <^k_p (d_1; K_1, \ldots, K_t)$ for all K_{t+1}, $(\hat{\mathcal{L}}_p^{\restriction k+b}(d_1; K_1, \ldots, K_{t-p}, K_{t+1}, \ldots, K_t); K_1, \ldots, K_{t-p}, K_{t+1}, K_{t+2}, \ldots, K_t) < (d_1; K_1, \ldots, K_{t-p}, K_{t+1}, \ldots, K_t)$ where $b = 0$ if $k \leq t - p$ and $b = 1$ if $k > t - p$, etc.

We also define $<^0_p$ to the ordering generated by the relations:

(1) $(d_1; K_1, \ldots, K_t) <^0_p (d_2; K_1, \ldots, K_t)$ for $d_1 < d_2$.

(2) $(\hat{\mathcal{L}}^{\restriction 1}(d_1); K_1, \ldots, K_{t-p}, K_{t+1}, \ldots, K_t) <^0_p (d_1; K_1, \ldots, K_t)$, $(\hat{\mathcal{L}}^{\restriction 1}(d_1); K_1, \ldots, K_{t-p}, K_{t+1}, K_{t+2}, \ldots, K_t) <^0_p (d_1; K_1, \ldots, K_{t-p}, K_{t+1}, \ldots, K_t)$, etc. Here $t + 1$, $t + 2$, etc. precede K_1 if $t = p$.

We let $f^k_p(d; K_1, \ldots, K_t)$ for $\delta_d(a) < k \leq k(d)$ be the rank of $(d; K_1, \ldots, K_t)$ with respect to $<^k_p$, computed as $<_r$.

We will also define below an auxiliary function $\bar{f}^k_p(d; K_1, \ldots, K_t)$ for such d, k, and we will have $E(2n-1) < \bar{f}^k_p(d; K_1, \ldots, K_t)) < E(2n+3)$.

We assume (inductively on n) that the corresponding functions, which we denote by g^k_p, $\bar{g}^k_p$ have been defined for $d \in \mathcal{D}_{2n-1}$.

We introduce the following hypotheses concerning $\bar{f}^k_p$:

A^k_p: for d defined and satisfying C, A relative to $K_1, ..., K_t$ and $\delta_d(a) < k \leq k(d)$ (where $I_a = (f(\bar{K}_1); \bar{K}_2, \ldots, \bar{K}_a)$ is the index for (d), $\bar{f}^k_p(d; K_1, ..., K_t) = \bar{f}^{k+b}_p(d; K_1, \ldots, K_{t-p}, K_{t+1}, \ldots, K_t)$, where $b = 0$ if $k \leq t - p$ and $b = 1$ if $k > t - p$. Also, if $p = t$ and $k = 0$, then $\bar{f}^0_t(d_1; K_1, \ldots, K_t) = \bar{f}^1_t(d_1; K_{t+1}, K_1, \ldots, K_t)$.

B^k_p: for $d^{(I_a)}$, k as above, we assume d' is obtained by reindexing d. That is, for $I_a =$

$(f(\bar{K}_1); \bar{K}_2, \ldots, \bar{K}_a)$, $I'_{a'}$ the index of $d' = (f(\bar{K}_1); K'_2, \ldots, \bar{K}'_{a'})$, where $\delta_{d'}(K'_a) < k(d') = k(d)$, $\bar{K}_2, \ldots, \bar{K}_a$ is a subsequence of $K'_2, \ldots, K'_{a'}$, and $d = d'$ except that for each component tuple $d_i^{(I_{a_i})}$ of d, the corresponding component tuple $d_i'^{(I_{a_i})}$ of d' has index $(f(\bar{K}_1); K'_2, \ldots, K'_{a'}, \bar{K}_{a+1}, \ldots, \bar{K}_{a_i})$, where $f(\bar{K}_1); \bar{K}_2, \ldots, \bar{K}_{a_i})$ is the index I_{a_i} of d_i. We then require that $\bar{f}_p^k(d'; K_1, \ldots, K_t) = \bar{f}_p^k(d; K_1, \ldots, K_t)$.

R_p^k: if $d_1^{(I_a)}$, $d_2^{(I_a)}$ are define and satisfy C, A relative to $K_1, \ldots, K_t$, where $\delta_{d_1}(a) < k \leq k(d_1)$ and similarly for d_2, and $d_2 < d_1$, then $\bar{f}_p^{k+b}(d_2; K_1, \ldots, K_{t-p}, K_{t+1}, \ldots, K_t) < \bar{f}_p^k(d_1; K_1, \ldots, K_t)$, where $b = 0$ if $k \leq t - p$ and $b = 1$ if $k > t - p$, or $k = 0$.

R_p: if $d \in \mathcal{D}_{2n+1}^{l,m}$, then $\bar{f}_p^k(d; K_1, \ldots, K_t) < E(2n+2)$.

We assume inductively that A_p^k, B_p^k, R_p^k, R_p are satisfied for $d \in \mathcal{D}_{2n-1}$, and the corresponding function $\bar{g}_p^k$ on $\mathcal{D}_{2n-1}$.

Definition: For $d^{(I_a)}$ where $(I_a) = (f;)$, we set $\bar{f}_p = \bar{f}_p^1$, and similarly $\bar{g}_p = \bar{g}_p^1$ on $\mathcal{D}_{2n-1}$. We let $\mathcal{M}(K_1, \ldots, K_t)$ denote the sequence of measures $v(K_i)$ corresponding to those $K_i = S_{2n+1}^{l_i,m_i}$ for $l_i > 1$.

Assuming $\bar{f}_p^k$ defined we will extend it slightly to a function $\bar{\bar{f}}_p^k$ defined on objects of the form (d) or $(d)^s$, where d satisfies C, A relative to $K_1, \ldots, K_t$ (but not necessarily D). We do this as follows: For s not appearing, we set $\bar{\bar{f}}_p^k((d); K_1, \ldots, K_t) = \bar{\bar{f}}_p^k((d)^s; K_1, \ldots, K_t) + 1$, and also $\bar{\bar{f}}_p^k((d)^s; K_1, \ldots, K_t) = 2 \cdot \bar{f}_p^k(d; K_1, \ldots, K_t)$. If we consider the corresponding versions of A_p^k, B_p^k, R_p^k, for $\bar{\bar{f}}_p^k$, it is immediate that they are satisfied if they are satisfied for $\bar{f}_p^k$.

We now proceed to define $\bar{f}_p^k$. We proceed by reverse induction on k.

(I) $k = \infty$. Hence d basic of type 1 or 0.

(1) d basic of type 1, so $d^{(I_a)} = (\bar{d}_2)^{s(f:\bar{K}_2,\ldots,\bar{K}_a)}$ where s may or may not appear, $\bar{d}_2 \in \mathcal{D}_{2n-1}$, and $(\bar{d}_2)^s$ satisfies D, A relative to $\bar{K}_2, \ldots, \bar{K}_a$ (where s appears here iff s appears in d).

(a) If s appears we set $\bar{f}_p^\infty(d; K_1, \ldots, K_t) = E(2n-1) + 2 \cdot \bar{g}_q(\bar{d}_2; \mathcal{M}(K_1, \ldots, K_t)) + 1$.

(b) If s does not appear

$\bar{f}_p^\infty(d; K_1, \ldots, K_t) = E(2n-1) + 2 \cdot \bar{g}_q(\bar{d}_2, \mathcal{M}(K_1, \ldots, K_t)) + 2$, where q is minimal such that $\delta(\tilde{K}_{w-q}) \leq t - p$, where $\tilde{K}_1, \ldots, \tilde{K}_w$ enumerates $\mathcal{M}(K_1, \ldots, K_t)$ and $\delta(v(K_i)) = i$ for $v(K_i)$ in the sequence $\mathcal{M}(K_1, \ldots, K_t)$.

(2) d basic of type 0, so $d^{(I_a)} = (k)^{(fm;\bar{K}_2,\ldots,\bar{K}_a)}$, where $k \leq m$. We set

$$\bar{f}_p^\infty(d; K_1, \ldots, K_t) = E(2n-1) + m.$$

(II) $k < \infty,\ k < k(d),\ k \neq t-p$.

(a) $K_k = S_{2n+1}^{l_k, m_k}$, with $l_k > 1$. We set

$$\bar{f}_p^k(d; K_1, \ldots, K_t) = \sum_{\beta < f_p^{k+1}(d; K_1, \ldots, K_t)} 2 \cdot (\beta + 1) + 1.$$

(b) $K_k = S_{2n+1}^{1, m_k}$.

$$\bar{f}_p^k(d; K_1, \ldots, K_t) = \sum_{\beta < \bar{f}_p^{k+1}(d; K_1, \ldots, K_t)} (m_k + \tau_{m_k - 1}(\beta) + m_k) + 1,$$

where $\tau_m(\beta)$ is defined by recursion by $\tau_1(\beta) = 2 \cdot \beta$, $\tau_m(\beta) = (m + \tau_{m-1}(\beta) + (m) \cdot \beta + (m + \tau_{m-1}(\beta))$.

(c) $K_k = W_{2n+1}^{m_k}$.

$$\bar{f}_p^k(d : K_1, \ldots, K_t) = E(2n; m_k) + \bar{f}_p^{k+1}(d; K_1, \ldots, K_t).$$

(d) K_k not of these forms.

$$\bar{f}_p^k(d; K_1, \ldots, K_t) = \bar{f}_p^{k+1}(d; K_1, \ldots, K_t).$$

(e) If $k = 0$, we set

$$\bar{f}_p^k(d; K_1, \ldots, K_t) = \bar{f}_p^{k+1}(d; K_1, \ldots, K_t).$$

(III) $k < \infty,\ k < k(d),\ k = t - p$.

(a) $K_k = S_{2n+1}^{l_k, m_k}$, with $l_k > 1$. We set

$$\bar{f}_p^k(d; K_1, \ldots, K_t) = \sum_{\beta < \omega^{\omega^{\bar{f}_p^{k+1}(d; K_1, \ldots, K_t)}}} (\beta + 1) + 1.$$

(b) $K_k = S_{2n+1}^{1, m_k}$.

$$\bar{f}_p^k(d; K_1, \ldots, K_t) = \Big[\sum_{\beta < \omega^{\omega^{\bar{f}_p{}^{k+1}(d; K_1, \ldots, K_t)}}} (m_k + \tau_{m_{k-1}}(\beta) + m_k) \Big] + 1.$$

(c) $K_k = W_{2n+1}^{m_k}$.

$$\bar{f}_p^k(d; K_1, \ldots, K_t) = E(2n; m_k) + \omega^{\omega^{\bar{f}_p^{k+1}(d; K_1, \ldots, K_t)}}.$$

(d) K_k not of these forms.

$$\bar{f}_p^k(d; K_1, \ldots, K_t) = \omega^{\omega^{\bar{f}_p^{k+1}(d;K_1,\ldots,K_t)}}.$$

(e) If $k = 0$, we set $\bar{f}_p^k(d; K_1, \ldots, K_t) = \omega^{\omega^{\bar{f}_p^{k+1}(d;K_1,\ldots,K_t)}}$.

(IV) $k < \infty$, $k = k(d)$, $k \neq t - p$.

(1) $K_k = S_{2n+1}^{l_k, m_k}$, with $l_k > 1$. Hence $d^{(I_a)} = (k; d_2^{(I_a+1)})^{s(I_a)}$, where s may or may not appear.

$$\bar{f}_p^k(d; K_1, \ldots, K_t) = \sum_{\beta < \bar{f}_p^{k+1}(\hat{d}; K_1, \ldots, K_t)} 2 \cdot (\beta + 1) + 2 \cdot \bar{f}_p^{k+1}(d_2; K_1, \ldots, K_t)$$
$$+ (1 \text{ if } s \text{ appears, } 2 \text{ if } s \text{ does not appear}).$$

(2) $K_k = S_{2n+1}^{1, m_k}$. Here $d^{(I_a)} = (k; d_1^{(I_a)}, \ldots, d_r^{(I_a)})^{a(I_a)}$ where $r \leq m_k$ and s may or may not appear.

(a) $r = m_k$.

$$\bar{f}_p^k(d; K_1, \ldots, K_t) = \sum_{\beta < \bar{f}_p^{k+1}(d_1; K_1, \ldots, K_t)} (m_k + \tau_{m_k - 1}(\beta) + m_k) + ((m_k - 1)$$
$$+ \tau_{m_k - 2}(\bar{f}_p^{k+1}(d_1; K_1, \ldots, K_t)) + (m_k - 1)) \cdot \bar{f}_p^{k+1}(d_2; K_1, \ldots, K_t)$$
$$+ \cdots + (2 + \tau_1(\bar{f}_p^{k+1}(d_1; K_1, \ldots, K_t)) + 2) \cdot \bar{f}_p^{k+1}(d_{r-1}; K_1, \ldots, K_t)$$
$$+ \tau_1(\bar{f}_p^{k+1}(d_r; K_1, \ldots, K_t)) + (1 \text{ or } 2)$$

depending on whether s appears or not.

(b) $r < m_k$ and s appears.

$$\bar{f}_p^k(d; K_1, \ldots, K_t) = \sum_{\beta < \bar{f}_p^{k+1}(d_1; K_1, \ldots, K_t)} (m_k + \tau_{m_k - 1}(\beta) + m_k)$$
$$+ ((m_k - 1) + \tau_{m_k - 2}(\bar{f}_p^{k+1}(d_1; K_1, \ldots, K_t)) + (m_k - 1).$$
$$\bar{f}_p^{k+1}(d_2; K_1, \ldots, K_t) + \cdots + ((m_k - r + 1) + \tau_{m-r}(\bar{f}_p^{k+1}(d_1; K_1, \ldots, K_t))$$
$$+ (m_k - r + 1)) \cdot \bar{f}_p^{k+1}(d_r; K_1, \ldots, K_t) + 1.$$

(c) $r < m_k$ and s does not appear.

$$\bar{f}_p^k(d; K_1, \ldots, K_t) = \sum_{\beta < \bar{f}_p^{k+1}(d_1; K_1, \ldots, K_t)} (m_k + \tau_{m_k - 1}(\beta) + m_k) + ((m_k - 1)$$
$$+ \tau_{m_k - 2}(\bar{f}_p^{k+1}(d_1; K_1, \ldots, K_t)) + (m_k - 1)) \cdot \bar{f}_p^{k+1}(d_2; K_1, \ldots, K_t))$$
$$+ \cdots + ((m_k - r + 1) + \tau_{m_k - r}(\bar{f}_p^{k+1}(d_r; K_1, \ldots, K_t)) + (m_k - r + 1)) \cdot$$

$$\bar{f}_p^{k+1}(d_r; K_1, \ldots, K_t) + \tau_{m_k - r}(\bar{f}_p^{k+1}(d_1; K_1, \ldots, K_t)) + (m_k - r + 1).$$

(3) $K_k = W_{2n+1}^{m_k}$.

(a) $d = (k;\ )$. If $n > 0$,

$\bar{f}_p^k(d; K_1, \ldots, K_t) = E(2n; m_2)$ if $k < t - p$ and $= E(2n-1)$ if $k > t - p$. If $n = 0$, and $d = (k; r)$, where $r \leq m_k$, then $\bar{f}_p^k(d; K_1, \ldots, K_t) = r$.

(b) $d = (k; \bar{d}_2)^s$, where s may or may not appear.

$$\bar{f}_p^k(d; K_1, \ldots, K_t) = E(2n-1) + 2 \cdot (\bar{g}_q(\bar{d}_2; K_{b(k)}, \ldots, K_{b_u}, \mathcal{M}(K_1, \ldots, K_{k-1})) + (1 \text{ or } 2)$$

depending on whether s appears or not, where q is minimal such that $\delta(\tilde{K}_{w-q}) \leq t - p$ where $\tilde{K}_1, \ldots, \tilde{K}_w$ enumerates $K_{b(k)}, \ldots, K_{b_u}$, $\mathcal{M}(K_1, \ldots, K_{k-1})$ and $\delta(K_{b_j}) = b_j$, $\delta(v(K_i)) = i$ for $v(K_k)$ in the sequence $\mathcal{M}(K_1, \ldots, K_{k-1})$, if such a q exists, and otherwise $q = w$.

(V) $k < \infty$, $k = k(d)$, $k = t - p$.

(1) $K_k = S_{2n+1}^{l_k, m_k}$, where $l_k > 1$.

$$\bar{f}_p^k(d; K_1, \ldots, K_t) = \sum_{\beta < \omega^{\omega^{\bar{f}_p^{k+1}(\hat{d}; K_1, \ldots, K_t)}}} (\beta + 1) + \omega^{\omega^{\bar{f}_p^{k+1}(d_2; K_1, \ldots, K_t)}} + (1 \text{ or } 2)$$

depending on whether s appears or not.

(2) $K_k = S_{2n+1}^{1, m_k}$.

(a) $r = m_k$.

$$\begin{aligned} \bar{f}_p^k(d; K_1, \ldots, K_t) = & \sum_{\beta < \omega^{\omega^{\bar{f}_p^{k+1}(d_1; k_1, \ldots, k_t)}}} (m_k + \tau_{m_k - 1}(\beta) + m_k) + (m_k - 1)) \\ & + \tau_{m_k - 2}(\omega^{\omega^{\bar{f}_p^{k+1}(d_1; K_1, \ldots, K_t)} + (m_k - 1)}) \cdot \omega^{\omega^{\bar{f}_p^{k+1}(d_2; K_1, \ldots, K_t)}} \\ & + \cdots + (2 + \tau_1(\omega^{\omega^{\bar{f}_p^{k+1}(d_1; K_1, \ldots, K_t)}}) + 2) \cdot \omega^{\omega^{\bar{f}_p^{k+1}(d_{r-1}; K_1, \ldots, K_t)}} \\ & + \tau_1(\omega^{\omega^{\bar{f}_p^{k+1}(d_r; K_1, \ldots, K_t)}}) + (1 \text{ or } 2) \end{aligned}$$

depending on whether s appears or not.

(b) $r < m_k$ and s appears. Similarly to (a) above, $\bar{f}_p^k(d; K_1, \ldots, K_t)$ is obtained from the formula in IV.2.b by replacing terms of the form $\bar{f}_p^{k+1}(\ )$ by $\omega^{\omega^{\bar{f}_p^{k+1}(\ )}}$.

(c) $r < m_k$ and s does not appear. As above, using the formula from IV.2.c.

(3) $K_k = W_{2n+1}^{m_k}$.

(a) $d = (k;\ )$. $\bar{f}_p^k = E(2n; m_k)$.

(b) $d = (k; \bar{d}_2)^s$, where s may or may not appear.

$$\bar{f}_p^k(d; K_1, \ldots, K_t) = E(2n-1) + 2(\bar{g}_q(\bar{d}_2; K_{b(k)}, \ldots, K_{b_u}, \mathcal{M}(K_1, \ldots, K_{k-1}))) + (1 \text{ or } 2)$$

depending on whether s appears or not. Here q is as in IV.3.b (where here the first clause applies).

We now proceed to establish A_p^k, B_p^k, R_p^k for $d \in \mathcal{D}_{2n+1}$. We assume these for $d \in \mathcal{D}_{2n-1}$ and the corresponding functions $\bar{g}_q^k$. We proceed by simultaneous reverse induction on k.

We first consider A_p^k.

(I) $k = \infty$.

(1) d basic of type 1. A_p^k follows from A_q^1 for $\bar{d}_2 \in \mathcal{D}_{2n-1}$, with q as in I.1.a.

(2) d basic of type 0. immediate.

(II) $k < \infty$, $k < k(d)$, $k \neq t - p$. The result follows by induction, A_p^{k+1}, and the formulas for $\bar{f}_p^k$.

(III) $k < \infty$, $k < k(d)$, $k = t - p$. We require the following easy lemma:

LEMMA. *If $\alpha = \omega^{\omega^\gamma}$ for some $\gamma \in ON$, then $\sum_{\beta < \alpha} \beta^n = \alpha$ for all n.*

By induction and A_p^{k+1}, we have

$$\bar{f}_p^{k+1}(d; K_1, \ldots, K_t) = \bar{f}_p^{k+2}(d; K_1, \ldots, K_{t-p}, K_{t+1} \ldots, K_t).$$

Hence $\omega^{\omega^{\bar{f}_p^{k+1}(d;K_1,\ldots,K_t)}} = \omega^{\omega^{\bar{f}_p^{k+2}(d:K_1,\ldots,K_{t-p},K_{t+1},\ldots,K_t)}}$. From the lemma and the formulas for $\bar{f}_p^k$, it follows that $\bar{f}_p^k(d; K_1, \ldots, K_t) \quad = \omega^{\omega^{\bar{f}_p^{k+1}(d;K_1,\ldots,K_t)}}$, and $\bar{f}_p^{k+1}(d; K_1, \ldots, K_{t-p}, K_{t+1}, \ldots, K_t) = \omega^{\omega^{\bar{f}_p^{k+2}(d;K_1,\ldots,K_{t-p},K_{t+1},\ldots,K_t)}}$, hence

$$\bar{f}_p^k(d; K_1, \ldots, K_{t-p}, K_{t+1}, \ldots, K_t) = \omega^{\omega^{\bar{f}_p^{k+2}(d;K_1,\ldots,K_{t-p},K_{t+1},\ldots,K_t)}} = \bar{f}_p^k(d; K_1, \ldots, K_t)$$

follows from the lemma and the formulas for $\bar{f}_p^k$.

(IV) $k < \infty$, $k = k(d)$, $k \neq t - p$. The result follows by induction and A_p^{k+1}, and in Case 3 ($K_k = W_{2n+1}^{m_k}$), by A_q^1 or A_q^0 for $\bar{d}_2 = \mathcal{D}_{2n-1}$.

(V) $k < \infty$, $k = k(d)$, $k = t - p$.

A_p^k then follows from the formulas for $\bar{f}_p^k$, A_p^{k+1}, and the lemma. For example, in case V.2.a,

$$\bar{f}_p^k(d;K_1,\ldots,K_t) = \sum_{\beta<\omega^{\omega^{\bar{f}_p^{k+1}(d_1;K_1,\ldots,K_t)}}} (m_k + \tau_{m_k-1}(\beta) + m_k)$$
$$+ \cdots + \tau_{m_k-2}(\omega^{\omega^{\bar{f}_p^{k+1}(d_1;K_1,\ldots,K_t)}}) \cdots = \omega^{\omega^{\bar{f}_p^{k+1}(d_1;K_1,\ldots,K_t)}}$$
$$+ \cdots + \tau_{m_k-2}(\omega^{\omega^{\bar{f}_p^{k+1}(d_1;K_1,\ldots,K_t)}}) \ldots, \quad \text{and}$$

$$\bar{f}_p^k(d;K_1,\ldots,K_{t-p},K_{t+1},\ldots,K_t) = \sum_{\beta<\bar{f}_p^{k+1}(d_1;K_1,\ldots,K_{t-p},K_{t+1},\ldots,K_t)} 2\cdot(\beta+1)$$
$$+ \cdots + \tau_{m_k-2}(\bar{f}_p^{k+1}(d_1;K_1,\ldots,K_{t-p},\ldots,K_{t+1},\ldots,K_t))\cdots.$$

(We consider here the case $K_{t+1} = S_{2n+1}^{l_{t+1},m_{t+1}}$ where $l_{t+1} > 1$). Also,

$$\bar{f}_p^{k+1}(d_1;K_1,\ldots,K_{t-p},K_{t+1},\ldots,K_t) = \omega^{\omega^{\bar{f}_p^{k+2}(d_1;K_1,\ldots,K_{t-p},K_{t+1},\ldots,K_t)}}$$

using the lemma, $= \omega^{\omega^{\bar{f}_p^{k+1}(d_1;k_1,\ldots,k_t)}}$ by induction. The result then follows. In case 3, we again use A_q^0 on $\mathcal{D}_{2n-1}$.

This establishes A_p^k in all cases.

We now consider B_p^k.

(I) $k = \infty$.

(1) $d^{(I_a)}$ basic of type 1, so $d = (\bar{d}_2)^s$, where s may or may not appear, and $I_a = (f(\bar{K}_1); \bar{K}_2,\ldots,\bar{K}_a)$. We let $\bar{d}$ be a re-indexing of d as in the statement of B_p^k, so $\bar{d}^{(\bar{I}_a)} = (\bar{d}_2')^{(f(\bar{K}_1);\tilde{K}_2,\ldots,\tilde{K}_{\tilde{a}})}$, where $\bar{d}_2' = \bar{d}_2$, and $\bar{K}_2,\ldots,\bar{K}_a$ is a subsequence of $\tilde{K}_2,\ldots,\tilde{K}_{\tilde{a}}$. The formula for $\bar{f}_p^k$, however involves $\bar{g}_q(\bar{d}_2)$ computed relative to $\mathcal{M}(K_1,\ldots,K_t)$. Hence, $\bar{f}_p^k(d; K_1,\ldots,K_t) = \bar{f}_p^k(\bar{d};K_1,\ldots,K_t)$.

(2) d basic of type 0 immediate.

(II) $k < \infty$, $k < k(d)$, $k \neq t - p$ and

(III) $k < \infty$, $k < k(d)$, $k = t - p$.

The result follows immediately from induction and B_p^{k+1}.

(IV) $k < \infty$, $k = k(d)$, $k \neq t - p$, and

(V) $k < \infty$, $k = k(d)$, $k = t - p$.

Also immediate from induction and B_p^{k+1}, where in Case 3b, we use the fact that $\mathcal{M}(K_1,\ldots,K_{k-1})$ is the same for both d, $\bar{d}$.

This establishes B_p^k.

We consider R_p^k.

We let $d_1^{(I_1)}$, $d_2^{(I_a)}$ satisfy C, A relative to $K_1,\ldots,K_t$, and assume that $d_1 < d_2$, and establish that

$\bar{f}_p^{k+b}(d_1;K_1,\ldots,K_{t-p},K_{t+1},\ldots,K_t) < \bar{f}_p^k(d_2;K_1,\ldots,K_t)$, where $b=0$ if $k \le t-p$, and $b=1$ if $k > t-p$ or $k=0$. We recall that $\delta_{d_1}(\bar{K}_a) = \delta_{d_2}(\bar{K}_a) < k \le k(d_1)$ or $k(d_2)$.

I) $k(d_1) > k(d_2)$.

1) $k < k(d_2)$.

a) $k \ne t-p$. By induction, $\bar{f}_p^{k+1+b}(d_1;K_1,\ldots,K_{t-p},K_{t+1},\ldots,K_t) < \bar{f}_p^{k+1}(d_2;K_1,\ldots,K_t)$. From the formulas for $\bar{f}_p^k$, it follows that $\bar{f}_p^{k+b}(d_1;\ K_1,\ldots,K_{t-p},\ K_{t+1},\ldots,K_t) < \bar{f}_p^k(d_2;\ K_1,\ldots,K_t)$.

b) $k = t-p$. By induction, $\bar{f}_p^{k+2}(d_1;K_1,\ldots,K_{t-p},K_{t+1},\ldots,K_t) < \bar{f}_p^{k+1}(d_2;K_1,\ldots,K_t)$. We have from the formulas for $\bar{f}_p^k$ that $\bar{f}_p^k(d_2;K_1,\ldots,K_t) = \omega^{\omega^{\bar{f}_p^{k+1}(d_2;K_1,\ldots,K_t)}}$, and $\bar{f}_p^{k+1}(d_1;\ K_1,\ldots,K_{t-p},K_{t+1},\ldots,K_t) = \omega^{\omega^{\bar{f}_p^{k+2}(d_1;K_1,\ldots,K_{t-p},K_{t+1},\ldots,K_t)}}$. From the lemma and the formulas for $\bar{f}_p^k$, we have that $\bar{f}_p^k(d_2;K_1,\ldots,K_{t-p},K_{t+1},\ldots,K_t) = \omega^{\omega^{\bar{f}_p^{k+2}(d_1;K_1,\ldots,K_{t-p},K_{t+1},\ldots,K_t)}} < \omega^{\omega^{\bar{f}_p^{k+1}(d_2;K_1,\ldots,K_t)}} = \bar{f}_p^k(d_2;K_1,\ldots,K_t)$.

2) $k = k(d_2)$, $K_k = S_{2n+1}^{\ell_k,m_k}$, $\ell_k > 1$. Hence, $d_1 \le \hat{d}_2$.

a) $k \ne t-p$. By induction, $\bar{f}_p^{k+1+b}(d_1;K_1,\ldots,K_{t-p},K_{t+1},\ldots,K_t) \le \bar{f}_p^{k+1}(\hat{d}_2;K_1,\ldots,K_t)$. From the formulas for $\bar{f}_p^k$, it follows that $\bar{f}_p^{k+b}(d_1;K_1,\ldots,K_{t-p},\ K_{t+1},\ldots,K_t) < \bar{f}_p^k(d_2;K_1,\ldots,K_t)$, since $\bar{f}_p^{k+1}((d_2)_2;K_1,\ldots,K_t) \ge 1$.

b) $k = t-p$. By induction, $\bar{f}_p^{k+2}(d_2;K_1,\ldots,K_{t-p},K_{t+1},\ldots,K_t) \le \bar{f}_p^{k+1}(\hat{d}_2;K_1,\ldots,K_t)$. From the formulas for $\bar{f}_p^k$, we have that $\bar{f}_p^k(d_2;K_1,\ldots,K_t) = \omega^{\omega^{\bar{f}_p^{k+1}(\hat{d}_2;K_1,\ldots,K_t)}} + \omega^{\omega^{\bar{f}_p^{k+1}((d_2)_2;K_1,\ldots,K_t)}} + (1 \text{ or } 2)$, and $\bar{f}_p^{k+1}(d_1;K_1,\ldots,K_{t-p},K_{t+1},\ldots,K_t) = \omega^{\omega^{\bar{f}_p^{k+a}(d_1;K_1,\ldots,K_{t-p},K_{t+1},\ldots,K_t)}} \le \omega^{\omega^{\bar{f}_p^{k+1}(\hat{d}_2;K_1,\ldots,K_t)}} < \bar{f}_p^k(d_2;K_1,\ldots,K_t)$. The result then follows from the lemma.

3) $k = k(d_2)$, $K_k = S_{2n+1}^{1,m_k}$. Hence, $d_1 \le (d_2)_1$.

a) $k \neq t-p$. By induction $\bar{f}_p^{k+1+b}(d_1; K_1, \ldots, K_{t-p}, K_{t+1}, \ldots, K_t) \leq \bar{f}_p^{k+1}((d_2)_1; K_1, \ldots, K_t)$. From the formulas for $\bar{f}_p^k$ it follows that $\bar{f}_p^{k+b}(d_1; K_1, \ldots, K_{t-p}, K_{t+1}, \ldots, K_t) < \bar{f}_p^k(d_2; K_1, \ldots, K_t)$.

b) $k = t-p$. By induction, $\bar{f}_p^{k+2}(d_1; K_1, \ldots, K_{t-p}, K_{t+1}, \ldots, K_t) \leq \bar{f}_p^{k+1}((d_2)_1; K_1, \ldots, K_t)$. From the formulas for $\bar{f}_p^k$, we then have that $\bar{f}_p^k(d_2; K_1, \ldots, K_t) = \omega^{\omega^{\bar{f}_p^{k+1}((d_2)_1; K_1, \ldots, K_t)}} + (\alpha)$, where $\alpha \geq 1$, and $\bar{f}_p^{k+1}(d_1; K_1, \ldots, K_{t-p}, K_{t+1}, \ldots, K_t) = \omega^{\omega^{\bar{f}_p^{k+2}(d_1; K_1, \ldots, K_{t-p}, K_{t+1}, \ldots, K_t)}}$, and hence by the lemma $\bar{f}_p^k(d_1; K_1, \ldots, K_{t-p}, K_{t+1}, \ldots, K_t) = \omega^{\omega^{\bar{f}_p^{k+2}(d_1; K_1, \ldots, K_{t-p} K_{t+1}, \ldots, K_t)}} \leq \omega^{\omega^{\bar{f}_p^{k+1}((d_2)_1; K_1, \ldots, K_t)}} < \bar{f}_p^k(d_2; K_1, \ldots, K_t)$.

4) $k = k(d_2)$, $K_2 = W_{2n+1}^{m_k}$. This case can not arise from the definition of $<$.

II) $k(d_1) < k(d_2)$.

1) $k < k(d_1)$. We proceed as in I.1.a. or I.1.b. depending on whether $k \neq t-p$ or $k = t-p$.

2) $k = k(d_1)$, $K_k = S_{2n+1}^{\ell_k, m_k}$, $\ell_k > 1$. Hence $\hat{d}_1 < d_2$.

a) $k \neq t-p$. By induction $\bar{f}_p^{k+1+b}(\hat{d}_1; K_1, \ldots, K_{t-p}, K_{t+1}, \ldots, K_t) < \bar{f}_p^{k+1}(d_2; K_1, \ldots, K_t)$. From the formulas for $\bar{f}_p^k$, we have that $\bar{f}_p^{k+b}(d_1; K_1, \ldots, K_{t-p}, K_{t+1}, \ldots, K_t) = \left[\sum_{\beta < \bar{f}_p^{k+1+b}(\hat{d}_1; K_1, \ldots, K_{t-p}, K_{t+1}, \ldots, K_t)} 2\cdot(\beta+1)\right] + 2\cdot(\bar{f}_p^{k+1+b}((d_1)_2; K_1, \ldots, K_{t-p}, K_{t+1}, \ldots, K_t)) +$ (1 or 2), and $\bar{f}_p^k(d_2; K_1, \ldots, K_t) = \left[\sum_{\beta < \bar{f}_p^{k+1}(d_2; K_1, \ldots, K_t)} 2\cdot(\beta+1)\right] + 1$. By B_p^{k+1+b} we have that $\bar{f}_p^{k+1+b}(\hat{d}_1; K_1, \ldots, K_{t-p}, K_{t+1}, \ldots, K_t) = \bar{f}_p^{k+1+b}(\hat{d}_1^{(I_{a+1})}; K_1, \ldots, K_{t-p}, K_{t+1}, \ldots, K_t)$, where $\hat{d}_1^{(I_{a+1})}$ is obtained from $\hat{d}_1^{(I_a)}$ by re-indexing as in B, where $(I_{a+1}) = (I_a; v(K_k))$. We also have that $(d_1)_2^{(I_{a+1})} \leq \hat{d}_1^{(I_{a+1})}$, and hence, by induction, $\bar{f}_p^{k+1+b}(\hat{d}_1^{(I_{a+1})}; K_1, \ldots, K_{t-p}, K_{t+1}, \ldots, K_t) \geq \bar{f}_p^{k+1+b}((d_1)_2^{(I_{a+1})}; K_1, \ldots, K_{t-p}, K_{t+1}, \ldots, K_t)$. (We have used here the fact that for d satisfying C, A relative to $K_1, \ldots, K_t$, $\bar{f}_p^{k+b}(d; K_1, \ldots, K_{t-p}, K_{t+1}, \ldots, K_t) \geq \bar{f}_p^k(d; K_1, \ldots, K_t)$; applied to $(d_1)_2^{(I_{a+1})}$). Hence, $\bar{f}_p^{k+1+b}(\hat{d}_1; K_1, \ldots, K_{t-p}, K_{t+1}, \ldots, K_t) \geq \bar{f}_p^{k+1+b}((d_1)_2^{(I_{a+1})}; K_1, \ldots, K_{t-p}, K_{t+1}, \ldots, K_t)$. In fact, we have strict inequality here unless s appears in d_1 (i.e. $d_1 = (k; (d_1)_2)^s$. Hence, in either case, $\bar{f}_p^{k+b}(d_1; K_1, \ldots, K_{t-p}, K_{t+1}, \ldots, K_t) \leq \sum_{\beta \leq \bar{f}_p^{k+1+b}(\hat{d}_1; K_1, \ldots, K_{t-p}, K_{t+1}, \ldots, K_t)} 2\cdot(\beta+1) < \sum_{\beta < \bar{f}_p^{k+1}(d_2; K_1, \ldots, K_t)} 2\cdot(\beta+1)+1 = \bar{f}_p^k(d_2; K_1, \ldots, K_t)$.

b) $k = t-p$. By induction, $\bar{f}_p^{k+2}(\hat{d}_1; K_1, \ldots, K_{t-p}, K_{t+1}, \ldots, K_t) < \bar{f}_p^{k+1}(d_2; K_1, \ldots, K_t)$. From the formulas for $\bar{f}_p^k$, we have that $\bar{f}_p^k(d_1; K_1, \ldots, K_{t-p}, K_{t+1}, \ldots, K_t) = \sum_{\beta < \bar{f}_p^{k+1}(\hat{d}_1; K_1, \ldots, K_{t-p}, K_{t+1}, \ldots, K_t)} 2\cdot(\beta+1) + \bar{f}_p^{k+1}((d_1)_2; K_1, \ldots, K_{t-p}, K_{t+1}, \ldots, K_t) +$ (1 or 2).

We further have that $\bar{f}_p^{k+1}(\hat{d}_1; K_1, \ldots, K_{t-p}, K_{t+1}, \ldots, K_t) = \omega^{\omega^{\bar{f}_p^{k+2}(\hat{d}_1; K_1, \ldots, K_{t-p}, K_{t+1}, \ldots, K_t)}}$, and $\bar{f}_p^{k+1}((d_1)_2; K_1, \ldots, K_{t-p}, K_{t+1}, \ldots, K_t) = \omega^{\omega^{\bar{f}_p^{k+2}((d_1)_2; K_1, \ldots, K_{t-p}, K_{t+1}, \ldots, K_t)}}$, hence $\bar{f}_p^k(d_1; K_1, \ldots, K_{t-p}, K_{t+1}, \ldots, K_t) = \omega^{\omega^{\bar{f}_p^{k+2}(\hat{d}_1; K_1, \ldots, K_{t-p}, K_{t+1}, \ldots, K_t)}} + \omega^{\omega^{\bar{f}_p^{k+2}((d_1)_2; K_1, \ldots, K_{t-p}, K_{t+1}, \ldots, K_t)}} + (1 \text{ or } 2)$. We also have that $\bar{f}_p^k(d_2; K_1, \ldots, K_t) = \omega^{\omega^{\bar{f}_p^{k+1}(d_2; K_1, \ldots, K_t)}} + 1$. It also follows from B_p^{k+2}, as in the previous case, that $\bar{f}_p^{k+2}((d_1)_2^{(I_a+1)}; K_1, \ldots, K_{t-p}, K_{t+1}, \ldots, K_t) \leq \bar{f}_p^{k+2}(\hat{d}_1; K_1, \ldots, K_{t-p}, K_{t+1}, \ldots, K_t)$. Hence $\bar{f}_p^k(d_1; K_1, \ldots, K_{t-p}, K_{t+1}, \ldots, K_t) \leq \omega^{\omega^{\bar{f}_p^{k+2}(\hat{d}_1; K_1, \ldots, K_{t-p}, K_{t+1}, \ldots, K_t)}} \cdot 2 + (1 \text{ or } 2) < \omega^{\omega^{\bar{f}_p^{k+1}(d_2; K_1, \ldots, K_t)}} < \bar{f}_p^k(d_2; K_1, \ldots, K_t)$.

3) $k = k(d_1)$, $K_k = S_{2n+1}^{1,m_k}$. Hence $(d_1)_1 < d_2$ (where $d_1 = (k; (d_1)_1, (d_1)_2, \ldots, (d_1)_r)^s$, where s may or may not appear).

a) $k \neq t-p$. By induction, $\bar{f}_p^{k+1+b}((d_1)_1; K_1, \ldots, K_{t-p}, K_{t+1}, \ldots, K_t) < \bar{f}_p^{k+1}(d_2; K_1, \ldots, K_t)$. From the formulas for $\bar{f}_p^k$ we have that $\bar{f}_p^{k+b}(d_1; K_1, \ldots, K_{t-p}, K_{t+1}, \ldots, K_t) =$

$$\sum_{\beta < \bar{f}_p^{k+1+b}((d_1)_1; K_1, \ldots, K_{t-p}, K_{t+1}, \ldots, K_t)} (m_k + \tau_{m_k-1}(\beta) + m_k) + \alpha,$$

where $\alpha \leq \tau_{m_k-1}(\bar{f}_p^{k+1+b}((d_1)_1; K_1, \ldots, K_{t-p}, K_{t+1}, \ldots, K_t)) + m_k$. This follows from the formulas for $\bar{f}_p^k$, the fact that $\bar{f}_p^{k+1}(d_i; K_1, \ldots, K_{t-p}, K_{t+1}, \ldots, K_t) < \bar{f}_p^{k+1}(d_1; K_1, \ldots, K_{t-p}, K_{t+1}, \ldots, K_t)$, and the fact that $((m-1) + \tau_{m-2}(\alpha) + (m-1)) \cdot \alpha + ((m-2) + \tau_{m-3}(\alpha) + (m-1)) \cdot \alpha + \cdots + ((m-r+1) + \tau_{m-r}(\alpha) + (m-r+1) \cdot \alpha + (m-r+1) + \tau_{m-r}(\alpha) = \tau_{m-1}(\alpha)$ for all α. We also have that

$$\bar{f}_p^k(d_2; K_1, \ldots, K_t) = \sum_{\beta < \bar{f}_p^{k+1}(d_2; K_1, \ldots, K_t)} (m_k + \tau_{m_k-1}(\beta) + m_k) + 1,$$

and hence $\bar{f}_p^{k+b}(d_1; K_1, \ldots, K_{t-p}, K_{t+1}, \ldots, K_t) < \bar{f}_p^k(d_2; K_1, \ldots, K_t)$.

b) $k = t-p$. We have that $\bar{f}_p^k(d_1; K_1, \ldots, K_{t-p}, K_{t+1}, \ldots, K_t) =$

$$\sum_{\beta < \bar{f}_p^{k+1}((d_1)_1; K_1, \ldots, K_{t-p}, K_{t+1}, \ldots, K_t)} \alpha \leq \tau_{m_k-1}(\bar{f}_p^{k+1}((d_1)_1; K_1, \ldots, K_{t-p}, K_{t+1}, \ldots, K_t)) + m_k.$$

Also $\bar{f}_p^{k+1}((d_1)_1; K_1, \ldots, K_{t-p}, K_{t+1}, \ldots, K_t) = \omega^{\omega^{\bar{f}_p^{k+2}((d_1)_1; K_1, \ldots, K_{t-p}, K_{t+1}, \ldots, K_t)}} < \omega^{\omega^{\bar{f}_p^{k+1}(d_1; K_1, \ldots, K_t)}}$ and $\bar{f}_p^k(d_2; K_1, \ldots, K_t) = \omega^{\omega^{\bar{f}_p^{k+1}(d_2; K_1, \ldots, K_t)}} + 1$, from the lemma, hence $\bar{f}_p^k(d_1; K_1, \ldots, K_{t-p}, K_{t+1}, \ldots, K_t) \leq \omega^{\omega^{\bar{f}_p^{k+2}((d_1)_1; K_1, \ldots, K_{t-p}, K_{t+1}, \ldots, K_t)}} \cdot m$, for some m, $< \omega^{\omega^{\bar{f}_p^{k+1}(d_2; K_1, \ldots, K_t)}} = \bar{f}_p^k(d_2; K_1, \ldots, K_t)$.

4) $k = k(d_1)$, $K_k = W_{2n+1}^{m_k}$.

a) $k \neq t-p$. From the formulas for $\bar{f}_p^k$ and R_q on $\mathcal{D}_{2n-1}$, we have that $\bar{f}_p^{k+b}(d_1; K_1, \ldots, K_{t-p}, K_{t+1}, \ldots, K_t) \leq E(2n; m_k)$. Also, $\bar{f}_p^k(d_2; K_1, \ldots, K_t) = E(2n; m_k) + \alpha$, where $\alpha \geq 1$ (in fact $\alpha \geq E(2n-1)$), and hence the result follows.

b) $k = t - p$. Follows as in the previous case.

III) $k(d_1) = k(d_2)$.

1) $k < k(d_1) = k(d_2)$.

a) $k \neq t - p$ and b) $k = t - p$. We proceed as in I.1.a. and I.1.b.

2) $k = k(d_1) = k(d_2)$, $K_k = S_{2n+1}^{\ell_k, m_k}$, $\ell_k > 1$. Hence $d_1 = (k; \bar{d}_{2,1}^{(I_a+1)})^s$, $d_2 = (k; \bar{d}_{2,2}^{(I_a+1)})^s$ where s may or may not appear in d_1, d_2.

i) $\bar{d}_{2,1} < \bar{d}_{2,2}$

a) $k \neq t - p$. We have by induction that $\bar{f}_p^{k+b+1}(\bar{d}_{2,1}; K_1, \ldots, K_{t-p}, K_{t+1}, \ldots, K_t) < \bar{f}_p^{k+1}(\bar{d}_{2,2}, K_1, \ldots, K_t)$. Since $\bar{d}_{2,1} < \bar{d}_{2,2}$, it follows that $\hat{d}_1 \leq \hat{d}_2$, and hence by induction, $\bar{f}_p^{k+b+1}(\hat{d}_1; K_1, \ldots, K_{t-p}, K_{t+1}, \ldots, K_t) \leq \bar{f}_p^{k+1}(\hat{d}_2; K_1, \ldots, K_t)$, where we use A_p^{k+1} in case $\hat{d}_1 = \hat{d}_2$. $\mathcal{R}_p^k$ then follows from the formulas for $\bar{f}_p^k$.

b) $k = t - p$. Similar to the above case.

ii) $\bar{d}_{2,1} = \bar{d}_{2,2}$, s appears in d_1 and not in d_2. R_p^k is immediate from the formulas for $\bar{f}_p^k$.

3) $k = k(d_1) = k(d_2)$, $K_k = S_{2n+1}^{1,m_k}$. Here $d_1 = (k; d_{1,1}, d_{2,1}, \ldots, d_{r_1,1})^s$, $d_2 = (k; d_{1,2}, d_{2,2}, \ldots, d_{r_2,2})^s$, where s may or may not appear in d_1, d_2.

i) $d_{1,1} < d_{1,2}$.

a) $k \neq t-p$. By induction, $\bar{f}_p^{k+b+1}(d_{1,1}; K_1, \ldots, K_{t-p}, K_{t+1}, \ldots, K_t) < \bar{f}_p^{k+1}(d_{1,2}; K_1, \ldots, K_t)$. From the formulas for $\bar{f}_p^k$, $\bar{f}_p^k(d_2; K_1, \ldots, K_t) > \sum_{\beta < \bar{f}_p^{k+1}(d_{1,2}; K_1, \ldots, K_t)} (m_k + \tau_{m_k - 1}(\beta) + m_k)$, and $\bar{f}_p^{k+b}(d_1; K_1, \ldots, K_{t-p}, K_{t+1}, \ldots, K_t) = [\sum_{\beta < \bar{f}_p^{k+b+1}(d_{1,1}; K_1, \ldots, K_{t-p}, K_{t+1}, \ldots, K_t)} (m_k + \tau_{m_k-1}(\beta) + m_k)] + \alpha$, where $\alpha \leq \tau_{m_k-1}(\bar{f}_p^{k+b+1}(d_{1,1}; K_1, \ldots, K_{t-p}, K_{t+1}, \ldots, K_t)) + m_k$. Hence $\bar{f}_p^{k+b}(d_1; K_1, \ldots, K_{t-p}, K_{t+1}, \ldots, K_t) < \bar{f}_p^k(d_2; K_1, \ldots, K_t)$ follows.

b) $k = t-p$. By induction, $\bar{f}_p^{k+2}(d_{1,1}; K_1, \ldots, K_{t-p}, K_{t+1}, \ldots, K_t) < \bar{f}_p^{k+1}(d_{1,2}; K_1, \ldots, K_t)$. Also, $\bar{f}_p^k(d_2; K_1, \ldots, K_t) > \omega^{\omega^{\bar{f}_p^{k+1}(d_{1,2}; K_1, \ldots, K_t)}}$ and $\bar{f}_p^k(d_1; K_1, \ldots, K_{t-p}, K_{t+1}, \ldots, K_t) \leq \omega^{(\omega^{\bar{f}_p^{k+2}(d_{1,1}; K_1, \ldots, K_{t-p}, K_{t+1}, \ldots, K_t)} \cdot m)}$ for some m, hence is $< \bar{f}_p^k(d_2; K_1, \ldots, K_t)$.

ii) There is an $2 \leq r \leq \min\{r_1, r_2\}$ such that $d_{i,1} = d_{i,2}$ for $1 \leq i \leq r - 1$, and $d_{r,1} < d_{r,2}$.

a) $k \neq t - p$. By induction, $\bar{f}_p^{k+b+1}(d_{r,1}; K_1, \ldots, K_{t-p}, K_{t+1}, \ldots, K_t) < \bar{f}_p^{k+1}(d_{r,2}; K_1, \ldots, K_t)$. From the formulas for $\bar{f}_p^k$, we also have that $\bar{f}_p^k(d_2; K_1, \ldots, K_t) \geq$

$\sum_{\beta<\bar{f}_p^{k+1}(d_{1,2};K_1,\ldots,K_t)} (m_k+\tau_{m_k-1}(\beta)+m_k)+\cdots+((m_k-r+2)+\tau_{m_k-r+1}(\bar{f}_p^{k+1}$ $(d_{1,2};K_1,\ldots,K_t))+(m_k-r+2))\cdot \bar{f}_p^{k+1}(d_{r-1,2};K_1,\ldots,K_t)+((m_k-r+1)+\tau_{m_k-r}(\bar{f}_p^{k+1}$ $(d_{1,2};K_1,\ldots,K_t))+(m_k-r+1))\cdot \bar{f}_p^{k+1}(d_{r,2};K_1,\ldots,K_t)+1$. We further have that $\bar{f}_p^{k+b}(d_1;$ $K_1,\ldots,K_{t-p},K_{t+1},\ldots,K_t)\leq \sum_{\beta<\bar{f}_p^{k+b+1}(d_{1,1};K_1,\ldots,K_{t-p},K_{t+1},\ldots,K_t)} (m_k+\tau_{m_k-1}(\beta)+m_k)+\cdots+$ $((m_k-r+2)+\tau_{m_k-r+1}(\bar{f}_p^{k+b+1}(d_{1,1};K_1,\ldots,K_{t-p},K_{t+1},\ldots,K_t))+(m_k-r+2))\cdot \bar{f}_p^{k+b+1}$ $(d_{r-1,1};K_1,\ldots,K_{t-p},K_{t+1},\ldots,K_t)+((m_k-r+1)+\tau_{m_k-r}(\bar{f}_p^{k+b+1}(d_{1,1}K_1,\ldots,K_{t-p},$ $K_{t+1},\ldots,K_t))+(m_k-r+1))\cdot(\bar{f}_p^{k+b+1}(d_{r,1};K_1,\ldots,K_{t-p},K_{t+1},\ldots,K_t)+1)$. Hence, it follows from A_p^k that R_p^k is satisfied.

b) $k=t-p$. Similar to the above case.

iii) $r_1<r_2,\ d_{i,1}=d_{i,2}$ for $1\leq i\leq r_1$, and s appears in d_1.

a) $k\neq t-p$. This case follows readily from A_p^k, the formulas for $\bar{f}_p^k$, and the fact that $\bar{f}_p^{k+1}(d_{r_1,2};K_1,\ldots,K_t)\geq 1$.

b) $k=t-p$. Similar to the above case.

iv) $r_1>r_2,\ d_{i,1}=d_{i,2}$ for $1\leq i\leq r_2$, and s does not appear in d_2.

a) $k\neq t-p$. From the formulas for $\bar{f}_p^k$, $\bar{f}_p^{k+b}(d_1;K_1,\ldots,K_{t-p},K_{t+1},\ldots,K_t)=$ $\sum_{\beta<\bar{f}_p^{k+b+1}(d_{1,1};K_1,\ldots,K_{t-p},K_{t+1},\ldots,K_t)} (m_k+\tau_{m_k-1}(\beta)+m_k)+\cdots+((m_k-r_2+1)+$ $\tau_{m_k-r_2}(\bar{f}_p^{k+b+1}(d_{1,1};K_1,\ldots,K_{t-p},K_{t+1},\ldots,K_t)+(m_k-r_2+1))\cdot \bar{f}_p^{k+b+1}(d_{r_2,1};K_1,\ldots,K_{t-p},$ $K_{t+1},\ldots,K_t)+\alpha$, where $\alpha\leq((m_k-r_2)+\tau_{m_k-r_2-1}(\bar{f}_p^{k+b+1}(d_{1,1};K_1,\ldots,K_{t-p},K_{t+1},\ldots,K_t))+$ $(m_k-r_2))\cdot(\bar{f}_p^{k+b+1}(d_{1,1};K_1,\ldots,K_{t-p},K_{t+1},\ldots,K_t)+1)$. We also have that $\bar{f}_p^k(d_2;K_1,\ldots,K_t)=$ $\sum_{\beta<\bar{f}_p^{k+1}(d_{1,2};K_1,\ldots,K_t)} (m_k+\tau_{m_k-1}(\beta)+m_k)+\cdots+((m_k-r_2+1)+\tau_{m_k-r_2}(\bar{f}_p^{k+1}(d_{1,2};K_1,\ldots,K_t))+$

$(m_k-r_2+1))\cdot(\bar{f}_p^{k+1}(d_{R_2,2};K_1,\ldots,K_t)+1)$. R_p^k then follows form A_p^k and the fact that $(m+\tau_{m-1}(\beta)+m)\cdot(\beta+1)<((m+1)+\tau_m(\beta)+(m+1))$ for all $m>1,\ \beta$.

b) $k=t-p$. Similar to the above case.

v) $r_1=r_2,\ d_{i,1}=d_{i,2}$ for $1\leq i\leq r_1$, s appears in d_1 and not in d_2. R_p^k follows immediately from the formulas for $\bar{f}_p^k$.

4) $k=k(d_1)=k(d_2),\ K_k=W_{2n+1}^{m_k}$.

i) $d_1=(k;\bar{d}_{2,1})^s$, where s may or may not appear, and $d_2=(k;\ )$.

a) $k \neq t-p$. Then $\bar{f}_p^k(d_2; K_1, \ldots, K_t) = E(2n; m_k)$. Also, by R_q and induction, $\bar{g}_q(\bar{d}_{2,1}; K_{b(k)}, \ldots, K_{b_u}, \mathcal{M}(K_1, \ldots, K_{k-1})) < E(2n; m_k)$, where q, etc. refer to the sequence $K_1, \ldots, K_{t-p}, K_{t+1}, \ldots, K_t$. R_p^k then follows.

b) $k = t-p$. Similar to the above.

ii) $d_1 = (k; \bar{d}_{2,1})^s$, $d_2 = (k; \bar{d}_{2,2})^s$, where $\bar{d}_{2,1} < \bar{d}_{2,2}$, and s may or may not appear in d_1, d_2.

a) $k \neq t-p$. If $q, K_{b(k)}, \ldots, K_{b_u}$, $\mathcal{M}(K_1, \ldots, K_{k-1})$ are as in the definition of $\bar{f}_p^k$ corresponding to $K_1, \ldots, K_t$, and $q', \ldots, K', \ldots, \mathcal{M}'$ are the values with respect to $K_1, \ldots, K_{t-p}$, $K_{t+1}, \ldots, K_t$, (and $k' = k + b$) then it follows from the definition of q that $q' = q$. Hence, it follows from induction and R_q^1 on $\mathcal{D}_{2n-1}$ that $\bar{g}_q(\bar{d}_{2,1}; \ldots K' \ldots, \mathcal{M}') < \bar{g}_q(\bar{d}_{2,2}; \ldots, K, \ldots \mathcal{M})$, and R_p^k follows.

b) $k = t-p$. Similar to the above.

iii) As above, where $\bar{d}_{2,1} = \bar{d}_{2,2}$ and s appears in d_1 and not in d_2. This case is immediate from the formulas for $\bar{f}_p^k$.

R_p^k has now been established in all cases. R_p now follows by reverse induction on k, R_p^k, the formulas for $\bar{f}_p^k$, and the fact that if $\alpha < E(2n+2; m_k)$ then $\alpha^n < E(2n+2; m_k)$ for all n.

It now follows from R_p^k that for any $(d^{(\mathcal{I}_a)})^s$, where s may not appear, and $\mathcal{I}_a = (f(\bar{K}_1);\ )$ and d is defined and satisfies C, A relative to $K_1, \ldots, K_t$, that $f((d)^s; K_1, \ldots, K_t) < E(2n+3)$.

VIII). The upper bound for $\boldsymbol{\delta}^1_{2n+5}$.

We are now in a position to collect the results of the previous sections and obtain the upper bound for $\boldsymbol{\delta}^1_{2n+5}$. We recall we are assuming I_{2n+1} and K_{2n+3}.

THEOREM. $\boldsymbol{\delta}^1_{2n+5} \leq \aleph_{E(2n+3)+1}$.

Proof. We recall (see [Kel]) that $\boldsymbol{\delta}^1_{2n+5} = [\sup_j j(\boldsymbol{\delta}^1_{2n+3})]^+$, where the sup ranges over embeddings from the ultra powers by the measures in $\cup_{m \leq n} S_{2m+1} \cup \cup_{m \leq n} W_{2m+3}$, these being, without loss of generality the most general measures arising from the homogeneous tree construction on a complete $\boldsymbol{\Pi}^1_{2n+3}$ set. From the global embedding theorem, we may restrict our attention to the measures $\mu_{2n+3}^{S^{\ell,m}_{2n+1}}$, where $\ell = 2^{n+1} - 1$ is maximal. We fix m, and let $F : \boldsymbol{\delta}^1_{2n+3} \to \boldsymbol{\delta}^1_{2n+3}$ be given, representing $[F]$ w.r.t. $\mu^{S^{\ell,m}_{2n+3}}$. From the weak partition relation on $\boldsymbol{\delta}^1_{2n+3}$, there is a $g : \boldsymbol{\delta}^1_{2n+3} \to \boldsymbol{\delta}^1_{2n+3}$ such that for almost all $f : \theta(S^{\ell,n}_{2n+1}) \to \boldsymbol{\delta}^1_{2n+3}$, $F([f]) < g(\sup_{\text{a.e.}} f)$. By the argument of the main theorem of section VI, it follows that

F represents an ordinal $\leq [\sup_{K_1}(\text{id}; f; d; K_1)]^+$, where $d = (\ )$. Repeating the argument gives $(\text{id}; f; d; K_1) \leq [\sup_{K_2}(\text{id}; f; (\tilde{d})^s; K_1, K_2)]^+$, where $\tilde{d}^{(I_a)}$ (where $I_a = (f(\bar{K}_1);\ ),\ \bar{K}_1 = v(S^{\ell,m}_{2n+1})$) is the maximal tuple relative to K_1, K_2. From the main theorem of section VI and R_p (where $p = 0$), it follows that F represents an ordinal $< \aleph_{E(2n+2)+2}$, and the result follows.

IX). A Lower Bound for f_p.

Our goal in this section is to obtain a lower bound for a certain rank function.

We first define some auxiliary measures. For each regular cardinal $\kappa < \boldsymbol{\delta}^1_{2n+1}$, we let M_κ be defined using the strong partition relation on $\boldsymbol{\delta}^1_{2n+1}$, functions $F : \boldsymbol{\delta}^1_{2n+1}$ of the correct type, and the normal measure on $\boldsymbol{\delta}^1_{2n+1}$ concentrating in points of cofinality κ. We let $N = M_{\kappa_1} \times \cdots \times M_{\kappa_p}$, where $\kappa_1, \ldots, \kappa_p$ enumerate the regular cardinals $< \boldsymbol{\delta}^1_{2n+1}$. We let $B^m_{2n+1} = (S^{1,m}_{2n+1} \times N \times \Pi_{\ell=2}^{2^{n+1}-1} \tilde{S}^{\ell,m}_{2n+1})^m$, where $\tilde{S}^{\ell,m}_{2n+1} = S^{\ell,m}_{2n+1}$ for $\ell \neq 2$, and for $\ell = 2$, we define $\tilde{S}^{\ell,m}_{2n+1}$ as follows: We let $k(m) = 1 + m + m(m-1) + m(m-1)(m-2) + \cdot\cdot + m!$ be the number of sequences $\pi = \langle i_1, i_2, \ldots, i_\ell \rangle$, where $0 \leq \ell \leq m$, each $1 \leq i_j \leq m$, and are distinct. For such a fixed π, we say a funciton $f : \omega_1^m \to \boldsymbol{\delta}^1_{2n+1}$ is of π type if $f(\alpha_1, \ldots, \alpha_m) < f(\beta_1, \ldots, \beta_m)$ whenever $(\alpha_{i_1}, \ldots, \alpha_{i_\ell}) <^L (\beta_{i_1}, \ldots, \beta_{i_\ell})$, and $f(\vec{\alpha}) = f(\vec{\beta})$ if $(\alpha_{i_1}, \ldots, \alpha_{i_\ell}) = (\beta_{i_1}, \ldots, \beta_{i_\ell})$. It is easy to see (using the weak partition relation on ω_1) that for any $f : \omega_1^m \to \boldsymbol{\delta}^1_{2n+1}$, there is a π and a measure one set A w.r.t. W_1^m such that $f \restriction A$ is of π type. We define $\tilde{S}^{2,m}_{2n+1}$ to be the measure on $k(m)$ tuples of ordinals defined by: A has measure one if there is a c.u.b. $C \subseteq \boldsymbol{\delta}^1_{2n+1}$ such that for all $F : \boldsymbol{\delta}^1_{2n+1} \to \boldsymbol{\delta}^1_{2n+1}$ of the correct type $(\ldots, \alpha_\pi, \ldots) \in A$, where for a fixed π, α_π is represented with respect $\mathcal{V} \equiv \mu^\mu_{2n+1} \equiv$ the measure on $\boldsymbol{\delta}^1_{2n+1}$ induced by the weak partition relation on $\boldsymbol{\delta}^1_{2n+1}$ and functions $f : \omega_1^m \to \boldsymbol{\delta}^1_{2n+1}$ of π-type and non-normal of uniform cofinality ω, by $\alpha_\pi([f]) = F([f])$.

We modify some of the previous definitions as follows:

We consider sequences of measures $K'_1, \ldots, K'_t$, where each $K'_k = B^{m_k}_{2n+1}$. We let $K^k_1, \ldots, K^k_{c_k}$ enumerate the measures in the product measure $B^{m_k}_{2n+1} = K'_k$. We can then consider $d \in \mathcal{D}_{2n+1}$ to be defined relative to $K'_1, \ldots, K'_t$ if d is defined relative to $K^1_1, \ldots, K^1_{c_1}, \ldots, K^t_1, \ldots, K^t_{c_t}$. We let $k(K^a_b)$ be the integer k such that K^a_b is the kth measure in the sequence $K^1_1, \ldots, K^t_{c_t}$. We let K_k denote the kth element of the sequence $K^1_1, \ldots, K^t_{c_t}$.

We allow now $\mathcal{D}_{2n+1}$ to further contain descriptions of the form $d^{(I_a)} = (k; d_2^{(I_a)})$, where $K_k = M_\kappa$ for some κ.

We define d to satisfy C relative to $K'_1, \ldots, K'_t$ if d satisfies C relative to $K^1_1, \ldots, K^t_{c_t}$, where we remove the restriction $r > 1$ if s appears for $K_k = s^{1,m_k}_{2n+1}$, and we define $(k; d_2)$ to satisfy C, where $K_k = M_\kappa$ provided for almost all $h_1^1, \ldots, h_{c_t}^t, f, \bar{h}_2, \ldots, \bar{h}_a$, $\operatorname{cof}(H(d_2; h_1^1, \ldots, h_{c_t}^t, f, \bar{h}_2, \ldots, \bar{h}_a)) = \kappa$. Also, if $d = (k; d_2)^s$ where $K_k = \tilde{S}^{2,m}_{2n+1}$, then condition C imposes no restriction if s appears, and if s does not appear, we define d to satisfy C if for almost all $h_1^1, \ldots, h_{c_t}^t$, $f, \bar{h}_2, \ldots, \bar{h}_a$, the function $g : \omega_1^m \to \boldsymbol{\delta}^1_{2n+1}$ given by $g([\bar{h}_{a+1}]) = H(d_2; h_1^1, \ldots, h_{c_t}^t, f, \bar{h}_2, \ldots, \bar{h}_{a+1})$ is such that for some measure one set A w.r.t. W_1^m, $g \restriction A$ is non-normal of uniform cofinality ω.

Conditions D, A are as before.

We define modified operations $\mathcal{L}_M, \hat{\mathcal{L}}_M, \hat{\mathcal{L}}_M^{\restriction k}$ on descriptions defined and satisfying C relative to $K_1', \ldots, K_t'$. The definition of $\hat{\mathcal{L}}_M^{\restriction k}$ proceeds as the definition of $\hat{\mathcal{L}}_M^{\restriction k}$, with the following modifications: (cases numbered as in the definition of $\hat{\mathcal{L}}_M^{\restriction k}$).

Case 3). If s appears and d_2 not minimal w.r.t $\hat{\mathcal{L}}_M^{\restriction k+1}$, then we set $\hat{\mathcal{L}}_M^{\restriction k}(d) =$ $(k; \hat{\mathcal{L}}_M^{\restriction k+1}(d_2)^{(I_a+1)})$ if this tuple satisfies C, and otherwise $= (k; \hat{\mathcal{L}}_M^{\restriction k+1}(d_2)^{(I_a+1)})^s$. This case includes now the case $K_k = \tilde{S}^{2,m}_{2n+1}$. If d_2 is minimal wiht respect to $\hat{\mathcal{L}}_M^{\restriction k+1}$ then d is minimal w.r.t. $\hat{\mathcal{L}}_M^{\restriction k}$.

We also add the following cases:

8) $k < \infty$, $k = k(d)$, $K_k = M_k$. Hence, $d^{(I_a)} = (k; d_2^{(I_a)})$. If d_2 is minimal w.r.t. $\hat{\mathcal{L}}_M^{\restriction k+1}$, then d is minimal w.r.t. $\hat{\mathcal{L}}_M^{\restriction k}$. Otherwise, we set $\hat{\mathcal{L}}_M^{\restriction k}(d) = (k; \hat{\mathcal{L}}_M^{\restriction k+1}(d_2))$ if for almost all $h_1^1, \ldots, h_{c_t}^t, f, \bar{h}_2, \ldots, \bar{h}_a$, $\operatorname{cof} H(\hat{\mathcal{L}}_M^{\restriction k+1}(d_2);\ h_1^1, \ldots, h_{c_t}^t,\ f, \bar{h}_2, \ldots, \bar{h}_a) = \kappa$, and otherwise $\hat{\mathcal{L}}_M^{\restriction k}(d) = d_2$.

9) $k < \infty$, $k < k(d)$, $K_k = M_\kappa$. If d is minimal w.r.t. $\hat{\mathcal{L}}_M^{\restriction k+1}$, then d is minimal w.r.t. $\hat{\mathcal{L}}_M^{\restriction k}$. Otherwise, we set $\hat{\mathcal{L}}_M^{\restriction k}(d) = (k; \hat{\mathcal{L}}_M^{\restriction k+1}(d))$ if $\operatorname{cof} H(\hat{\mathcal{L}}_M^{\restriction k+1}(d); h_1^1, \ldots, \bar{h}_a) = \kappa$ almost everywhere, and otherwise $\hat{\mathcal{L}}_M^{\restriction k}(d) = \hat{\mathcal{L}}_M^{\restriction k+1}(d)$.

We set $\hat{\mathcal{L}}_M(d) = \hat{\mathcal{L}}_M^{\restriction 1}(d)$ for $d^{(I_a)}$ with $I_a = (f(\bar{\kappa}_1);\ )$.

If d satisfies C relative to $K_1', \ldots, K_t'$, it is easy to see that $\hat{\mathcal{L}}_M(d)$ also satisfies C. Also, property F^2_{2n+1} is still satisfied for d satisfying C.

If d is defined and satisfied C relative to $K_1', \ldots, K_t'$ and $G : \boldsymbol{\delta}^1_{2n+3} \to \boldsymbol{\delta}^1_{2n+3}$, then we define $(\mathrm{id}; f; (d)^s; K_1', \ldots, K_t')$ and $(G; f; (d)^s; K_t', \ldots, K_t')$ as before, where $f : \theta(S^{\ell,m}_{2n+1}) \to \boldsymbol{\delta}^1_{2n+3}$. If s does not appear here, then we required (d) satisfy D. From Lemma 11 of section V, it follows that if s appears, then $(d)^s$ satisfies D, so these are well defined.

We will establish in part 2 that the ordinals $(\mathrm{id}; d; K_1', \ldots, K_t')$, for d satisfying C, are all cardinals.

We let $<_M$ denote the ordering on tuples $(d; K'_1, \ldots, K'_t)$, where d is defined and satisfies C relative to $K'_1, \ldots, K'_t$, generated by the relation:

1) $(\hat{\mathcal{L}}_M(d; K'_1, \ldots, K'_t);\ K'_1, \ldots, K'_t, K'_{t+1}) <_M (d; K'_1, \ldots, K'_t)$ for all K'_{t+1}.

We let $f(d; K'_1, \ldots, K'_t)$ denote the rank of $(d; K'_1, \ldots, K'_t)$ in the ordering $<_M$. It then follows from the above remark (granting the unproven assertion above) that

$$\boldsymbol{\delta}^1_{2n+5} \geq \aleph_{[\sup_{d, K'_1, \ldots, K'_t} f(d; K'_1, \ldots, K'_t)] + 1}.$$

We now proceed to establish that $\sup_{d, K'_1, \ldots, K'_t} f(d; K'_1, \ldots, K'_t) \geq E(2n+3)$, yielding the lower bound.

We define two auxiliary orderings $<_q$, $<^k_q$, where $q \geq 1$ and $1 \leq k \leq c$, and $c \equiv c_1 + c_2 + \cdots + c_t$, as follows:

$<_1$ is generated by the relation:

1) $(\hat{\mathcal{L}}^{(q)}_M(d; K'_1, \ldots, K'_t); K'_1, \ldots, K'_t, K'_{t+1}) <_q (d; K'_1, \ldots, K'_t)$ for all K'_{t+1}.

$<^k_q$ is generated by the relation:

$(\hat{\mathcal{L}}^{\lceil k(q)}_M(d; K'_1, \ldots, K'_t); K'_1, \ldots, K'_t, K'_{t+1}) <^k_q (d; K'_1, \ldots, K'_t)$. Here $\delta_d(\bar{K}_a) < k \leq k(d)$, d satisfies C relative to $K'_1, \ldots, K'_t$ and $\hat{\mathcal{L}}^{(q)}_M$ denotes the qth iterate of $\hat{\mathcal{L}}_M$.

We let f_q, f^k_q denote the ranks of $(d; K'_1, \ldots, K'_t)$ in these orderings. We let g_q, g^k_q denote the corresponding rank functions on $\mathcal{D}_{2n-1}$.

We define an auxiliary function $\tilde{f}^k_q$ for $1 \leq k \leq c$, on d satisfying C relative to $K'_1, \ldots, K'_t$, where $\delta_d(\bar{K}_a) < k \leq k(d)$ as follows:

I) $k(d) = \infty,\ k = c$.

1) d basic of type 1, so $d = (d_2)^s$, where s may or may not appear. We set $\tilde{f}^k_q(d; K'_1, \ldots, K'_t) = \omega^{\omega^{g_{q\cdot 4}(d_2; \bar{K}_2, \ldots, \bar{K}_a)}} + 1$, where $I_a = (f; \bar{K}_2, \ldots, \bar{K}_a)$ as usual, for $g_{q\cdot 4}(d_2; \bar{K}_2, \ldots, \bar{K}_a) > 0$, and otherwise $\tilde{f}^k_q(d; K'_1, \ldots, K'_t) = 0$.

2) d basic of type 0. $\tilde{f}^k_q(K'_1, \ldots, K'_t) = 0$.

II) $k < c,\ k < k(d)$. We set $\tilde{f}^k_q(d; K'_1, \ldots, K'_t) = \tilde{f}^{k+1}_q(d; K'_1, \ldots, K'_t)$.

III) $k < c,\ k = k(d)$.

1) $K_k = S^{1,m_k}_{2n+1}$. Hence, $d = (k; d_1, \ldots, d_r)^s$, where $r \leq m_k$, and s may or may not appear.

We let $\triangleleft$ be the ordering on tuples $(\alpha_1,\alpha_j)^s$, where $j = 1$ or 2, $\alpha_2 \le \alpha_1$, s may or may not appear, defined by: $(\alpha_1,\alpha_{j_1})^s \triangleleft (\beta_1,\beta_{j_2})^s$ if one of the following holds:

i) $\alpha_1 < \beta_1$

ii) $\alpha_1 = \beta_1$, $j_1 = j_2 = 2$, and $\alpha_2 < \beta_2$.

iii) $\alpha_1 = \beta_1$ and

a) $j_1 = 1$, $j_2 = 2$, $(\vec{\alpha})$ involves s,

or b) $j_1 = 2$, $j_2 = 1$, $(\vec{\beta})$ does not involve s,

or c) $j_1 = j_2 = 1$, $(\vec{\alpha})$ involves s and $(\vec{\beta})$ does not, where we identify (α_1,α_2) with $(\alpha_1,\alpha_2)^s$, $(\alpha,0)$ with $(\alpha)^s$, and $(0,0)$ with $(0)^s$. We then let $\tilde{f}^k_q(d;K'_1,\ldots,K'_t) = r((\tilde{f}^{k+1}_q(d_1;$ $K_1,\ldots,K'_t)$, $\tilde{f}^{k+1}_q(d_2;K'_1,\ldots,K'_t))^s) \equiv$ the rank of $(\tilde{f}^{k+1}_q(d_1;K'_1,\ldots,K'_t)$, $\tilde{f}^{k+1}_q(d_2;K'_1,\ldots,K'_t))^s$ in $\triangleleft$.

2) $K_k = S^{\ell_k,m_k}_{2n+1}$, with $\ell_k > 1$. Here $d = (k;d_2)^s$, where s may or may not appear. We set $\tilde{f}^k_q(d;K'_1,\ldots,K'_t) = \tilde{f}^{k+1}_q(d_2;K'_1,\ldots,K'_t)$.

3) d basic of type -1. We set $\tilde{f}^k_q(d;K'_1,\ldots,K'_t) = 0$.

IV) $k = c$, $k = k(d)$. As in III) where we replace $\tilde{f}^{k+1}_q(\ )$ by $\omega^{\omega^{g_{q\cdot 4}(\ )}}$ in cases 1,2, and retain 3. (Actually only case 2 arises).

We introduce the following hypothesis:

R^k_q: for d satisfying C relative to $K'_1,\ldots,K'_t$, where $\delta_d(\bar{K}_a) < k \le k(d)$, $f^k_q(d;K'_1,\ldots,K'_t) \ge$

$\tilde{f}^k_q(d;K'_1,\ldots,K'_t)$.

We establish R^k_q by reverse induction on k, and for fixed k by induction on $<$.

We let $\tau(\beta) =$ the rank of (β) in $\triangleleft$. We consider the following cases:

I) $k(d) = \infty$, $k = c$. If d is basic of type 0, the result is immediate. Hence, $d = (d_2)^s$, where we may assume s appears. We have that $f^c_q(d;K'_1,\ldots,K'_t) \ge \sup_{K'_{t+1}} [f^c_q((\mathcal{L}^{(q)}(d_2;$ $\bar{K}_2,\ldots,\bar{K}_a))^s;\ K'_1,\ldots,K'_t,K'_{t+1}) + 1] \ge \sup_{K'_{t+1}} [\ \sup_{K'_{t+2}} f^c_q((k_1;(k_2;\cdots(k_u;(k_v;(\mathcal{L}^{(q\cdot 2)}(d_2;$ $\bar{K}_2,\ldots,\bar{K}_a))^s)))\cdots));\ K'_1,\ldots,K'_t,\ K'_{t+1},K'_{t+2}) + 1]$, where $k_1,\ldots,k_u$ enumerate the components of K'_{t+1} of the form $S^{1,m}_{2n+1}$, and K_{k_v} is the component of K'_{t+1} of the form M_κ such that the above satisfies C. This is $\ge$

$\sup_m \tau^{(m)}(f_q^c((\mathcal{L}^{(q\cdot 2)}(d_2; \bar{K}_2, \ldots, \bar{K}_a))^s; K_1', \ldots, K_t'))$, where $\tau^{(m)}$ denotes the mth iterate of τ.

We also have that $f_q^c((\mathcal{L}^{(q\cdot 2)}(d_2; \bar{K}_2, \ldots, \bar{K}_a)^s);\ K_1', \ldots, K_t') \geq \sup_{K_{t+3}'} [f_q^c((\mathcal{L}^{(q\cdot 3)}\ (d_2;$ $\bar{K}_2, \ldots, \bar{K}_a))^s; K_1', \ldots, K_t',\ K_{t+3}')] \geq \sup_{K_{t+3}'} [f_q^c((k_1; (k_2, \ldots, (k_u; (\mathcal{L}^{(q\cdot 4)}(d_2$ $\bar{K}_2, \ldots, \bar{K}_a))^{s(I_a+u)})^{s(I_a+u-1)} \ldots)^{s(I_a+1)})^{s(I_a)}; K_1', \ldots, K_t', K_{t+3}')]$, where here $k_1, \ldots, k_u$ enumerate the components of K_{t+3}' of the form $S_{2n+1}^{\ell_k, m_k}$ with $\ell_k > 1$. By induction this is $\geq \sup_{K_{t+3}'} [\tilde{f}_q^{c'}$ $((\mathcal{L}^{(q\cdot 4)}(d_2; \bar{K}_2, \ldots, \bar{K}_a))^{s(I_a+u)}; K_1', \ldots, K_t', K_{t+3}']$, where c' is the value of c corresponding to the sequence $K_1', \ldots, K_t',\ K_{t+3}'$. This is $\geq \sup_{\bar{K}_{a+1}} \omega^{\omega^{[g_{q\cdot 4}(\mathcal{L}^{(q\cdot 4)}(d_2; \bar{K}_2, \ldots, \bar{K}_a); \bar{K}_2, \ldots, \bar{K}_a, \bar{K}_{a+1})]}}$ $= \omega^{\omega^{[\sup_{\bar{K}_{a+1}} (g_{q\cdot 4}(\mathcal{L}^{(q\cdot 4)}(d_2; \bar{K}_2, \ldots, \bar{K}_a); \bar{K}_2, \ldots, \bar{K}_{a+1}))]}} \geq \omega^{\omega^{g_{q\cdot 4}(d_2; \bar{K}_2, \ldots, \bar{K}_a) - 1}}$, where $\lambda - 1 = \lambda$ if λ is a limit ordinal. But, for all ordinals β, $\sup_m \tau^{(m)}(\omega^{\omega^\beta}) = \omega^{\omega^{\beta+1}}$, hence $f_q^c(d; K_1', \ldots, K_t') \geq$ $\omega^{\omega^{g_{q\cdot 4}(d_2; \bar{K}_2, \ldots, \bar{K}_a)}} = \tilde{f}_q^c(d; K_1', \ldots, K_t')$.

II) $k < c,\ k < k(d)$.

1) $k(d) < \infty$. For $\delta_d(\bar{K}_a) < k_1 < k_2 \leq k(d)$, it follows readily that $f_q^{k_1}(d; K_1', \ldots, K_t') \geq$ $f_q^{k_2}(d; K_1', \ldots, K_t')$. Hence, $f_q^k(d; K_1', \ldots, K_t') \geq f_q^{k(d)}(d; K_1', \ldots, K_t') \geq$ (by induction) $\tilde{f}_q^k(d;$ $K_1', \ldots, K_t')$.

2) $k(d) = \infty$. $f_q^k(d; K_1', \ldots, K_t') \geq f_q^c(d; K_1', \ldots, K_t') \geq \tilde{f}_q^c(d; K_1', \ldots, K_t')$ by case I, $=$ $\tilde{f}_q^k(d; K_1', \ldots, K_t')$.

III) $k < c,\ k = k(d)$.

1) $K_k = M_\kappa$. Hence, $d = (k; d_2)$. Then, $f_q^k(d; K_1', \ldots, K_t') \geq f_q^{k+1}(d_2; K_1', \ldots, K_t') \geq$ $\tilde{f}_q^{k+1}(d_2; K_1', \ldots, K_t') = \tilde{f}_q^k(d; K_1', \ldots, K_t')$.

2) $K_k = S_{2n+1}^{\ell_k, m_k}$, where $\ell_k > 1$. Here $d^{(I_a)} = (k; d_2^{(I_a+1)})^s$, where we may assume s appears.

a) d_2 is minimal w.r.t. $\hat{\mathcal{L}}_M^{\restriction k+1(q)}$, that is, $\hat{\mathcal{L}}_M^{\restriction k+1(q)}(d_2)$ is not defined. Then by induction, $\tilde{f}_q^{k+1}(d_2; K_1', \ldots, K_t') = 0$, hence $\tilde{f}_q^k(d; K_1', \ldots, K_t') = \tilde{f}_q^{k+1}(d_2; K_1', \ldots, K_t') = 0$, so R_q^k is immediate.

b) $\hat{\mathcal{L}}_M^{\restriction k+1(q)}(d_2)$ is defined.

We first establish that $f_q^k(d; K_1', \ldots, K_t') \geq f_q^{k+1}(d_2; K_1', \ldots, K_t')$ for d of this form by induction w.r.t. the ordering $<$. We have that $f_q^k(d; K_1', \ldots, K_t') = \sup_{K_{t+1}'} [f_q^k(\hat{\mathcal{L}}_M^{\restriction k(q)}(d; K_1', \ldots, K_t');$

$K'_1, \ldots, K'_t, K'_{t+1})+1] \geq \sup_{K'_{t+1}} [f_q^k((k; \hat{\mathcal{L}}_M^{\restriction k+1(q)}\ (d_2; K'_1, \ldots, K'_t))^s; K'_1, \ldots, K'_{t+1})+1] \geq \sup_{K'_{t+1}} [f_q^{k+1}$ $(\hat{\mathcal{L}}_M^{\restriction k+1(q)}(d_2; K'_1, \ldots, K'_t); K'_1, \ldots, K'_{t+1}) + 1]$ by induction, $= f_q^{k+1}(d_2; K'_1, \ldots, K'_t)$.

It then follows that $f_q^k(d; K'_1, \ldots, K'_t) \geq f_q^{k+1}(d_2; K'_1, \ldots, K'_t) \geq \tilde{f}_q^{k+1}(d_2; K'_1, \ldots, K'_t) = \tilde{f}_q^k(d; K'_1, \ldots, K'_t)$.

3) $K_k = S_{2n+1}^{1,m_k}$. Hence, $d = (k; d_1, \ldots, d_r)^s$, where we may assume s appears.

a) $r > 2$. We have that $f_q^k(d; K'_1, \ldots, K'_t) \geq f_q^k((k; d_1, d_2)^s;\ K'_1, \ldots, K'_t) \geq \tilde{f}_q^k((k; d_1, d_2)^s;$ $K'_1, \ldots, K'_t) = \tilde{f}_q^k(d; K'_1, \ldots, K'_t)$. We use here the fact that if $d' \leq d$ then $f_q^k(d'; K'_1, \ldots, K'_t) \leq f_q^k(d; K'_1, \ldots, K'_t)$, established by an easy induction.

b) $r = 2$.

i) $\hat{\mathcal{L}}_M^{\restriction k+1(q)}(d_2; K'_1, \ldots, K'_t)$ is defined. We have that $f_q^k(d; K'_1, \ldots, K'_t) \geq \sup_{K'_{t+1}} f_q^k(\hat{\mathcal{L}}_M^{\restriction k(q)}$ $(d; K'_1, \ldots, K'_t);\ K'_1, \ldots, K'_{t+1})+1 \geq \sup_{K'_{t+1}} f_q^k((k; d_1, \hat{\mathcal{L}}_M^{\restriction k+1(q)}\ (d_2; K'_1, \ldots, K'_t))^s; K'_1, \ldots, K'_{t+1})+$ 1, by cases on whether or not $r = m_k$. By induction, this is $\geq \sup_{K'_{t+1}} r(f_q^{k+1}(d_1; K'_1, \ldots, K'_{t+1}),$ $f_q^{k+1}(\hat{\mathcal{L}}_M^{\restriction k+1(q)}(d_2; K'_1, \ldots, K'_t);\ K'_1, \ldots, K'_{t+1})) + 1 \geq r(f_q^{k+1}(d_1; K'_1, \ldots, K'_t), \sup_{K'_{t+1}} f_q^{k+1}$ $(\hat{\mathcal{L}}_M^{\restriction k+1(q)}(d_2; K'_1, \ldots, K'_t);\ K'_1, \ldots, K'_{t+1})+1)$. Hence, $f_q^k(d; K'_1, \ldots, K'_t) \geq r(f_q^{k+1}(d_1; K'_1, \ldots, K'_t),$ $f_q^{k+1}(d_2; K'_1, \ldots, K'_t)) = \tilde{f}_q^k(d; K'_1, \ldots, K'_t)$.

ii) $\hat{\mathcal{L}}_M^{\restriction k(q)}(d_2; K'_1, \ldots, K'_t)$ is not defined. In this case, $f_q^k(d; K'_1, \ldots, K'_t) \geq f_q^k((k; d_1)^s;$ $K'_1, \ldots, K'_t) \geq r((f_q^{k+1}(d_1; K'_1, \ldots, K'_t))$, by induction, $= r(f_q^{k+1}(d_1;\ K'_1, \ldots, K'_t), 0) =$ $\tilde{f}_q^k(d; K'_1, \ldots, K'_t)$, since $f_q^{k+1}(d_2; K'_1, \ldots, K'_t) = 0$.

c) $r = 1$ and s appears.

i) $\hat{\mathcal{L}}_M^{\restriction k+1(q)}(d_1; K'_1, \ldots, K'_t)$ is defined. We have that $f_q^k(d; K'_1, \ldots, K'_t) \geq \sup_{K'_{t+1}} [f_q^k(\hat{\mathcal{L}}_M^{\restriction k+1(q)}$ $(d; K'_1, \ldots, K'_t); K'_1, \ldots, K'_{t+1}) + 1] \geq \sup_{K'_{t+1}} [f_q^k(k; \hat{\mathcal{L}}_M^{\restriction k+1(q)}(d_1; K'_1, \ldots, K'_t);\ K'_1, \ldots, K'_{t+1}) + 1]$ $\geq \sup_{K'_{t+1}} [r(f_q^{k+1}\ (\hat{\mathcal{L}}_M^{\restriction k+1(q)}(d_1; K'_1, \ldots, K'_t);\ K'_1, \ldots, K'_{t+1})) + 1] = r(f_q^{k+1}(d_1; K'_1, \ldots, K'_{t+1}))^s$, from the definition of r, $= \tilde{f}_q^k(d; K'_1, \ldots, K'_t)$.

ii) $\hat{\mathcal{L}}_M^{\restriction k+1(q)}(d_1; K'_1, \ldots, K'_t)$ is not defined. In this case, $\tilde{f}_q^{k+1}(d_1; K'_1, \ldots, K'_t) = 0$ by induction, so R_q^k is immediate.

d) $r = 1$ and s does not appear.

i) $\hat{\mathcal{L}}_M^{\restriction k+1(q)}(d_1; K'_1, \ldots, K'_t)$ is defined. Proceeding as above, $f_q^k(d; K'_1, \ldots, K'_t) \geq \sup_{K'_{t+1}}$

$$[f_q^k((k;d_1,\hat{\mathcal{L}}_M^{\restriction k+1(q)}(d_1;K'_1,\ldots,K'_t))^s;K'_1,\ldots,K'_{t+1})+1]\geq \sup_{K'_{t+1}}[r(f_q^{k+1}(d_1;K'_1,\ldots,K'_t),$$
$$f_q^{k+1}(\hat{\mathcal{L}}_M^{\restriction k+1(q)}(d_1;K'_1,\ldots,K'_t);K'_1,\ldots,K'_{t+1}))+1]\geq r(f_q^{k+1}(d_1;K'_1,\ldots,K'_t),(\sup_{K'_{t+1}}$$
$$f_q^{k+1}(\hat{\mathcal{L}}_M^{\restriction k+1(q)}(d_1;K'_1,\ldots,K'_t);K'_1,\ldots,K'_{t+1})+1))=\tilde{f}_q^k(d;K'_1,\ldots,K'_t).$$

ii) $\hat{\mathcal{L}}_M^{\restriction k+1(q)}(d_1;K'_1,\ldots,K'_t)$ not defined. By induction $\tilde{f}_q^{k+1}(d_1;K'_1,\ldots,K'_{t+1})=0$, so R_q^k is immediate.

IV) $k=c,\ k=k(d)$. This case is similar to the previous cases (in fact by a trivial change in the definition of B_{2n+1}^m, so that its last component is in $\cup_{m<n}R_{2m+1}$, we may assume that this case does not arise).

This establishes R_q^k in all cases.

If we assume inductively that $\sup_{\bar{d},\bar{K}_1,\ldots,\bar{K}_t} g_q(\bar{d};\bar{K}_1,\ldots,\bar{K}_t)=E(2n+1)$ for all q, then it follows from R_q and the formulas for $\tilde{f}_q^k$ that $\sup_{d,K'_1,\ldots,K'_t} f_q(d;K'_1,\ldots,K'_t)=E(2n+3)$ for all q.

References

[J] Jackson, Steve. *A Calculation of $\boldsymbol{\delta}_5^1$*. Ph.D. Thesis, UCLA, 1983.

[Ke] Kechris, A. S. *Homogeneous Trees and Projective Scales.* Cabal Seminar 77-79, p. 33-73. Springer-Verlag, vol. 839.

[Ma] Martin, D. A. *AD and the Normal Measures on $\boldsymbol{\delta}_3^1$*. To appear.

[Mo] Moschovakis, Y. N., *Descriptive Set Theory, Studies in Logic*, vol. 100, North Holland, Amsterdam, 1980.

APPENDIX: VICTORIA DELFINO PROBLEMS II

At the "Very Informal Gathering" of January 1984, the Cabal announced the addition of seven problems to the Victoria Delfino list. We are happy (and not at all embarrassed) to report that since then four of these problems have been solved. Below we list the new problems, beginning with #6 since there were five problems on the original list. For each we describe briefly what was known when it was added to the list, and what has been its fate since.

#6. The extent of definable scales. Assume $\mathbf{\Pi}^1_1 - \mathrm{AD}^{\Sigma_3}$. Do all $\Game^{\Sigma_2}\mathbf{\Pi}^1_1$ sets admit $\mathrm{HOD}(\mathbf{R})$ scales?

The terminology is explained in Steel's paper "Long Games" in this volume. The strongest result in this direction had been Martin's theorem that for $\lambda < \omega_1$ a limit ordinal, $\mathbf{\Pi}^1_1 - \mathrm{AD}^\lambda$ implies all $\Game^\lambda\mathbf{\Pi}^1_1$ sets admit $\Game^\lambda\mathbf{\Pi}^1_1$ scales ([5]). Work of Woodin and Steel had shown that a positive answer to #6 implies that some form of definable determinacy (i.e. $\mathbf{\Pi}^1_1 - \mathrm{AD}^{\Sigma_3}$) yields an inner model of $\mathrm{AD}_{\mathbf{R}}$.

Steel obtained a positive answer to #6 in February 1984; his results in this area are described in "Long Games".

#7. The Kleene ordinal. Let κ be the least ordinal not the order type of a prewellordering of $\mathbf{R}$ recursive in Kleene's ${}^3\mathrm{E}$ and a real. Assume $\mathrm{AD}^{L(\mathbf{R})}$. Is κ the least weakly inaccessible cardinal?

That the answer is positive is an old conjecture of Moschovakis, who had shown that κ is a regular limit of Suslin cardinals ([7], [8]). Steel showed in [9] that κ is the least regular limit of Suslin cardinals. Thus the problem amounted to bounding the growth of the Suslin cardinals below κ. Building on work of Kunen and Martin, Jackson had done this for the first ω Suslin cardinals; this work is described in his long paper in this volume.

In the Fall of 1985, Jackson obtained a positive answer to #7. His new work extends the theory presented in his paper in this volume. Because of its length and complexity, as of now no one but Jackson has been through this new work.

#8. Regular cardinals in $L(\mathbf{R})$. Assume $\mathrm{AD} + V = L(\mathbf{R})$. Are all regular cardinals below θ measurable?

Moschovakis and Kechris had shown, in $\mathrm{ZFC}+\mathrm{AD}^{L(\mathbf{R})}$, that if κ is regular (in V, where AC holds!) and $\kappa < \theta^{L(\mathbf{R})}$, then $L(\mathbf{R}) \models \kappa$ is measurable. This led them to conjecture a positive answer to #8. Jackson's detailed analysis of cardinals and measures had verified the conjecture for κ below the sup of the first ω Suslin cardinals (cf. his paper in this volume).

The only progress on this problem since its addition to the list is Jackson's new work cited above, which presumably yields a positive answer to #8 for κ below the Kleene ordinal.

#9. Large cardinals implying determinacy. Does the existence of a nontrivial, elementary $j: V_{\lambda+1} \rightarrow V_{\lambda+1}$ imply $\mathbf{\Pi}^1_3$ determinacy?

The world view embodied in the statements of this and the succeeding problem was seriously mistaken. That view was inspired by Martin's result ([4]) that the existence of a nontrivial, Σ_1-elementary $j: V_{\lambda+1} \rightarrow V_{\lambda+1}$ implies that $\mathbf{\Pi}^1_2$ determinacy, together with work of Mitchell [6] which promised to lead to a proof that nothing much weaker than the existence of such an embedding would imply $\mathbf{\Pi}^1_2$ determinacy. Martin naturally conjectured that a nontrivial, fully elementary $j: V_{\lambda+1} \rightarrow V_{\lambda+1}$ would yield PD; hence the inclusion of #9 on our list.

Partly because this view was so mistaken, progress in this area since 1984 has been dramatic. In February - April of 1984 Woodin showed that the existence of a nontrivial, elementary $j: L(V_{\lambda+1}) \rightarrow L(V_{\lambda+1})$ implies PD and in fact $\mathrm{AD}^{L(\mathbf{R})}$. This was still consistent with the view underlying #9, and in spirit was a positive answer, although even for $\mathbf{\Pi}^1_3$ determinacy Woodin's result required a hypothesis slightly stronger than that allowed in #9. However, at about the same time Foreman, Magidor, and Shelah ([3]) developed a powerful new technique for producing generic elementary embeddings under relatively "weak" large cardinal hypotheses such as the existence of supercompact cardinals. Woodin realized at once the potential in their technique and used it to show, in May 1984, that the existence of a supercompact cardinal implies all projective sets of reals are Lebesgue measurable. Immediately thereafter, Shelah and Woodin improved this to include all sets in $L(\mathbf{R})$.

If the relationship between large cardinals and determinacy were to exhibit anything like the pattern it had previously, supercompact cardinals had to imply $\mathrm{AD}^{L(\mathbf{R})}$. In September 1985, Martin and Steel showed that in fact they do (thereby answering #9 positively). (Their proof of PD is self-contained. Their proof of $\mathrm{AD}^{L(\mathbf{R})}$ requires work done by Woodin using the generic embedding techniques.) The Martin-Steel theorem required much less than supercompactness; e.g., for $\mathbf{\Pi}^1_{n+1}$ determinacy it required the existence of n "Woodin cardinals" with a measurable above them all. [The notion of a "Woodin cardinal" had been isolated by Woodin in his work on generic embeddings; it is a refinement of a notion due to Shelah.] In May - July of 1986, Martin and Steel pushed the theory of inner models for large cardinals far enough to show that the hypothesis of their theorem was best possible: the existence of n Woodin cardinals does not imply $\mathbf{\Pi}^1_{n+1}$ determinacy. More recently, Woodin has obtained relative consistency results in this direction by a different method; cf. #10 below.

Unfortunately, with the exception of [3], none of this recent work has been published.

#10. Supercompacts in $\mathrm{HOD}^{L(\mathbf{R})}$. Assume $\mathrm{AD}^{L(\mathbf{R})}$. Does $\mathrm{HOD}^{L(\mathbf{R})}$ satisfy the sentence "$\exists\kappa$ (κ is 2^κ-supercompact)"?

Becker and Moschovakis ([2]) had shown that that $\mathrm{HOD}^{L(\mathbf{R})} \models \exists\kappa(O(k) = \kappa^+)$. Martin (unpublished) then showed $\mathrm{HOD}^{L(\mathbf{R})} \models \exists\kappa$ (κ is μ-measurable.). Steel (unpublished) then showed $\mathrm{HOD}^{L(\mathbf{R})} \models \exists\kappa$ (κ is λ-strong, where $\lambda > \kappa$ is measurable). Inspired by these results, the Cabal conjectured that $\mathrm{HOD}^{L(\mathbf{R})}$ satisfies all large cardinal hypotheses weaker than that which implies $\mathrm{AD}^{L(\mathbf{R})}$ (which is false in $\mathrm{HOD}^{L(\mathbf{R})}$). Problem #10 resulted from our mistaken guess as to what these hypotheses are.

The Woodin-Shelah theorem that the existence of supercompacts implies all sets in $L(\mathbf{R})$ are Lebesgue measurable settles #10 negatively, since, assuming $\mathrm{AD}^{L(\mathbf{R})}$, $\mathrm{HOD}^{L(\mathbf{R})} \models$ "There is a wellorder of $\mathbf{R}$ in $L(\mathbf{R})$". However, except for the mistake about the cardinals involved, the answer to #10 is positive. Woodin has recently (February 1987) shown that, assuming $\mathrm{AD}^{L(\mathbf{R})}$, $\mathrm{HOD}^{L(\mathbf{R})} \models \exists\kappa$ (κ is a Woodin cardinal), and under the same assumption found a natural submodel of $\mathrm{HOD}^{L(\mathbf{R})}$ satisfying "There are ω Woodin cardinals". The work of Martin, Steel, and Woodin referred to in the discussion of #9, together with further work of Woodin reducing its large cardinal hypothesis, shows that $\mathrm{AD}^{L(\mathbf{R})}$ follows from the existence of ω Woodin cardinals with a measurable above them all, so that Woodin's recent work is in spirit a positive answer to #10.

#11. The GCH in $\mathrm{HOD}^{L(\mathbf{R})}$. Assume $\mathrm{AD}^{L(\mathbf{R})}$. Does $\mathrm{HOD}^{L(\mathbf{R})}$ satisfy the GCH?

Becker ([1]) had shown that, assuming $\mathrm{AD}^{L(\mathbf{R})}$, $\mathrm{HOD}^{L(\mathbf{R})} \models 2^\kappa = \kappa^+$ for many cardinals κ.

There has been little progress on this question since January 1984. Woodin's recent work on large cardinals in $\mathrm{HOD}^{L(\mathbf{R})}$ does show that, letting $\delta = (\boldsymbol{\delta}^2_1)^{L(\mathbf{R})}$ and $\theta = \theta^{L(\mathbf{R})}$, $\mathrm{HOD}^{L(\mathbf{R})} \models \delta$ is θ-strong. It follows by an easy reflection argument that if $\mathrm{HOD}^{L(\mathbf{R})}$ satisfies the GCH below $(\boldsymbol{\delta}^2_1)^{L(\mathbf{R})}$, then it satisfies the GCH.

#12. Projective uniformization, measure, and category. Does the theory ZFC + "Every projective relation can be uniformized by a projective function" + "Every projective set is Lebesgue measurable and has the property of Baire" prove PD?

Woodin ([10]) showed that the theory in question proves $\forall x \subseteq \omega(x^\dagger$ exists) and more in this direction, together with some other consequences, of PD, and conjectured a positive answer to #12.

There has been no direct progress on this problem since 1984.

REFERENCES

[1] H. Becker, *Thin collections of sets of projective ordinals and analogs of* L, Annals of Math. Logic **19** (1980), pp. 205-241.

[2] H. Becker and Y. Moschovakis, *Measurable cardinals in playful models,* Cabal Seminar 77-79, Springer Lecture notes in Mathematics **839**, pp. 203-215.

[3] M. Foreman, M. Magidor, and S. Shelah, *Martin's maximum, saturated ideals, and non-regular ultrafilters,* to appear in Annals of Mathematics.

[4] D. Martin, *Infinite games,* Proceedings of the International Congress of Mathematicians, Helsinki, 1978.

[5] D. Martin, *The real game quantifier propagates scales,* Cabal Seminar 79-81, Springer Lecture notes in Mathematics, v. **1019**, pp. 157-172.

[6] W. Mitchell, *Hypermeasurable cardinals,* Logic Colloquium '78, M. Boffa, D. van Dalen and K. McAloon, eds., North Holland, Amsterdam, 1979.

[7] Y. Moschovakis, *Determinacy and prewellorderings of the continuum,* Math. Logic and Foundations of Set Theory, Y. Bar Hillel ed., North Holland, Amsterdam, 1970, pp. 24-62.

[8] Y. Moschovakis, *Inductive scales on inductive sets,* Cabal seminar 76-77, Springer Lecture notes in Mathematics, v. **689** (1978), pp. 185-192.

[9] J. Steel, *Closure properties of pointclasses,* Cabal Seminar 77-79, Springer Lecture notes in Mathematics, v. **839**, pp. 147-165.

[10] H. Woodin, *On the consistency strength of projective uniformization,* Proceedings of the Herbrand Symposinon, Logic Colloquium '81, North Holland, v. **107**.

Vol. 1173: H. Delfs, M. Knebusch, Locally Semialgebraic Spaces. XVI, 329 pages. 1985.

Vol. 1174: Categories in Continuum Physics, Buffalo 1982. Seminar. Edited by F.W. Lawvere and S.H. Schanuel. V, 126 pages. 1986.

Vol. 1175: K. Mathiak, Valuations of Skew Fields and Projective Hjelmslev Spaces. VII, 116 pages. 1986.

Vol. 1176: R.R. Bruner, J.P. May, J.E. McClure, M. Steinberger, H_∞ Ring Spectra and their Applications. VII, 388 pages. 1986.

Vol. 1177: Representation Theory I. Finite Dimensional Algebras. Proceedings, 1984. Edited by V. Dlab, P. Gabriel and G. Michler. XV, 340 pages. 1986.

Vol. 1178: Representation Theory II. Groups and Orders. Proceedings, 1984. Edited by V. Dlab, P. Gabriel and G. Michler. XV, 370 pages. 1986.

Vol. 1179: Shi J.-Y. The Kazhdan-Lusztig Cells in Certain Affine Weyl Groups. X, 307 pages. 1986.

Vol. 1180: R. Carmona, H. Kesten, J.B. Walsh, École d'Été de Probabilités de Saint-Flour XIV – 1984. Édité par P.L. Hennequin. X, 438 pages. 1986.

Vol. 1181: Buildings and the Geometry of Diagrams, Como 1984. Seminar. Edited by L. Rosati. VII, 277 pages. 1986.

Vol. 1182: S. Shelah, Around Classification Theory of Models. VII, 279 pages. 1986.

Vol. 1183: Algebra, Algebraic Topology and their Interactions. Proceedings, 1983. Edited by J.-E. Roos. XI, 396 pages. 1986.

Vol. 1184: W. Arendt, A. Grabosch, G. Greiner, U. Groh, H.P. Lotz, U. Moustakas, R. Nagel, F. Neubrander, U. Schlotterbeck, One-parameter Semigroups of Positive Operators. Edited by R. Nagel. X, 460 pages. 1986.

Vol. 1185: Group Theory, Beijing 1984. Proceedings. Edited by Tuan H.F. V, 403 pages. 1986.

Vol. 1186: Lyapunov Exponents. Proceedings, 1984. Edited by L. Arnold and V. Wihstutz. VI, 374 pages. 1986.

Vol. 1187: Y. Diers, Categories of Boolean Sheaves of Simple Algebras. VI, 168 pages. 1986.

Vol. 1188: Fonctions de Plusieurs Variables Complexes V. Séminaire, 1979–85. Edité par François Norguet. VI, 306 pages. 1986.

Vol. 1189: J. Lukeš, J. Malý, L. Zajíček, Fine Topology Methods in Real Analysis and Potential Theory. X, 472 pages. 1986.

Vol. 1190: Optimization and Related Fields. Proceedings, 1984. Edited by R. Conti, E. De Giorgi and F. Giannessi. VIII, 419 pages. 1986.

Vol. 1191: A.R. Its, V.Yu. Novokshenov, The Isomonodromic Deformation Method in the Theory of Painlevé Equations. IV, 313 pages. 1986.

Vol. 1192: Equadiff 6. Proceedings, 1985. Edited by J. Vosmansky and M. Zlámal. XXIII, 404 pages. 1986.

Vol. 1193: Geometrical and Statistical Aspects of Probability in Banach Spaces. Proceedings, 1985. Edited by X. Femique, B. Heinkel, M.B. Marcus and P.A. Meyer. IV, 128 pages. 1986.

Vol. 1194: Complex Analysis and Algebraic Geometry. Proceedings, 1985. Edited by H. Grauert. VI, 235 pages. 1986.

Vol. 1195: J.M. Barbosa, A.G. Colares, Minimal Surfaces in $\mathbb{R}^3$. X, 124 pages. 1986.

Vol. 1196: E. Casas-Alvero, S. Xambó-Descamps, The Enumerative Theory of Conics after Halphen. IX, 130 pages. 1986.

Vol. 1197: Ring Theory. Proceedings, 1985. Edited by F.M.J. van Oystaeyen. V, 231 pages. 1986.

Vol. 1198: Séminaire d'Analyse, P. Lelong – P. Dolbeault – H. Skoda. Seminar 1983/84. X, 260 pages. 1986.

Vol. 1199: Analytic Theory of Continued Fractions II. Proceedings, 1985. Edited by W.J. Thron. VI, 299 pages. 1986.

Vol. 1200: V.D. Milman, G. Schechtman, Asymptotic Theory of Finite Dimensional Normed Spaces. With an Appendix by M. Gromov. VIII, 156 pages. 1986.

Vol. 1201: Curvature and Topology of Riemannian Manifolds. Proceedings, 1985. Edited by K. Shiohama, T. Sakai and T. Sunada. VII, 336 pages. 1986.

Vol. 1202: A. Dür, Möbius Functions, Incidence Algebras and Power Series Representations. XI, 134 pages. 1986.

Vol. 1203: Stochastic Processes and Their Applications. Proceedings, 1985. Edited by K. Itô and T. Hida. VI, 222 pages. 1986.

Vol. 1204: Séminaire de Probabilités XX, 1984/85. Proceedings. Edité par J. Azéma et M. Yor. V, 639 pages. 1986.

Vol. 1205: B.Z. Moroz, Analytic Arithmetic in Algebraic Number Fields. VII, 177 pages. 1986.

Vol. 1206: Probability and Analysis, Varenna (Como) 1985. Seminar. Edited by G. Letta and M. Pratelli. VIII, 280 pages. 1986.

Vol. 1207: P.H. Bérard, Spectral Geometry: Direct and Inverse Problems. With an Appendix by G. Besson. XIII, 272 pages. 1986.

Vol. 1208: S. Kaijser, J.W. Pelletier, Interpolation Functors and Duality. IV, 167 pages. 1986.

Vol. 1209: Differential Geometry, Peñíscola 1985. Proceedings. Edited by A.M. Naveira, A. Ferrández and F. Mascaró. VIII, 306 pages. 1986.

Vol. 1210: Probability Measures on Groups VIII. Proceedings, 1985. Edited by H. Heyer. X, 386 pages. 1986.

Vol. 1211: M.B. Sevryuk, Reversible Systems. V, 319 pages. 1986.

Vol. 1212: Stochastic Spatial Processes. Proceedings, 1984. Edited by P. Tautu. VIII, 311 pages. 1986.

Vol. 1213: L.G. Lewis, Jr., J.P. May, M. Steinberger, Equivariant Stable Homotopy Theory. IX, 538 pages. 1986.

Vol. 1214: Global Analysis – Studies and Applications II. Edited by Yu.G. Borisovich and Yu.E. Gliklikh. V, 275 pages. 1986.

Vol. 1215: Lectures in Probability and Statistics. Edited by G. del Pino and R. Rebolledo. V, 491 pages. 1986.

Vol. 1216: J. Kogan, Bifurcation of Extremals in Optimal Control. VIII, 106 pages. 1986.

Vol. 1217: Transformation Groups. Proceedings, 1985. Edited by S. Jackowski and K. Pawalowski. X, 396 pages. 1986.

Vol. 1218: Schrödinger Operators, Aarhus 1985. Seminar. Edited by E. Balslev. V, 222 pages. 1986.

Vol. 1219: R. Weissauer, Stabile Modulformen und Eisensteinreihen. III, 147 Seiten. 1986.

Vol. 1220: Séminaire d'Algèbre Paul Dubreil et Marie-Paule Malliavin. Proceedings, 1985. Edité par M.-P. Malliavin. IV, 200 pages. 1986.

Vol. 1221: Probability and Banach Spaces. Proceedings, 1985. Edited by J. Bastero and M. San Miguel. XI, 222 pages. 1986.

Vol. 1222: A. Katok, J.-M. Strelcyn, with the collaboration of F. Ledrappier and F. Przytycki, Invariant Manifolds, Entropy and Billiards; Smooth Maps with Singularities. VIII, 283 pages. 1986.

Vol. 1223: Differential Equations in Banach Spaces. Proceedings, 1985. Edited by A. Favini and E. Obrecht. VIII, 299 pages. 1986.

Vol. 1224: Nonlinear Diffusion Problems, Montecatini Terme 1985. Seminar. Edited by A. Fasano and M. Primicerio. VIII, 188 pages. 1986.

Vol. 1225: Inverse Problems, Montecatini Terme 1986. Seminar. Edited by G. Talenti. VIII, 204 pages. 1986.

Vol. 1226: A. Buium, Differential Function Fields and Moduli of Algebraic Varieties. IX, 146 pages. 1986.

Vol. 1227: H. Helson, The Spectral Theorem. VI, 104 pages. 1986.

Vol. 1228: Multigrid Methods II. Proceedings, 1985. Edited by W. Hackbusch and U. Trottenberg. VI, 336 pages. 1986.

Vol. 1229: O. Bratteli, Derivations, Dissipations and Group Actions on C*-algebras. IV, 277 pages. 1986.

Vol. 1230: Numerical Analysis. Proceedings, 1984. Edited by J.-P. Hennart. X, 234 pages. 1986.

Vol. 1231: E.-U. Gekeler, Drinfeld Modular Curves. XIV, 107 pages. 1986.

Vol. 1232: P.C. Schuur, Asymptotic Analysis of Soliton Problems. VIII, 180 pages. 1986.

Vol. 1233: Stability Problems for Stochastic Models. Proceedings, 1985. Edited by V.V. Kalashnikov, B. Penkov and V.M. Zolotarev. VI, 223 pages. 1986.

Vol. 1234: Combinatoire énumérative. Proceedings, 1985. Edité par G. Labelle et P. Leroux. XIV, 387 pages. 1986.

Vol. 1235: Séminaire de Théorie du Potentiel, Paris, No. 8. Directeurs: M. Brelot, G. Choquet et J. Deny. Rédacteurs: F. Hirsch et G. Mokobodzki. III, 209 pages. 1987.

Vol. 1236: Stochastic Partial Differential Equations and Applications. Proceedings, 1985. Edited by G. Da Prato and L. Tubaro. V, 257 pages. 1987.

Vol. 1237: Rational Approximation and its Applications in Mathematics and Physics. Proceedings, 1985. Edited by J. Gilewicz, M. Pindor and W. Siemaszko. XII, 350 pages. 1987.

Vol. 1238: M. Holz, K.-P. Podewski and K. Steffens, Injective Choice Functions. VI, 183 pages. 1987.

Vol. 1239: P. Vojta, Diophantine Approximations and Value Distribution Theory. X, 132 pages. 1987.

Vol. 1240: Number Theory, New York 1984–85. Seminar. Edited by D.V. Chudnovsky, G.V. Chudnovsky, H. Cohn and M.B. Nathanson. V, 324 pages. 1987.

Vol. 1241: L. Gårding, Singularities in Linear Wave Propagation. III, 125 pages. 1987.

Vol. 1242: Functional Analysis II, with Contributions by J. Hoffmann-Jørgensen et al. Edited by S. Kurepa, H. Kraljević and D. Butković. VII, 432 pages. 1987.

Vol. 1243: Non Commutative Harmonic Analysis and Lie Groups. Proceedings, 1985. Edited by J. Carmona, P. Delorme and M. Vergne. V, 309 pages. 1987.

Vol. 1244: W. Müller, Manifolds with Cusps of Rank One. XI, 158 pages. 1987.

Vol. 1245: S. Rallis, L-Functions and the Oscillator Representation. XVI, 239 pages. 1987.

Vol. 1246: Hodge Theory. Proceedings, 1985. Edited by E. Cattani, F. Guillén, A. Kaplan and F. Puerta. VII, 175 pages. 1987.

Vol. 1247: Séminaire de Probabilités XXI. Proceedings. Edité par J. Azéma, P.A. Meyer et M. Yor. IV, 579 pages. 1987.

Vol. 1248: Nonlinear Semigroups, Partial Differential Equations and Attractors. Proceedings, 1985. Edited by T.L. Gill and W.W. Zachary. IX, 185 pages. 1987.

Vol. 1249: I. van den Berg, Nonstandard Asymptotic Analysis. IX, 187 pages. 1987.

Vol. 1250: Stochastic Processes – Mathematics and Physics II. Proceedings 1985. Edited by S. Albeverio, Ph. Blanchard and L. Streit. VI, 359 pages. 1987.

Vol. 1251: Differential Geometric Methods in Mathematical Physics. Proceedings, 1985. Edited by P.L. García and A. Pérez-Rendón. VII, 300 pages. 1987.

Vol. 1252: T. Kaise, Représentations de Weil et GL_2 Algèbres de division et GL_n. VII, 203 pages. 1987.

Vol. 1253: J. Fischer, An Approach to the Selberg Trace Formula via the Selberg Zeta-Function. III, 184 pages. 1987.

Vol. 1254: S. Gelbart, I. Piatetski-Shapiro, S. Rallis. Explicit Constructions of Automorphic L-Functions. VI, 152 pages. 1987.

Vol. 1255: Differential Geometry and Differential Equations. Proceedings, 1985. Edited by C. Gu, M. Berger and R.L. Bryant. XII, 243 pages. 1987.

Vol. 1256: Pseudo-Differential Operators. Proceedings, 1986. Edited by H.O. Cordes, B. Gramsch and H. Widom. X, 479 pages. 1987.

Vol. 1257: X. Wang, On the C*-Algebras of Foliations in the Plane. V, 165 pages. 1987.

Vol. 1258: J. Weidmann, Spectral Theory of Ordinary Differential Operators. VI, 303 pages. 1987.

Vol. 1259: F. Cano Torres, Desingularization Strategies for Three-Dimensional Vector Fields. IX, 189 pages. 1987.

Vol. 1260: N.H. Pavel, Nonlinear Evolution Operators and Semi-groups. VI, 285 pages. 1987.

Vol. 1261: H. Abels, Finite Presentability of S-Arithmetic Groups. Compact Presentability of Solvable Groups. VI, 178 pages. 1987.

Vol. 1262: E. Hlawka (Hrsg.), Zahlentheoretische Analysis II. Seminar, 1984–86. V, 158 Seiten. 1987.

Vol. 1263: V.L. Hansen (Ed.), Differential Geometry. Proceedings, 1985. XI, 288 pages. 1987.

Vol. 1264: Wu Wen-tsün, Rational Homotopy Type. VIII, 219 pages. 1987.

Vol. 1265: W. Van Assche, Asymptotics for Orthogonal Polynomials. VI, 201 pages. 1987.

Vol. 1266: F. Ghione, C. Peskine, E. Sernesi (Eds.), Space Curves. Proceedings, 1985. VI, 272 pages. 1987.

Vol. 1267: J. Lindenstrauss, V.D. Milman (Eds.), Geometrical Aspects of Functional Analysis. Seminar. VII, 212 pages. 1987.

Vol. 1268: S.G. Krantz (Ed.), Complex Analysis. Seminar, 1986. VII, 195 pages. 1987.

Vol. 1269: M. Shiota, Nash Manifolds. VI, 223 pages. 1987.

Vol. 1270: C. Carasso, P.-A. Raviart, D. Serre (Eds.), Nonlinear Hyperbolic Problems. Proceedings, 1986. XV, 341 pages. 1987.

Vol. 1271: A.M. Cohen, W.H. Hesselink, W.L.J. van der Kallen, J.R. Strooker (Eds.), Algebraic Groups Utrecht 1986. Proceedings. XII, 284 pages. 1987.

Vol. 1272: M.S. Livšic, L.L. Waksman, Commuting Nonselfadjoint Operators in Hilbert Space. III, 115 pages. 1987.

Vol. 1273: G.-M. Greuel, G. Trautmann (Eds.), Singularities, Representation of Algebras, and Vector Bundles. Proceedings, 1985. XIV, 383 pages. 1987.

Vol. 1274: N. C. Phillips, Equivariant K-Theory and Freeness of Group Actions on C*-Algebras. VIII, 371 pages. 1987.

Vol. 1275: C.A. Berenstein (Ed.), Complex Analysis I. Proceedings, 1985–86. XV, 331 pages. 1987.

Vol. 1276: C.A. Berenstein (Ed.), Complex Analysis II. Proceedings, 1985–86. IX, 320 pages. 1987.

Vol. 1277: C.A. Berenstein (Ed.), Complex Analysis III. Proceedings, 1985–86. X, 350 pages. 1987.

Vol. 1278: S.S. Koh (Ed.), Invariant Theory. Proceedings, 1985. V, 102 pages. 1987.

Vol. 1279: D. Ieşan, Saint-Venant's Problem. VIII, 162 Seiten. 1987.

Vol. 1280: E. Neher, Jordan Triple Systems by the Grid Approach. XII, 193 pages. 1987.

Vol. 1281: O.H. Kegel, F. Menegazzo, G. Zacher (Eds.), Group Theory. Proceedings, 1986. VII, 179 pages. 1987.

Vol. 1282: D.E. Handelman, Positive Polynomials, Convex Integral Polytopes, and a Random Walk Problem. XI, 136 pages. 1987.

Vol. 1283: S. Mardešić, J. Segal (Eds.), Geometric Topology and Shape Theory. Proceedings, 1986. V, 261 pages. 1987.

Vol. 1284: B.H. Matzat, Konstruktive Galoistheorie. X, 286 pages. 1987.

Vol. 1285: I.W. Knowles, Y. Saitō (Eds.), Differential Equations and Mathematical Physics. Proceedings, 1986. XVI, 499 pages. 1987.

Vol. 1286: H.R. Miller, D.C. Ravenel (Eds.), Algebraic Topology. Proceedings, 1986. VII, 341 pages. 1987.

Vol. 1287: E.B. Saff (Ed.), Approximation Theory, Tampa. Proceedings, 1985–1986. V, 228 pages. 1987.

Vol. 1288: Yu. L. Rodin, Generalized Analytic Functions on Riemann Surfaces. V, 128 pages, 1987.

Vol. 1289: Yu. I. Manin (Ed.), K-Theory, Arithmetic and Geometry. Seminar, 1984–1986. V, 399 pages. 1987.